Adaptive predictive control

Prentice Hall International
Series in Systems and Control Engineering

M. J. Grimble, Series Editor

BAGCHI, A., *Optimal Control of Stochastic Systems*
BENNETT, S., *Real-time Computer Control: An introduction*, second edition
BITMEAD, R. R., GEVERS, M. and WERTZ, V., *Adaptive Optimal Control*
BROWN, M. and HARRIS, C., *Neurofuzzy Adaptive Modelling and Control*
COOK, P. A., *Nonlinear Dynamical Systems*, second edition
GRIMBLE, M. J., *Robust Industrial Control*
ISERMANN, R., LACHMANN, K. H. and MATKO, D., *Adaptive Control Systems*
KUCERA, V., *Analysis and Design of Discrete Linear Control Systems*
MARTINS DE CARVALHO, J. L., *Dynamical Systems and Automatic Control*
MATKO, D., ZUPANCIC, B. and KARBA, R., *Simulation and Modelling of Continuous Systems: A case study approach*
OLSSON, G. and PIANI, G., *Computer Systems for Automation and Control*
ÖZGÜLER, A. B., *Linear Multichannel Control*
PARKS, P. C. and HAHN, V., *Stability Theory*
PETKOV, P. H., CHRISTOV, N. D. and KONSTANTINOV, M. M., *Computational Methods for Linear Control Systems*
SÖDERSTRÖM, T. D., *Discrete-time Stochastic Systems*
SÖDERSTRÖM, T. D. and STOICA, P., *System Identification*
SOETERBOEK, A. R. M., *Predictive Control: A unified approach*
WATANABE, K., *Adaptive Estimation and Control*
WILLIAMSON, D., *Digital Control and Instrumentation*

Preface

Scope

The aim of this book is to give the reader a historical, didactic, comprehensive and unified presentation of adaptive predictive control systems (APCS), their application and its contribution to optimization in the operation of industrial plants.

The book starts from the concept of predictive control as it was originally proposed, and presents the formulation of control laws derived from this concept. Predictive control was originally proposed within the framework of adaptive control; moreover, adaptability plays an important role in the practical applications. So the book also focuses on adaptive systems from the perspective of predictive control, within a unified setting based on stability theory.

Progressing from theory to practice has never been a simple matter. That is why we include some of the first milestone applications of adaptive predictive control. The purpose of these applications is twofold: first, they serve as examples which illustrate how the theoretical concepts are applied and, second, they show the effectiveness of adaptive predictive control in processes with significant difficulties requiring to be controlled.

The book, on its journey from the theory to the industrial applications, goes on to describe some implementations in which adaptive predictive control is in continuous operation nowadays. These implementations serve to introduce and illustrate the optimization in the process operation derived from the application of adaptive predictive control.

Taking the preceding three paragraphs into account, the book is divided into three parts: 'The Theory', 'The Applications' and 'The Optimization'.

In each of these parts we have used, let us say, different 'languages'. While in the first part we have tried to use the language of the scientist with its mathematical rigor, in the second part we have used the language of the engineer trying to implement a technology practically. Finally, in the third part we have used the language of the industrial plant operators, who wish to obtain the best performance from the process operation. It will not be surprising that some readers will feel more comfortable with one of these languages than with the others, but the journey described in this book certainly requires communication at the three previously mentioned levels. We have tried to make such communication as easy as possible and if some reader considers that the contents of this book are simple, then this will mean that our objective has been reached in this case.

In the following we will briefly advise the reader of the contents of the different parts of this book.

Contents

Chapter 1 discusses the requirements for an advanced industrial control system and introduces in an intuitive manner the basic concepts for discrete time modelling, predictive and adaptive predictive control. Also, this chapter sets the bases for the theoretical and methodological developments presented in the first part of the book and describes its overall structure.

Chapter 2 focuses on the mathematical formulation of the principle of predictive control and on its basic strategy, and introduces the concepts of projected desired trajectory (PDT) and driving desired trajectory (DDT). These new concepts are of paramount importance in the unification of stability theory within the context of adaptive predictive control and, as a result, in the theoretical analysis presented in this book. The drawback of the basic strategy when controlling processes with unstable inverse is illustrated by means of an example that shows the need for an extended strategy.

Chapter 3 develops the extended strategy of predictive control which is based on the evaluation of the predictive process input/output sequences within a multistep prediction horizon that is redefined at each sampling instant. As examples for this evaluation, lineal quadratic and step-shaped indexes are considered. The capability of the extended strategy to control unstable inverse processes, as well as its stability and robustness properties in the presence of modelling errors, is analyzed using elemental tools and illustrated by means of several examples. The role of the main design parameters, such as the length of the prediction horizon, is also considered.

Chapter 4 presents the basic concepts of adaptive systems within the context of predictive control and defines different scenarios characterized by different hypotheses progressively approaching the real industrial environment. The principles for the design of adaptive systems are stated in the following terms: (i) the desired performance objectives that a global adaptive predictive system should achieve are defined on an intuitive basis, that is, when trying to approach the desired behaviour of predictive control when the process dynamics is known; (ii) these objectives are translated into classical stability concepts, establishing what will be referred to as *APCS stability*; (iii) specific conditions for the adaptive systems are stated by means of a conjecture such that, if they are satisfied, the desired objectives are achieved.

The remainder of this chapter solves the problem of synthesis of the adaptive system for the different scenarios considered. The properties derived for the adaptive system in these sections match the desired goals under the assumption that the input/output vector remains bounded. Later, in the next chapter, these properties are used to prove that this assumption is satisfied when the desired objective is physically realizable and predictive control is combined with the adaptive system. In this way, Chapters 4 and 5 prove that the overall adaptive predictive control system is globally stable, that is to say, it verifies the desired performance objective.

Chapter 5 analyzes and proves under what conditions predictive control and adaptive predictive control can guarantee stability. For predictive control the effect of incorrect modelling is analyzed and the limits of stability are established. For adaptive predictive control global stability is guaranteed if there is no difference in structure between the process and the predictive model. When there is a difference in structure, the limits to the result on stability and the restrictions on the control objectives are derived. It is proved that these limits and restrictions depend on the dynamic nature of the process.

The above stability results are based on the properties of the adaptive systems derived in Chapter 4, the stability nature of the process and the properties of the driving desired trajectory (DDT), that is, physical realizability and/or boundedness. The stability analysis takes into account the following three types of process progressively: (a) stable or unstable processes with a stable inverse; (b) stable processes with a stable or unstable inverse and (c) unstable processes with an unstable inverse. As previously mentioned, the DDT concept is used to obtain a general and unifying proof within the body of APCS stability theory presented in this book.

Chapter 6 is intended to supply the reader with a bridge between the theoretical presentations of the preceding part and their applications. With this purpose in mind, the chapter emphasizes the key issues involved in the practical application of APCS to real processes. These practical issues are illustrated by actual application to single-input/single-output and the multivariable control of a pilot scale binary distillation column. This project was carried out in the first months of 1976 at the Department of Chemical Engineering at the University of Alberta, Canada. The multivariable control of a distillation column was at that time a typical example of the difficulties found in the practical application of modern control theory, and, consequently, the application of APCS to this process was a challenging real project with which to assess its capabilities as an advanced control methodology.

The application of APCS to the design and evaluation of an automatic pilot for NASA's F-8 supersonic aircraft was the first important real time application of the new methodology. It was carried out in the spring of 1975 using the facilities of the Charles Stark Draper Laboratory, Cambridge, Massachusetts, USA, and in particular its high-fidelity hybrid simulation of the aircraft. This application is presented in Chapter 7, showing the link between the methodology and its use in a practical problem.

Chapters 6 and 7 present applications that were carried out in the context of a two-year research and development program (1974–1976), funded by the Juan March foundation of Spain and, in both cases, the basic strategy of predictive control was used.

Chapter 8 presents the first historical application of APCS to an industrial production process and illustrates the use of the extended strategy of predictive control, drawing the links between the theory presented in Chapter 3 and its implementation. In this application, carried out in 1984, APCS controlled the bleach plant of the pulp factory of CANFOR Ltd in Port Mellon, British Columbia, Canada. This project was the result of a joint effort between the Paper and Pulp Research Institute of Canada (PAPRICAN) and the Department of Chemical Engineering of the University of Alberta, and was carried out in the context of a five-year research and development programme supported by the Natural Sciences and Engineering Research Council (NSERC) of Canada.

One of the challenging applications of control theory in recent years has been the real time mitigation of vibrations of flexible structures subjected to dynamic actions. Examples can be found in aerospace and in mechanical and civil engineering. Chapter 9 describes the application of the extended strategy of predictive control, using a state space representation, to experimental models of building structures subjected to earthquake loads. The implementation was carried out in 1986 at the laboratories of the National Center for Earthquake Engineering Research in Buffalo, USA, and was one of the first experimental applications of control theory reported in this field.

A simple, systematic and generalized industrial implementation of APCS was the main objective of an intensive research and development effort initiated in 1986 in Spain and supported by the 'Centro para el Desarrollo Tecnológico Industrial (CDTI)', the 'Comisión Interministerial de Ciencia y Tecnología (CICYT)', the 'Dirección General de Nuevas Tecnologías del Ministerio de Industria' and the company SCAP Europa SA.

Chapter 10 presents a concept of optimization which is defined in the context of the application of APCS and the systems that were developed to make this concept a reality. These systems, which are known as APCS optimization systems, were the results of the above-mentioned research and development effort. From the APCS perspective, optimization is related to the real time evaluation of the performance of industrial processes and the experimental search of the operating points at which this performance is maximized. Also, this chapter presents, from the conceptual point of view, the application of APCS optimization systems to the processes of the cement industry, including sinterization and milling, and illustrates the strategies used to approach the concept of optimization.

Chapter 11 presents the generic application of an APCS optimization system to a cement kiln. From the optimization criteria defined in Chapter 10, centred basically on searching for the lowest possible energy consumption and the maximum production, it describes the general configuration of the system and the adaptive predictive control loops. It also analyzes the results obtained for the stabilization of the kiln, as well as for the application of the optimization strategies.

In a similar way, Chapter 12 presents the detailed application to a cement milling process. It considers the control and optimization strategies used in order to maximize the mill production with the simultaneous minimization of energy consumption, and evaluates the benefits derived from the application.

Four appendices have been included in the book. Appendix A summarizes the main basic tools for describing linear systems in both continuous and discrete time, which will be used throughout the book. Appendix B extends the results of Chapter 3, developing and discussing other predictive control laws derived using different forms of predictive models and performance indices. Appendix C states an input/output property that characterizes the class of *stable and linear nature processes* and plays an important role in the APCS stability theory. Finally, Appendix D presents an alternative proof of APCS global stability for processes of a stable and linear nature, the inverse of which may be unstable, which does not use the assumption of physical realizability of the driving desired trajectory (DDT) required in the general and unifying proof of APCS global stability.

Readership

This book is addressed to undergraduate and postgraduate students of sciences and engineering, academics, engineers in industry and researchers whose interest may lie in the area of automatic control and, particularly, in advanced applications within this area. Without relaxing the scientific rigor, we have tried to avoid the complexity of books that are strictly devoted to control theory and have aimed for simplicity of exposition, with the purpose of making its contents interesting and understandable for a wide spectrum of potential readers, from the theoretically oriented to the practitioners.

Acknowledgements

This book contains the results of research and development work carried out over more than two decades by the authors and their colleagues. These colleagues deserve a first place in our recognition of their help and contribution to this book. Thus, our gratitude is to Sirish L. Shah and Grant Fisher from the Department of Chemical Engineering at the University of Alberta (Canada), William Cluett from the University of Toronto (Canada), Guy Dumont from the University of British Columbia and the Paper and Pulp Research Institute of Canada (PAPRICAN), Tsu T. Soong and Andrei Reinhorn from the National Center for Earthquake Engineering Research at the State University of New York at Buffalo (USA).

Our gratitude is also extended to Manuel García Gil de Bernabé who, being President of the Spanish Institute of Engineering, decided to support the adaptive predictive control methodology personally and founded the company SCAP

Europa SA in order to promote its industrial application. Also, this book, and particularly its optimization part, contains the results of projects developed by this company. We must thank SCAP Europa and its staff for their contribution to the book and for making possible what most people believed was impossible.

The writing of this book has been a shared activity for the authors, who have enjoyed the friendly and motivating atmosphere at the Department of Energy Systems at the School of Mining Engineering of the Universidad Politécnica de Madrid (Spain) and the Department of Applied Mathematics III at the School of Civil Engineering of the Universitat Politècnica de Catalunya, Barcelona (Spain).

Our thanks also go to those who contributed to the theoretical developments from which our research work began, particularly Ioan D. Landau and Karl J. Aström, and the research community as a whole, since without this community any advance is practically impossible these days.

Finally we owe a special debt of gratitude to our wives and families, who provided unwavering support during the writing of this book.

Juan M. Martín Sánchez
José Rodellar

Madrid–Barcelona, Spain, July 1995

Chapter 1

INTRODUCTION

1.1 THE CONCEPT OF THE CONTROL LOOP IN AN INDUSTRIAL PROCESS

Nature always appears in the form of processes with a basic logic of cause-effect. The history of science is, in large measure, the history of the knowledge of nature's processes, while the history of technology concerns the use and practical application of this knowledge to the creation of other artificial processes that are capable of producing goods and services for mankind. The productive processes which we will be referring to in this book may be represented by a basic scheme such as that shown in Figure 1.1.

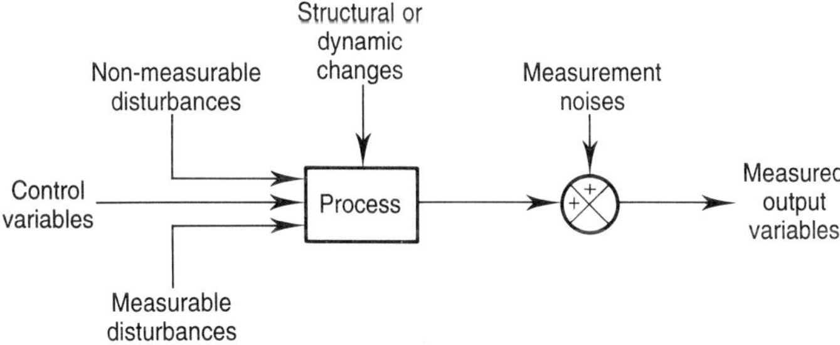

Fig. 1.1. Basic representation of a process.

Processes evolve over time following their own particular dynamics, which in general is subject to change, and responding to external actions. The reaction of the process may be characterized by the so-called output variables. On the one hand, these variables are generally measurable and are normally affected by so-called measurement noise, while, on the other hand, the actions on the process are represented using two types of input variable, defined, respectively, as control variables and disturbances. The essential difference between these two types is that only the first can be manipulated. Among the disturbances we may distinguish between the measurable and the non-measurable disturbances.

The control of a process is generally taken to mean the continuous or dynamic manipulation of its control variables, using the available output variables and disturbances measured in real time, with the final objective of making these output variables, or at least a subset of them, reach and maintain an assigned value or setpoint. If this manipulation is carried out by a system in the absence of a human operator, the control performed is termed automatic.

Figure 1.2 represents the basic elements of an automatic control system within a configuration referred to as a closed control loop. The output variables and the measurable disturbances of the process are measured by sensors and are used by a device, known as a controller, to generate the control signal or signals which, through a set of actuators, become the control actions on the process. The controller is designed on the basis of a methodology of control systems, which determines the generation of the aforementioned control signal or signals and, in summary, the driving of the process to certain specifications. The concept of a closed loop, as defined in this way includes a multivariable character capable of taking the interactions between the different process variables into account.

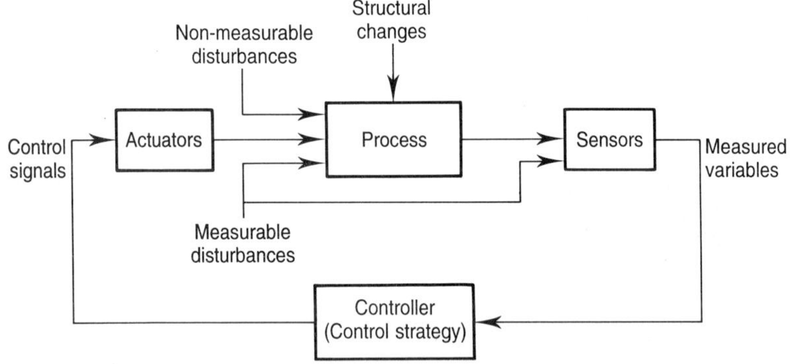

Fig. 1.2. Representation of a closed control loop.

As is well known, the controller may either be made from analog devices or based on a digital computer. In the first case, the control action is generated continuously in real time. In the second case, the control action may only be modified in discrete time instants, whose interval is the so-called control period.

The computation of the latter control action is performed by a program based on the control methodology chosen.

Over the last two decades, the process control field has shown a clear and progressive preference for the use of computers that has been mainly motivated by the availability of fast and powerful microprocessors at low prices, which introduce important advantages over analog hardware. Of all these advantages, perhaps the most important is the possibility of implementing advanced control strategies in a relatively easy manner. Within the practical applications of predictive and adaptive predictive control presented in this book, we will consider control loops similar to that shown in Figure 1.2. Nevertheless, in general, the complexity of the process, and the convenience of using the control methodology in a simple manner, make it advisable to consider the process as being divided into subprocesses, with different control loops assigned to them and interrelated. This way of using the methodology will be illustrated in this book.

1.2 REQUIREMENTS FOR AN ADVANCED INDUSTRIAL CONTROL METHODOLOGY

On the one hand, the implementation of control loops obviously requires the availability of adequate technology for the sensors, actuators and devices that make up the control system. On the other hand, in order to achieve a satisfactory performance in its operation, the selected control methodology or system should be able to solve the problems inherent in the process dynamics and its environment. This satisfactory performance would demand an 'ideal' control methodology able to satisfy the following requirements:

(a) to guarantee the stability and, particularly, to be robust, as explained below;

(b) to be as efficient as possible, as required by the desired performance criteria;

(c) to be easy to implement and operate in real time through the use of digital computers.

Furthermore, in order to consider this a 'general purpose' control methodology, these properties must be valid for a wide variety of processes and operating conditions.

Stability is a basic property that is essential in the application of an automatic control system. The control system has to ensure that the output variable will reach a band around the setpoint value. If, due to circumstances such as changes in the process dynamics or disturbances acting on the process, the output variable moves away from the setpoint, the control system should drive the output variable back into the required band, all this via the application of a control action that is compatible with the physical restrictions of the process and its actuators.

A discrete time control system, that is to say, a control system developed for implementation through the use of a digital computer, is generally based on a mathematical model of the process under control that may include the dynamics of the associated instrumentation, that is, sensors and actuators. In this context, the control system may be formulated and tested on the proposed model, using it as an ideal representation of the process under consideration. A control system is said to be robust if its performance and, fundamentally, its stability are maintained when applied to the real process in spite of the differences that may exist between the process and the proposed model.

The robustness problem is especially important in the control of industrial processes, where their dynamic nature is complex, of non-linear character and varying over time during operation. This renders the linear models generally used in this context approximations only of the process behaviour under certain operating conditions.

The efficiency requirement has to be understood in the sense that the system should be based on, or should respond to, a criterion with a physical meaning that is as intuitive as possible, and that is able to guarantee optimization in the process operation.

The simplicity of the implementation and operation is an important property for a control methodology to be practicable. The mathematical formulation should be simple and the design parameters small in number and meaningful. Mathematical simplicity implies shorter calculation time, which is important for a real time control system, especially in processes with fast dynamics, where the control period has to be short. A small number of simple and meaningful design parameters help the operator during startup and maintenance of the control system, avoiding the necessity of a profound knowledge of the theory involved.

In the industrial field there is a wide variety of processes with their own particular dynamic characteristics [Shi88]. The dynamics of industrial processes have a multivariable, interactive character and may be of completely different orders and vary from a very fast nature (time constants of the order of milliseconds) to a very slow one (of the order of minutes or hours). Likewise, within a control loop there may exist non-linearities associated with the process itself (pH control processes are a typical example) or with the actuators (for instance saturation, dead zones, etc.).

Another frequent characteristic is the presence of dead times, especially in processes involving matter or energy transportation. Such transportation introduces a time delay between the control action and its effect on the process output variable that could lead to instability or lack of robustness if the proper control method were not applied, especially if the time delay were unknown or variable with time, as occurs in many cases. Likewise, unstable processes may be found which the control system has to stabilize.

Furthermore, in many cases, processes have an unstable inverse nature that will be analyzed later in this book. This becomes an additional problem since, for instance, a process of this kind may require an unbounded control action (obviously impossible to produce in practice) to drive its output variable through a predetermined trajectory to the setpoint.

In this context, a control methodology that attempts to be general purpose has to be capable of being equally applicable to processes which include such diverse characteristics as those mentioned above.

In attempts to solve the problems described in this section, different methodological solutions have been proposed over the last few decades. Obviously, an objective evaluation of these contributions to control theory is difficult in the short term. However, the reality of the industrial applications is an objective criterion in itself. Indeed, only two methodologies have produced a significant industrial impact: the proportional, integral and derivative (PID) negative feedback methodology and the predictive control methodology. The first one introduced, half a century ago, automatic control in the industrial field, but the problems described above were left unsolved. The second one represents a new approach in which these problems find an adequate solution.

1.3 NEGATIVE FEEDBACK METHODOLOGY

Figure 1.3 presents the well-known scheme of PID control systems based on negative feedback methodology. The output variable under control is fed back and subtracted from the setpoint (desired value) in order to obtain an error signal which is used to generate the control action. The way to generate the control signal is extremely simple. The error is multiplied by a constant to obtain a *proportional* control action. Likewise, the integration of the error and its multiplication by a constant may produce an *integral* control action. Finally, based on the derivation of the error, a *derivative* control action may also be generated. The control signal applied to the process may result from the addition of these three actions, and the corresponding controller is usually called the PID.

As mentioned previously, negative feedback methodology heralded the technological era of automatic control. For the first time, productive processes could use a methodological solution for automatic control, of a general character, easy to implement and with satisfactory results in many cases. Where it could be applied, this solution allowed the autocorrection of the process operation, maintaining the output variable around the desired setpoint.

Negative feedback methodology and, more precisely, PID controllers, were an important step forward at the time, becoming widely used, even to the present

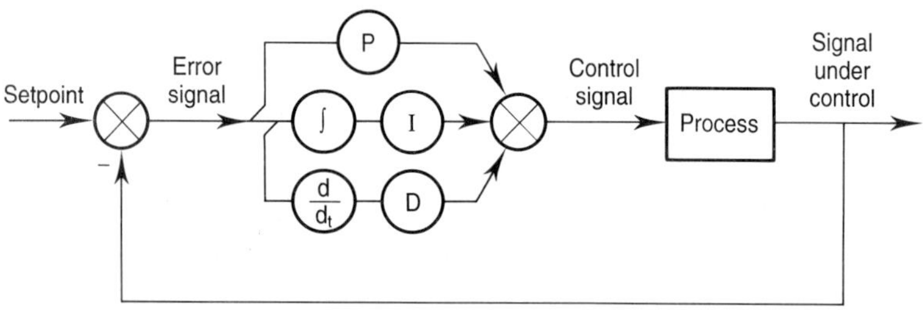

Fig. 1.3. General scheme of PID control systems.

day. As already indicated, their main advantage was their extreme simplicity, which allowed them to be implemented with the technology available at the time, that is to say, pneumatic technology at first and, later, analog electronic technology. This was certainly an important reason for their success.

We may now consider that even if a more advanced control theory, obviously requiring more sophisticated computations, had been developed in the 1940s or 1950s, its practical application would have been impossible due to the limitations of the supporting technologies available at the time.

However, the restrictions of negative feedback methodology were, and are, important. In particular, its application involves, among others, the well-known inherent problem of stability, which we are going to discuss now in a completely intuitive way. The corresponding rigorous analysis of this stability problem, inherent in the application of PID controllers, is a fundamental element in the textbooks used nowadays in technical studies for courses on process control [Oga70, Dor80, Kuo91].

The need for control arises from the fact that the variables we wish to control tend to move away from their desired operation points and, in general, oscillate. These oscillations of the output variable appear, due to the negative feedback strategy, to be inverted at the level of the error signal, leading to the dilemma described in the following paragraphs.

If we want the control system to react with a rapid response, that is, to generate a significant corrective control signal when a certain error is produced, the PID controller will have to produce a control signal that will, in a certain way, result from the amplification of the error signal, amplifying its oscillations at the same time. The amplification of these oscillations at the control signal level may cause their reproduction in the process output, which may lead to such an undesirable effect as the approach of the process to the well-known phenomenon of resonance. To avoid this, a slower response is desirable, at the cost of lower efficiency.

A descriptive example may help to illustrate the possible practical consequences of this problem. Let us consider the case of a climatization process in a building. The temperature in a specific room may be found, for energy saving reasons, to be around 12 °C at seven o'clock in the morning. On the arrival of the employees at eight o'clock, this temperature should have reached a comfortable level. For this purpose, if we install a PID control system for the room, the setpoint could be 20 °C. This control system should have a fast action in order to avoid a low temperature during working hours. In these conditions, the action of a PID control system will lead to the oscillations represented by graph 1 of Figure 1.4. These oscillations around 20 °C will produce frequent and undesirable periods of low temperature.

The immediate solution to avoid these periods of low temperature consists of increasing the setpoint, for instance to 24 °C. However, this produces overheating, as represented by graph 2 of Figure 1.4. These periods of overheating, reaching temperatures of 28 °C, result in, firstly, excessive energy consumption; secondly, this heat excess inside the room results in dry off of the air and thus in a decrease in the comfort of the environment. The problem is very familiar in large office buildings where PID digital systems are installed for climatization control.

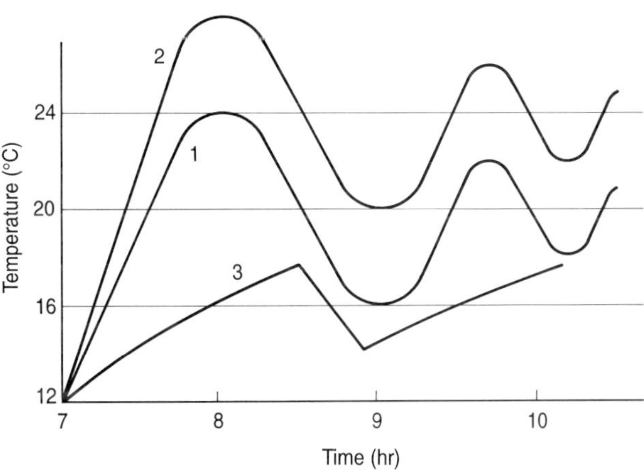

Fig. 1.4. Climatization example.

Oscillations may be reduced by requiring a slower reaction from the PID controller. In this case, the temperature would tend to reach its setpoint slowly but, as represented by graph 3 of Figure 1.4, opening a window would be enough to result in a significant deviation of the variable from its setpoint, leading to prolonged recuperation periods.

The climatization problem is a clear example of the insufficiency of systems based on negative feedback.

Textbooks in the field of classical automatic control point out that the best performance achievable with PID negative feedback control is generally a compromise between rapidity of response and level of oscillation. In this way, they reflect the inherent stability problem that we refer to here.

1.4 COMPUTERS AND PREDICTIVE CONTROL

The limitations of PID methodology justified new developments in control theory which began to intensify from the end of the 1950s. These developments were generally based on mathematical theories that described and acted on the processes in continuous time, that is to say, within the context of analog systems, the optimal control theory being the most representative [AF66, BH69, Ros70, KS72, SW77]. Additionally, the implementation of these developments was limited by the existing analog technology that allowed the application of PIDs.

The introduction of digital computers represented a revolution in the control field similar to those implemented in other areas of human activity. In fact, it offered a suitable tool for the implementation of new theoretical developments that were potentially capable, through more complex computations, of performing more efficient control than that obtained with PIDs. In this context, the main advantage of the controllers, their simplicity, lost its relevance.

On the one hand, already existing theories were translated into algorithmic form [RF58, AW84, FPW90, Kuo92]. On the other hand, research work in continuous time was continued and new developments for discrete time systems were formulated [Lan74, Mon74, NV78]. However, these developments were mostly based on concepts derived within the context of continuous time systems.

This research effort led to an extensive and complex body of control theory, with the potential to be implemented using computers, which, however, turned out to be unable to satisfy the requirements expected from an advanced industrial control strategy. In some cases, the complexity of the theory was excessive, thereby limiting its practical applicability. In other cases, the hypotheses necessary to guarantee stability and robustness were not realistic. Finally, the simplicity sought in some solutions limited or decreased practical performance [Fos73, LW76].

The characteristic common to all these theoretical developments, as already mentioned, was their origin in concepts formulated in continuous time and thus in principle their suitability for implementation using analog technology.

Discrete time modelling of processes introduced, for the first time in control theory, the possibility of predicting in real time the future evolution of process outputs. This capability was of paramount importance in filtering and prediction theory [Kal60]. Later, it was used by the so-called minimum variance control strategy [Ast70], derived in the context of optimal control theory. This strategy can be considered as the actual precursor to the methodology of predictive control, which for the first time fully exploited the possibilities of real time prediction for control purposes.

Before defining the principle of predictive control formally, let us introduce, with the help of a simple example, the concept of the model and show how the use of this concept along with a digital computer will help us to define the mathematical relation between the inputs and outputs of a process and to predict the evolution of the output variables.

1.5 BASIC CONCEPT OF DISCRETE TIME MODEL AND GENERALIZATIONS

If we are asked what the temperature of the room that we are in will be in five minutes, the answer does not seem to be completely obvious. Nevertheless, if we measure the present temperature with a thermometer, reading, for instance, 20 °C, we will be able to deduce that the temperature in five minutes will be lower than 25 °C and higher than 15 °C, because the room's dynamics will not allow the temperature to change any more. For now, we know that the temperature in five minutes will be around 20 °C, but the next question is related to whether it will be above or below this level. If we note that the temperature five minutes ago was 18 °C, the tendency to increase, deduced from it, will lead us to expect the temperature of the room to reach 22 °C in five minutes if this tendency is maintained. This extrapolation is represented in Figure 1.5.

In Figure 1.5, the present instant is given the value 0 and the corresponding temperature is $t(0)$. In the same way, the time instant five minutes ago is given the value -5 and the corresponding temperature is $t(-5)$. Finally, the temperature estimated in $+5$ minutes is $\hat{t}(+5)$. According to the extrapolation shown in Figure 1.5, the temperature $\hat{t}(+5)$ is equal to the temperature $t(0)$ added to the temperature increment between -5 and 0. This relation may be expressed by the equation

$$\hat{t}(+5) = t(0) + t(0) - t(5) = 2t(0) - t(-5) \tag{1.1}$$

However, the hypothesis of a direct extrapolation, such as the one considered above, does not seem to be very satisfactory. If we desire a precise estimation,

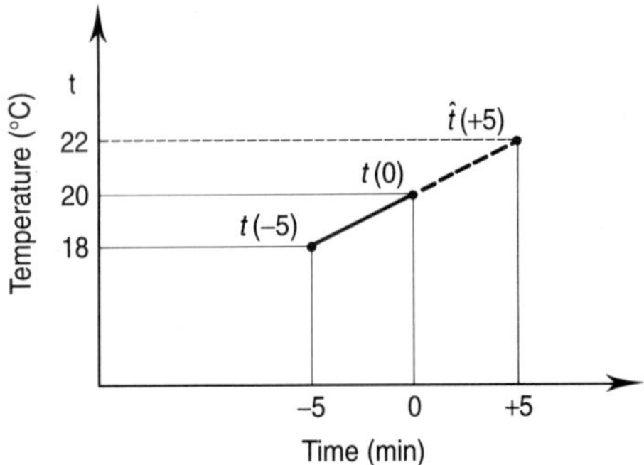

Fig. 1.5. Basic concept of model.

other factors which influence the temperature will need to be taken into account. For instance, if the room is being climatized, the temperature variation will depend on the cool air flow at the present instant and on the flow five minutes ago. If 5 l/min were entering five minutes ago and, at the present moment, the flow is 10 l/min, we may expect a reduction in the tendency of the temperature to increase and, as a result, the temperature will not reach 22 °C but will probably be around 21 °C, as represented in Figure 1.6.

In Figure 1.6, the cool air flow entering five minutes ago is given the value $q(-5)$ and the flow entering at the present instant is $q(0)$. In summary, the temperature in five minutes will depend on (among other things, but still fundamentally) the variables $t(0)$, $t(-5)$, $q(0)$ and $q(-5)$. If we assign weight parameters to these variables in order to evaluate their influence on the estimated temperature, and sum all of them subsequently, we obtain the following temperature estimation:

$$\hat{t}(+5) = a_1 t(0) + a_2 t(-5) + b_1 q(0) + b_2 q(-5) \tag{1.2}$$

where a_1, a_2, b_1 and b_2 are the aforementioned weight parameters. If these parameters are well chosen, the equation given above, even if not exact, will give us an estimation of the temperature in five minutes. Practical experience shows that, in order to maximize the precision of such an estimation, the values of the temperature t and the flow q in equation (1.2) should be chosen as increments or deviations with respect to the values of a steady state operation.

This example has been useful for illustrating the concept of the model in discrete time. In effect, equation (1.2) describes the relation between the flow of cool air and the temperature of the room, which are in this case the output and input variables respectively of the climatization process. Therefore, this equation is a model that describes the behaviour of the process at consecutive time instants.

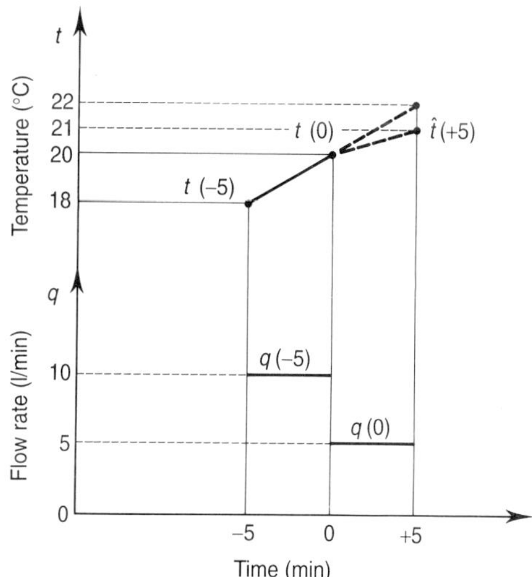

Fig. 1.6. Basic concept of model.

Obviously, this simple example allows generalizations such as those described in the following. In all these generalizations, choice of the variables of the model as increments with respect to steady state values is advisable.

Firstly, in order to generalize the present instant (instant 0) of the proposed model, we may designate it as instant k, where k is an integer ($k = 0, 1, 2, \ldots$) representing this present instant, and in a generic form the different sampling instants too, which are considered to be spaced every five minutes. In this way, instant 0 of the example now becomes instant k, instant + 5 becomes $k + 1$ and instant − 5 becomes $k - 1$. Equation (1.2) may then be written as follows:

$$\hat{t}(k+1) = a_1 t(k) + a_2 t(k-1) + b_1 q(k) + b_2 q(k-1) \tag{1.3}$$

Moreover, we may consider values of temperature and flow corresponding to previous instants, such as − 10 or − 15. Consequently, a model such as (1.3) may be extended thus:

$$\begin{aligned} \hat{t}(k+1) &= a_1 t(k) + a_2 t(k-1) + a_3 t(k-2) + \cdots \\ &\quad + b_1 q(k) + b_2 q(k-1) + b_3 q(k-2) + \cdots \end{aligned} \tag{1.4}$$

which may be expressed as follows:

$$\hat{t}(k+1) = \sum_{i=1}^{n} a_i t(k+1-i) + \sum_{j=1}^{m} b_j q(k+1-j) \tag{1.5}$$

where the coefficients n and m determining the number of terms of temperature and flow are to be taken into account in the model. These coefficients determine the so-called order of the model.

However, it seems obvious that the older the values for temperature and flow, the less will be their influence on the estimated value and, as a result, the most relevant values are those already given in equation (1.2).

Likewise, the influence of other variables, such as the external temperature or the heat given off by electrical appliances, may be considered by the model. The influence of these variables could be taken into account in the same manner as was the flow of cool air. For instance, a model that would consider the influence of the external temperature (t_e) could be expressed as follows:

$$\hat{t}(k+1) = \sum_{i=1}^{n} a_i t(k+1-i) + \sum_{j=1}^{m} b_j q(k+1-j) + \sum_{h=1}^{p} c_h t_e(k+1-h) \tag{1.6}$$

where the coefficient p determines the number of external temperature values to be taken into account in the model and the parameters c_h are the corresponding factors that weight the influence of these values on the estimation made by the model.

This type of model could be used to represent not only climatization processes, but also processes in any other industrial area, as will be described in the following chapters in this book. In this case, the output variable of the process is denoted by y, while the input variable uses the notation u. If the influence of a disturbance is also taken into account, this variable is normally denoted by w. Consequently, a generalized expression of this type of model would be the following:

$$\hat{y}(k+1) = \sum_{i=1}^{n} a_i y(k+1-i) + \sum_{j=1}^{m} b_j u(k+1-j) + \sum_{h=1}^{p} c_h w(k+1-h) \tag{1.7}$$

On the other hand, this type of expression could progress from relating purely scalar variables to relating variables of the vectorial type, leading us to a so-called multivariable model, which is usually represented as follows:

$$\hat{Y}(k+1) = \sum_{i=1}^{n} A_i Y(k+1-i) + \sum_{j=1}^{m} B_j U(k+1-j) + \sum_{h=1}^{p} C_h W(k+1-h) \tag{1.8}$$

where Y, U and W are the output, input and disturbance vectors, respectively, and A_i, B_j and C_h are the matrices that assign weights to the influence of each vector on the estimation given by the model.

So far, we have been considering a linear representation of the process dynamic behaviour because, although processes are basically non-linear, under certain operating conditions they may be approximated locally by linear equations with constant parameters, such as those given above. Different operating conditions for the same real process will lead to different local linear representations, but the set of all of them can be described globally by such equations as those given previously if the weighting parameters or matrices are variable over time, being represented as a function of the sampling time k. Likewise, a term $\Delta(k)$, known as the perturbation vector or signal, could be included in the model in order to match the estimated output $\hat{Y}(k+1)$ with the measured output $Y(k+1)$. This perturbation vector includes the influence of any kind of non-measurable perturbation and noise. Consequently, we may finally allocate the description of the process dynamics to a general expression such as

$$\begin{aligned} Y(k+1) = & \sum_{i=1}^{n} A_i(k) Y(k+1-i) + \sum_{j=1}^{m} B_j(k) U(k+1-j) \\ & + \sum_{h=1}^{p} C_h(k) W(k+1-h) + \Delta(k) \end{aligned} \tag{1.9}$$

We can see in (1.9) that there is one sampling period of time delay between the process input U and the process output Y. This time delay is due solely to the discrete time description of the process dynamics. However, in many processes we may find additional time delays due to their dynamic nature, typical examples being processes involving transportation times. This is why equation (1.9) may be further extended in the following form:

$$\begin{aligned} Y(k+1) = & \sum_{i=1}^{n} A_i(k) Y(k+1-i) + \sum_{j=1}^{m} B_j(k) U(k+1-j-r) \\ & + \sum_{h=1}^{p} C_h(k) W(k+1-h-r_1) + \Delta(k) \end{aligned} \tag{1.10}$$

where r is the time delay (in sampling periods) related to the process input U and r_1 is the time delay related to the disturbance vector W.

The types of model that have been considered in this section are usually called *difference equation models*. They will be used extensively in this book and some basic tools related to them are given in Appendix A.

One finds different names for this kind of model in the literature, such as ARMA (autoregressive moving average), ARMAX (ARMA with exogenous input) and CARMA (controlled ARMA), as well as, considering incremental input output variables in those models, ARIMA, ARIMAX and CARIMA.

In control theory there are other types of discrete time model that may also be used for the prediction of the process output at consecutive time instants. We will consider this issue in more detail in other chapters of this book, and particularly in Appendix B, where the state space representation is considered together with impulse and step response models.

More recently, the representation of dynamic systems by means of neural networks has also been a subject of attention [HSW89, GM90, NP90, HKP91]. The aim in neural network modelling is to find a parameterized structure able to emulate non-linear dynamic process behaviour. Local linear representations have also been considered within this framework in order to obtain a global description of the process dynamics [JF93, ZHDM94], thus providing model predictions of the process in a similar way to those considered for the above difference equation models. This book mainly considers difference equation models, but any other appropriate representation of the process dynamics could be used as a vehicle tool for prediction [SBM91] in the developments that follow.

At this introductory stage, in the following section we will use model (1.2) to illustrate the principle of predictive control and its application.

1.6 PREDICTIVE CONTROL, ORIGIN AND BASIC CONCEPTS

The methodology of predictive control was introduced in 1974 in a doctoral thesis [Mar74]. Subsequently, the original basic principle was formally defined in a US patent [Mar76a]. This principle may be expressed in the following way: 'Based on a model of the process, predictive control is the one that makes the predicted process dynamic output equal to a desired dynamic output conveniently predefined'. Defined in this way, predictive control is straight common sense and its objective of control has a clear physical meaning.

In the context of the climatization example considered previously, the application of the predictive control principle implies that the cool air flow rate to be introduced into the room has to be chosen in order to make the predicted temperature in five minutes $\hat{t}(+5)$ equal to the desired temperature for this future instant. Using the notation $t_d(+5)$ for this desired temperature, we may conclude that predictive control results in

$$\hat{t}(+5) = t_d(+5) \tag{1.11}$$

In order to compute the predictive control signal we may use the model defined by equation (1.2) which determines the predicted temperature within five minutes. Substituting $t_d(+5)$ for $\hat{t}(+5)$ in equation (1.2) and solving for $q(0)$, we obtain the computation of predictive control as follows:

$$q(0) = \frac{t_d(+5) - a_1 t(0) - a_2 t(-5) - b_2 q(-5)}{b_1} \tag{1.12}$$

If the prediction made by model (1.2) were correct, the application of predictive control (1.12) every five minutes would make the sequence of temperatures, measured every five minutes, equal to a sequence of desired values, that is to say, the measured temperatures would follow a desired trajectory. Obviously, this desired trajectory may be chosen conveniently.

The predictive control strategy described above may be generalized and implemented through a *predictive model* and a *driver block*, as presented in the block diagram of Figure 1.7.

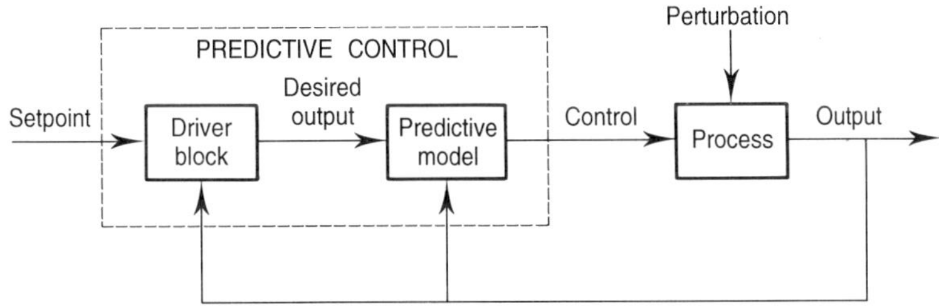

Fig. 1.7. Basic block diagram for predictive control.

As already described, the predictive model generates, from the previous input and output (I/O) process variables, the control signal that makes the predicted process output equal to the desired output. This will be generated by the driver block taking its physical feasibility and the desired dynamics into account. In this way, the desired output will belong to a trajectory that will drive the process output to the setpoint in a satisfactory, fast and smooth manner, without any offset or excessive control action.

1.7 THE NEED FOR ADAPTATION

Obviously, the performance of the predictive control system will depend significantly on the precision of the prediction made by the model. Indeed, if the prediction of the model is correct, when using predictive control the evolution of the process variables should be driven perfectly.

A model capable of making exact predictions would be able to match the process outputs if it received the same inputs as the process. When the predictions are not satisfactory, due to 'unadjusted' model parameters, it would be convenient to have an *adaptation mechanism* able to adjust the parameters of the model from the error in the comparison between process and model outputs. The corresponding scheme is represented in Figure 1.8.

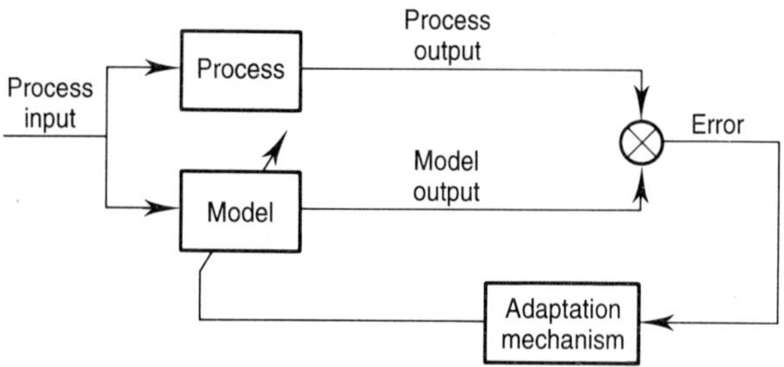

Fig. 1.8. Basic scheme for an adaptive system.

The problem of producing an adaptive system as presented in Figure 1.8 has generated considerable interest among researchers. Some authors have approached the problem from the perspective of optimization, that is to say, from the perspective of the minimization of an index or function, generally in quadratic form [AE71, Men73, Lju87], of the previously mentioned error. Another approach [Mar76b, Mar84, MSF84, Mar86, CMSF88] considers the solution from the perspective of stability and has led to the result summarized in the following paragraph.

After a given time instant, the adaptation mechanism ensures that the absolute value of the error considered in Figure 1.8 will be bounded by the smallest possible limit, according to the level of noise and perturbations acting on the process. If no noise or perturbations are involved, the error tends asymptotically towards zero.

1.8 ADAPTIVE PREDICTIVE CONTROL SYSTEM (APCS)

1.8.1 Block diagram

When the predictive control strategy described in Section 1.6 is combined with an adaptive system such as the one considered in the preceding section, the global adaptive predictive scheme represented in Figure 1.9 is obtained.

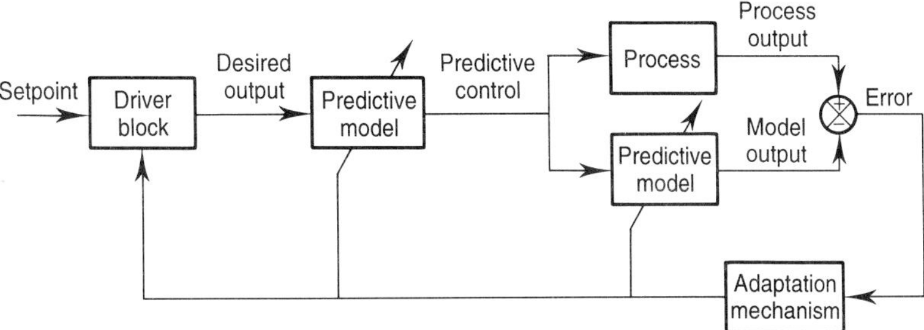

Fig. 1.9. Global adaptive predictive control scheme.

The predictive model, as observed in Figure 1.9, performs the following two roles simultaneously:

- On the one hand, it receives the same input signal as the process and generates the model output that, compared with the process output, allows the adaptation mechanism to adjust its parameters in order to obtain the stability result considered in the preceding section.

- On the other hand, the predictive model calculates the control signal from the desired output generated by the driver block. This computation, carried out according to the predictive control principle, renders the desired output equal to the previously considered model output, which is the predicted process output.

Consequently, when the adaptation mechanism makes the difference between the process and the model outputs tend towards zero, the difference between the process output and the desired output also tends towards zero. In this way global stability of the adaptive predictive control system is achieved. The mathematical theory that supports the stability results of APCS has appeared in a number of publications [Mar76b, Mar84, MSF84, Mar86, CMSF88].

The diagram given in Figure 1.9 may be simplified to at the one given in Figure 1.10, which is the diagram generally used to represent the methodology. The functional description of the blocks in this diagram may be summarized as follows:

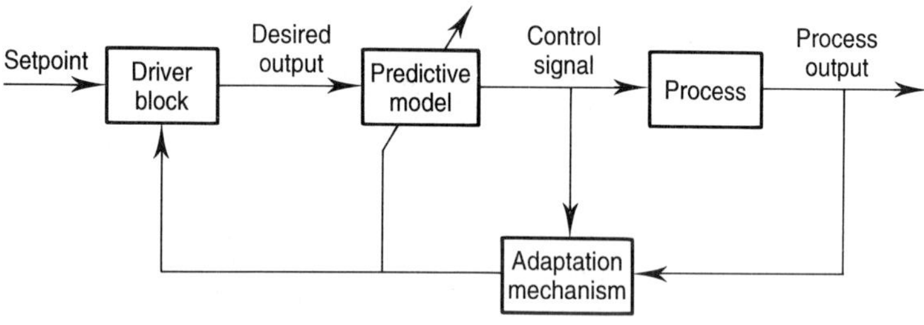

Fig. 1.10. Block diagram of the adaptive predictive control system.

- **The driver block**. Generates the desired output trajectory that will guide the process output to the setpoint in an optimal way.
- **The predictive model**. Calculates the control signal that ensures that the predicted process output is contained within the desired trajectory generated by the driver block.
- **The adaptation mechanism**. Adjusts the predictive model parameters from the prediction errors, in order to make these errors tend towards zero efficiently. Likewise, it informs the driver block of the deviation of its outputs with respect to the desired trajectory. In this way, the driver block may redefine the desired trajectory in a manner that is consistent with the actual process output.

1.8.2 Application example

In the following, we will describe, in an illustrative way and by means of Figure 1.11, the application of adaptive predictive control to the previously considered climatization problem.

At the initial control instant k (7 o'clock in the morning) of Figure 1.11, the driver block (represented in Figure 1.10) receives information on the one hand of the setpoint (20 °C) and, on the other, of the present value of the temperature,

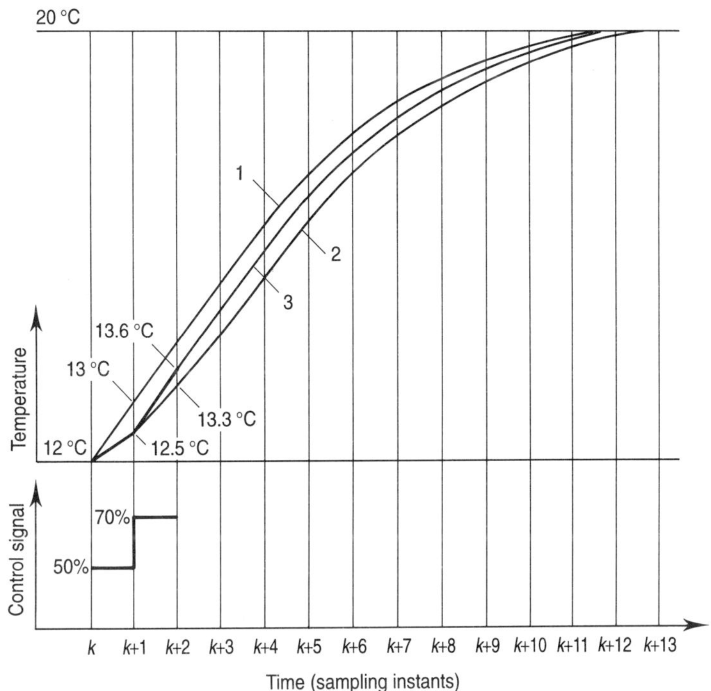

Fig. 1.11. Illustrative example of adaptive predictive control.

that is, 12 °C. From this information, the driver block generates the desired trajectory at instant k that the temperature under control should follow in order to attain the setpoint. This is shown by trajectory 1 in Figure 1.11.

According to this first desired trajectory, the temperature should increase within the next control period of five minutes from 12 °C to 13 °C. This last value is the desired output for the next control instant $k+1$, given by the driver block to the predictive model. From this desired output, the predictive model calculates the control signal at instant k that makes the predicted temperature at instant $k+1$ equal to 13 °C. Let us consider that this control signal corresponds to a warm air valve opening of 50%, as indicated in Figure 1.11.

Once the predictive control signal has been applied, the system waits until the next control instant, when the room temperature will be measured again. Let us assume that the measured temperature at instant $k+1$ is 12.5 °C instead of the desired 13 °C. Thus, there has been a prediction error of 0.5 °C.

In this situation, the adaptation mechanism reacts on two different levels. On the first level, it adjusts the parameters of the predictive model in order to cause the prediction error to tend towards zero. On the second level, it informs the driver block that the measured temperature is only 12.5 °C instead of 13 °C as expected according to the first desired trajectory.

Consequently, the driver block generates a second desired trajectory, taking the real evolution of the process variable into account. This second trajectory is represented by trajectory 2 of Figure 1.11 and, as can be seen, it starts from the last measured temperature value, that is, 12.5 °C.

According to this second desired trajectory, the temperature of the room should change from 12.5 °C to 13.3 °C during the new control period and the predictive model calculates the new control signal from this value. Let us assume that the predictive model, already adjusted at instant $k+1$, computes a control action this time corresponding to a valve opening of 70%, as indicated in Figure 1.11.

The system waits again for the process reaction and, more precisely, for the temperature achieved at the following control instant. Let us assume that the new temperature is 13.6 °C. Thus, this time, the measured value has exceeded the predicted one. The system will behave according to the same pattern: it will adjust the parameters of the predictive model, inform the driver block of the new measured temperature, and generate a new desired output trajectory (trajectory 3 of Figure 1.11) and a new desired output value, which will be used to compute a new predictive control action.

The result of the operation of the system at each control instant is that the prediction error tends rapidly towards zero and the different desired trajectories generated by the driver block converge into one unique trajectory. This unique trajectory is the one finally followed by the process variable, in this case the temperature, to achieve the setpoint of 20 °C.

1.9 FROM THE CONCEPTS TO THE PRACTICAL APPLICATION

The basic concepts of adaptive predictive control systems have been introduced in the preceding sections. The development of these concepts into an industrial system for widespread use in different areas has required intensive research and development effort. This effort has involved the following:

- theoretical developments in the stability, robustness, efficiency and generality of the methodology, as applied in an industrial environment;
- applications to simulations that reproduce complex industrial processes and

their challenging control problems;

- applications to industrial pilot plants;
- experimental applications to critical control loops of operational industrial plants;
- development, realization and exhaustive testing of prototypes that have been designed to satisfy the conversational and control requirements for an industrial control system;
- definitive implementation of APCS systems in industrial plants.

The following chapters of this book will present the most relevant results in the attempt to progress from theoretical developments to the optimization of industrial plants through a standard product.

Part 1

THE THEORY

Chapter 2

BASIC STRATEGY OF PREDICTIVE CONTROL

2.1 INTRODUCTION

With this chapter we open the part of the book that is devoted to the theoretical concepts of the methodology. Since the main purpose is to present these concepts in a didactic way, we will consider a single-input, single-output (SISO) formulation in order to simplify the mathematical developments. The theory for multivariable processes is essentially no more complicated than for the SISO case, being a natural extension, as illustrated in Appendix D. The multivariable case is also considered in Appendix B and applications are described in Chapters 6 and 9.

In this chapter we will focus on the mathematical formulation of the basic strategy of predictive control as it was originally proposed. Our primary aim will be to describe the role of the predictive model in the computation of the control signal and the concepts defining the driver block. These concepts will be developed introducing the *projected desired trajectory* and the *driving desired trajectory*, whose respective roles will be illustrated by an example. Another example will be used to illustrate the need to extend the basic predictive control strategy. We will develop this extension in the following chapter. A brief outline of the basic concepts of systems analysis used in this chapter and in Chapter 3 can be found in Appendix A.

2.2 BASIC STRATEGY OF PREDICTIVE CONTROL

2.2.1 The basic strategy

Predictive control, as presented in Chapter 1, consists of calculating the control action that will make the predicted output equal to a conveniently selected desired output. The *basic strategy of predictive control* implies the direct application of this principle in a single-step prediction. In order to do this and taking into account that predictive control is formulated in discrete time, at each sampling instant k the desired output for the next instant $k+1$, which is denoted by $y_d(k+1|k)$, is calculated. Once this value is available, the basic predictive control strategy can be summarized by the following condition:

$$\hat{y}(k+1|k) = y_d(k+1|k) \tag{2.1}$$

where $\hat{y}(k+1|k)$ is the output predicted at instant k for the next instant $k+1$. The control $u(k)$ to be applied at instant k is the one meeting the above condition. This control strategy will be realized by means of a predictive model and a driver block according to the predictive control principle.

Imposing condition (2.1) assumes that the control $u(k)$, applied at instant k, does not influence the output until the next time instant, $k+1$. This means that there exists a time delay equal to one sampling period. As is well known, when the process is under digital control, this basic time delay always exists and is often referred to as the *discretization delay* [FPW90, Kuo92]. In many other cases the process dynamics itself has a time delay, which can be represented by an integer number r of sampling periods and referred to as *pure delay*. In such a case, the total time delay is $r+1$ and the prediction is calculated at $k+r+1$ instead of at $k+1$, thus imposing the condition

$$\hat{y}(k+r+1|k) = y_d(k+r+1|k) \tag{2.2}$$

For the sake of simplicity, we will assume no time delay for the remainder of this chapter.

2.2.2 The predictive model

In order to predict the process output at each sampling instant, we can use a mathematical model describing the dynamics of the open loop that we wish to control. The basic components of a computer controlled open loop are the process, the sensors which measure the output, the signal converters (analog/digital and digital/analog) and the actuators applying the control action. If we consider these components to be connected as shown in Figure 2.1, we may include all of them in a block which operates at discrete time instants. As input it has the control $u(k)$ at instant k and as output it has the process output $y(k)$ measured at instant k.

As we have already discussed in Chapter 1, there exist different kinds of mathematical model which can be used to relate the output $y(k)$ to the input $u(k)$. In this chapter we will use the type of model that was considered in the preceding chapter, which can be written in the form

$$y(k) = \sum_{i=1}^{n} a_i y(k-i) + \sum_{i=1}^{m} b_i u(k-i) + \Delta(k) \tag{2.3}$$

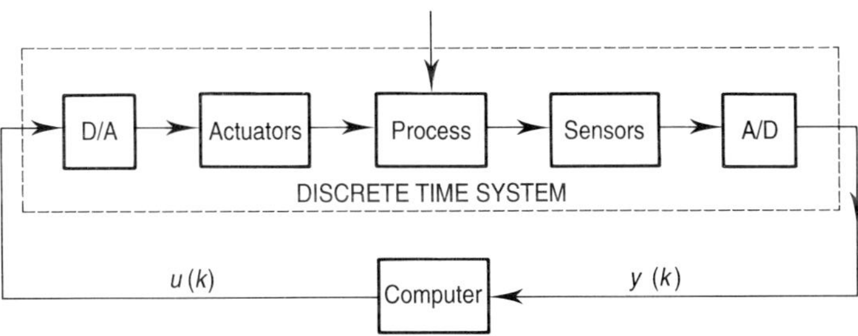

Fig. 2.1. Discrete time modelling of the control loop.

Note that, purely for the sake of simplicity, we do not include the presence of measurable disturbances in the model, but we do retain the perturbation signal $\Delta(k)$ in the description.

Since, in general, we cannot know *a priori* the precise dynamics of the open loop described in Figure 2.1, we will use an estimation of (2.3) in the following form as a predictive model:

$$\hat{y}(k+1|k) = \sum_{i=1}^{\hat{n}} \hat{a}_i y(k+1-i) + \sum_{i=1}^{\hat{m}} \hat{b}_i u(k+1-i) \tag{2.4}$$

where $\hat{y}(k+1|k)$ is the output predicted at instant k for the next instant $k+1$. As can be seen, the model of (2.4) does not include the perturbation signal $\Delta(k)$ since this is unknown. The number of parameters $\hat{a}_i$ and $\hat{b}_i$ and their values may differ from those given in equation (2.3). Moreover, the value of these parameters may be updated at each sampling instant k by means of the adaptation mechanism included in the APCS formulation. Another essential feature of the predictive model is that the prediction for instant $k+1$ is made using the information on the outputs $y(\cdot)$ and the inputs $u(\cdot)$ known at instant k and at preceding instants. Due to the effect of the disturbances and/or discrepancies between the parameters of the predictive model and those of the system under control, it is possible that the prediction $\hat{y}(k+1|k)$ will differ from the actual output, which will be measured later, at instant $k+1$. However, the new prediction at $k+1$ for instant $k+2$ will be made with the predictive model of (2.4) using the actual output measured at $k+1$ as the initial condition instead of the output that was predicted for this instant. This feedback of the measured output in the predictive model is essential to the effectiveness of predictive control, as the reader will see later on.

2.2.3 The driver block: concept of projected desired trajectory

Within the scope of the predictive control strategy, the driver block has the role of generating the desired output. As a design principle, we require the desired output to belong to a trajectory which, starting from the actual values of the process output, would cause it to reach the setpoint in a smooth manner without abrupt control actions but, at the same time, rapidly and without overshoots. This trajectory, which is redefined at each sampling instant k, is called the *projected desired trajectory* (PDT_k).

One very simple and efficient way of generating this kind of trajectory consists of using the output of a stable model having the setpoint as input and the actual process outputs as initial conditions. By way of example, we may consider a discrete time model of the form

$$y_d(k+j|k) = \sum_{i=1}^{p} \alpha_i y_d(k+j-i|k) + \sum_{i=1}^{q} \beta_i y_{sp}(k+j-i) \qquad (j = 1,2,3,\ldots) \tag{2.5}$$

where

$$y_d(k+1-i|k) = y(k+1-i) \qquad (i = 1,\ldots,p)$$

For instance, a reasonable way of choosing the parameters p, q, α_i and β_i is to make the response to a constant setpoint y_{sp}, starting from initial conditions $y(k)$ and $y(k-1)$, have the form shown in Figure 2.2. This kind of response is typical of a second-order model [Oga70] with critical damping and static gain equal to one. The time constant of this model will determine the velocity with which this trajectory tends to the setpoint.

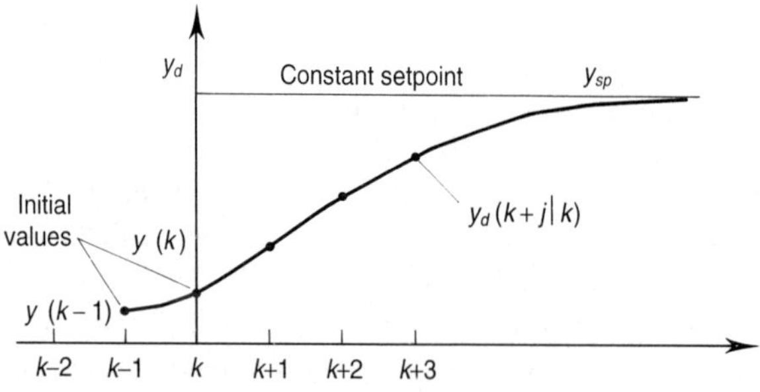

Fig. 2.2. Projected desired trajectory PDT_k.

At instant k the driver block selects the corresponding value of the PDT_k

$$y_d(k+1|k) = \sum_{i=1}^{p} \alpha_i y(k+1-i) + \sum_{i=1}^{q} \beta_i y_{sp}(k+1-i) \tag{2.6}$$

as the desired output for $k+1$.

The fact that the PDT_k is redefined at each sampling instant k introduces a level of feedback of the process output in the generation of the desired output which has been shown to be a key factor in practical applications such as those described in subsequent chapters of this book. In fact, the desired output trajectory could be calculated using just the model of (2.5) recursively from an initial instant $k = 0$ without any kind of redefinition at the following instants $k = 1,2,\ldots$ However, in such a case, the desired output would not take the actual process output into account. Therefore, due to a variety of factors, the desired output would diverge from the process output and become physically unrealizable, which could lead to instability. Such a design of the driver block would be similar to the concept of the reference model used in model reference adaptive systems [Lan74].

The driver block is an essential concept in predictive control. The lack of it would reduce the basic strategy of predictive control to the minimum variance control strategy [Ast70].

2.2.4 Control law of the basic strategy. Example and discussion

Once the predictive model and the driver block have been defined by equations (2.4) and (2.6), respectively, we only need to impose condition (2.1) to obtain the control $u(k)$ in the form

$$u(k) = \frac{y_d(k+1|k) - \sum_{i=1}^{\hat{n}} \hat{a}_i y(k+1-i) - \sum_{i=2}^{\hat{m}} \hat{b}_i u(k+1-i)}{\hat{b}_1} \tag{2.7}$$

Equations (2.6) and (2.7) describe the following two basic operations for the computation of the predictive control which are repeated at each sampling instant k:

1. Calculate the desired output $y_d(k+1|k)$ by means of (2.6).

2. Calculate the control $u(k)$ by means of (2.7).

Next we describe the application of this strategy to the control of a continuous time process whose transfer function is

$$G(s) = \frac{1}{(1+s)(1+4s)} \tag{2.8}$$

and whose step response is plotted in Figure 2.3.

The control law of (2.7) is applied with a sampling period of two seconds for which the difference equation describing the process in discrete time and obtained by the procedure outlined in Appendix A (Section A.1.4) takes the form

$$y(k) = a_1 y(k-1) + a_2 y(k-2) + b_1 u(k-1) + b_2 u(k-2) \tag{2.9}$$

with

$$\begin{array}{ll} a_1 = 0.7419 & a_2 = -0.0821 \\ b_1 = 0.2364 & b_2 = 0.1038 \end{array} \tag{2.10}$$

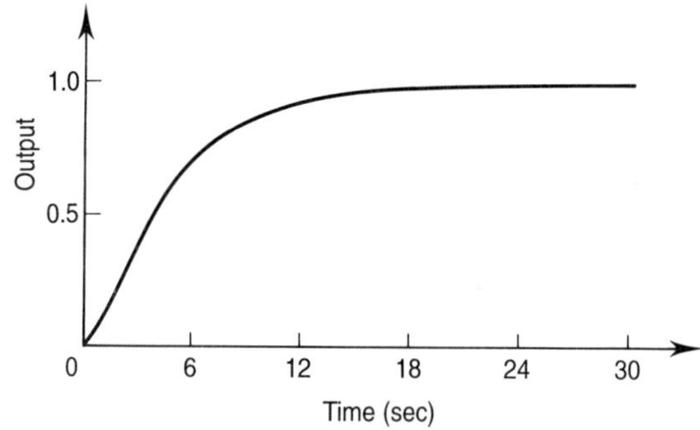

Fig. 2.3. Step response of process (2.8).

In this example, we will take as our predictive model the model of the process itself by assuming that there are no modelling errors, that is, $\hat{n} = \hat{m} = 2$ and parameters $\hat{a}_i$ and $\hat{b}_i$ in (2.4) are equal to those of process (2.10). However, we assume that the actuator limits the increments of the control action applied to the process. In order to obtain the desired output at each sampling instant k, we use a model such as that of (2.6), with the following parameters:

$$\begin{array}{ll} \alpha_1 = 0.667 & \alpha_2 = -0.111 \\ \beta_1 = 0.333 & \beta_2 = \ 0.111 \end{array} \tag{2.11}$$

Our intention now is to follow the steps involved in a few sampling instants to see how the control strategy operates. Assume that we are at the initial instant $k = 0$, the process being at equilibrium, with input and outputs equal to zero. From this situation, we apply the predictive control law of (2.7) with the purpose of driving the process output to the setpoint, which is assumed equal to one. The desired output for the next instant is calculated by (2.6), using the setpoint and the process outputs at instants $k = 0$ and $k = -1$ (both are 0), resulting in $y_d(1|0) = 0.333$. As we have already discussed, the value of $y_d(1|0)$ is on a projected desired trajectory evolving towards the setpoint according to the dynamics chosen when defining the parameters of equation (2.6). This trajectory is labelled as PDT_0 in Figure 2.4.

With the value of $y_d(1|0)$ and using the control law of (2.7), a control value $u(0) = 1.41$ results. However, the control actually applied to the process is not the computed value but, rather, is 0.4 since this is the value of the incremental saturation limit of the actuator. Knowing this limit, $u(0)$ is redefined as equal to 0.4 in order to be properly used in future calculations of the predictive control (2.7).

Obviously, if there were no control limits and the predictive model were an ideal representation of the process (as has been assumed in this example), the predicted output $\hat{y}(1|0)$, which had been set equal to the desired output $y_d(1|0)$, would be equal to the output $y(1)$ that would be measured at the next instant. However, in general, the prediction would not be correct due, among other things, to control limits and modelling errors. In such a case, the output $y(1)$ measured at $k = 1$ would not be equal to the desired output as had been planned. This happens in this example, where the output measured at instant 1 is $y(1) = 0.197$. If this error did not exist, we could assume a value over the same projected desired trajectory PDT_0 as the desired output $y_d(2|1)$ at the next instant. However, the existence of this error suggests leaving PDT_0 and redefining a new PDT_1 starting from the measured value $y(1)$, as can be observed in Figure 2.4. In PDT_1 we find the value of the desired output $y_d(2|1) = 0.576$, which is calculated analogously by means of (2.6). Using this value, equation (2.7) gives the next control action $u(1) = 1.73$. The control actually applied to the process, again due to the incremental limitations of the actuator, is only 0.8 and leads to an output $y(2) = 0.395$. This differs again from the desired output and requires redefinition of a new PDT_2 starting from $y(2)$, and so on at each consecutive sampling instant.

2.2.5 The concept of driving desired trajectory

From the behaviour of the driver block, as described above we can introduce the concept of the *driving desired trajectory* (DDT). This is produced from the first values of each of the projected desired trajectories that are defined at the consecutive control instants k, as illustrated in Figure 2.4.

Then the DDT will be generated point-by-point in real time and, from its values, the control action will be computed according to the predictive control principle. Consequently, this trajectory is the one which has to guide the process output to the setpoint in the desired way: rapidly, without oscillations and,

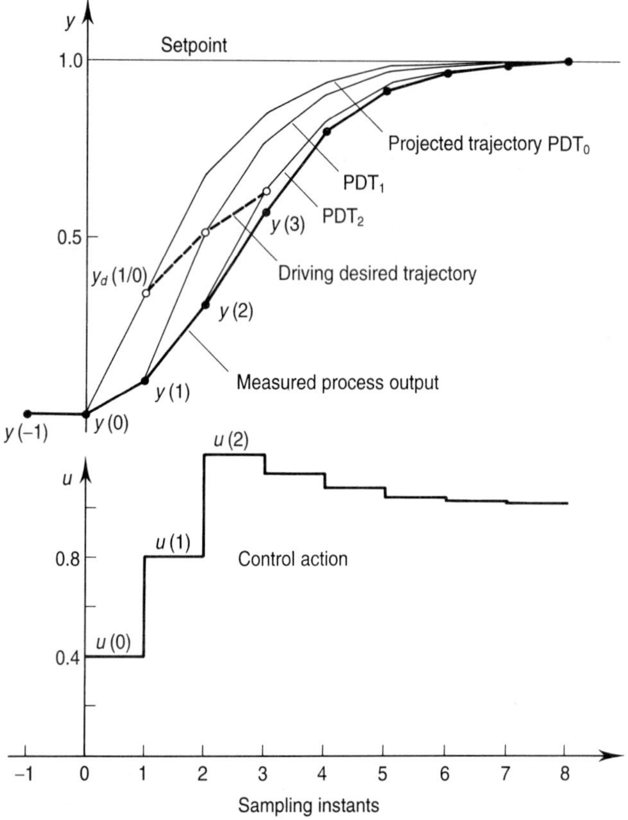

Fig. 2.4. Desired trajectories and process output under predictive control.

moreover, compatible with a bounded control action. Figure 2.4 shows the final evolution of the process output during consecutive control instants. Notice that the trajectory made by the measured values of the process output approaches the driving desired trajectory and finally converges to the setpoint. In the same way, we see the predictive control action approach a steady value with increments that are already compatible with the actuator limits.

The design of the projected desired trajectory at each instant k can be similar to the one proposed in Section 2.2.3 or it can accord with some other criteria, as we consider in the following chapter. But this design should always generate a driving desired trajectory that is able to guide the process output to the setpoint in a satisfactory manner.

2.3 NEED FOR AN EXTENSION TO THE BASIC STRATEGY

In the procedure described in the preceding section, we must point out that the projected desired trajectory at each instant k has been generated independently of the dynamic nature of the process. In fact, the parameters α_i and β_i, which define the form of the projected desired trajectory at each instant k, have been selected with no relation to the parameters a_i and b_i defining the process dynamics or to $\hat{a}_i$ and $\hat{b}_i$ defining the predictive model. In spite of this fact, the process used in the preceding example belongs to a class of processes for which, if the prediction is correct, the predictive control law of (2.7) generates a control sequence which is bounded, no matter what kind of trajectory the process is required to follow, the only condition being that this trajectory be bounded. However, another class of processes exists, for which the absolute value of the control sequence given by (2.7) can increase unboundedly over time for the process to follow a desired trajectory generated as we have done in the preceding section. Processes with an unstable inverse will behave in such a manner, that is, while the output is bounded, the input may be unbounded. We now illustrate this situation using an example.

Consider a continuous time process described by the transfer function

$$G(s) = \frac{1 - 4s}{(1 + s)(1 + 4s)} \tag{2.12}$$

whose step response is plotted in Figure 2.5.

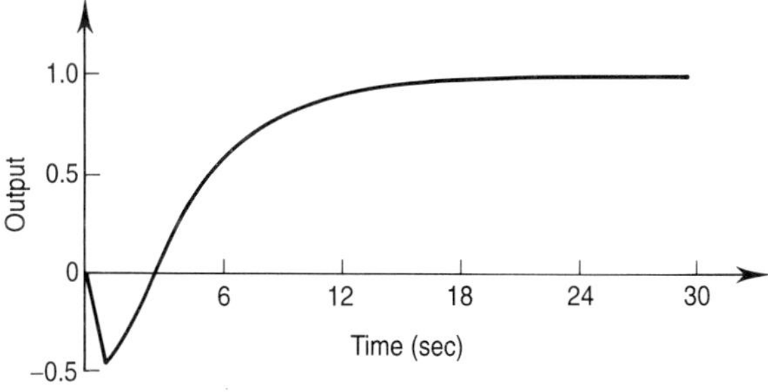

Fig. 2.5. Step response of process (2.12).

This process approaches the steady state in a similar way to that considered in Section 2.2. But, unlike the preceding case, the process here has an unstable inverse. The response of Figure 2.5 is representative of processes with this characteristic which have a response that is initially negative to a positive control action.

We have used a sampling period of two seconds for the application of the control law (2.7) and consecutive projected desired trajectories have been defined by equation (2.6) with parameters (2.11). Also in this example we assume that the predictive model is the same as the discrete time model of the process, which has the form of (2.9) with parameters

$$\begin{aligned} a_1 &= 0.7419 \qquad & a_2 &= -0.0821 \\ b_1 &= -0.391 \qquad & b_2 &= 0.7321 \end{aligned} \tag{2.13}$$

Moreover, we now consider that the actuator has no incremental limits, so the application of predictive control is made under ideal conditions. Figure 2.6 shows what happens in this case, starting from instant $k = 0$.

The application of predictive control under these ideal conditions makes all the consecutive projected desired trajectories coincide with the driving desired trajectory, as is illustrated in Figure 2.6. Unlike the example given in Section 2.2 in which the presence of control limits produced an error between the desired and the measured outputs at each time instant, in this case the desired output is equal to the measured output at each instant. However, it is interesting to follow the evolution of the control signal required to make the desired and measured outputs equal. At $k = 0$, the value $y_d(1|0) = 0.33$ is obtained by means of (2.6) and this value produces, by means of (2.7), a negative control value $u(0) = -0.851$ for the process output $y(1)$ to reach the value $y_d(1|0)$. The control required to obtain a value of $y_d(2|1) = 0.667$ at the next instant will also be negative, $u(1) = -2.659$, but greater in absolute value. The value required to obtain $y_d(3|2) = 0.852$ is $u(2) = -5.95$. Finally, we conclude that the control sequence is unbounded.

This feature can be interpreted if we look at Figure 2.6. It shows those trajectories predicted at each control instant that would be obtained if the control computed at the corresponding instant were applied indefinitely to the process. For example, the predicted trajectory 0 is the one obtained from instant 0 if the value $u(0) = -0.851$ were applied permanently to the process. Obviously, since $u(0)$ has been calculated from the basic predictive control law of (2.7), the predicted trajectory 0 will coincide with the driving desired trajectory at instant 1; but the trajectories separate from each other starting from instant 1. The predicted trajectory 1, defined from instant $k = 1$, is equal to the driving desired trajectory at $k = 2$ for the same reason as given above, but from $k = 2$ it differs, approaching a steady value that is further removed from the setpoint than is the steady value reached by the predicted trajectory 0. Similar comments can be made for the predicted trajectory at $k = 3$.

To summarize, the control signal given by the control law of (2.7) at each

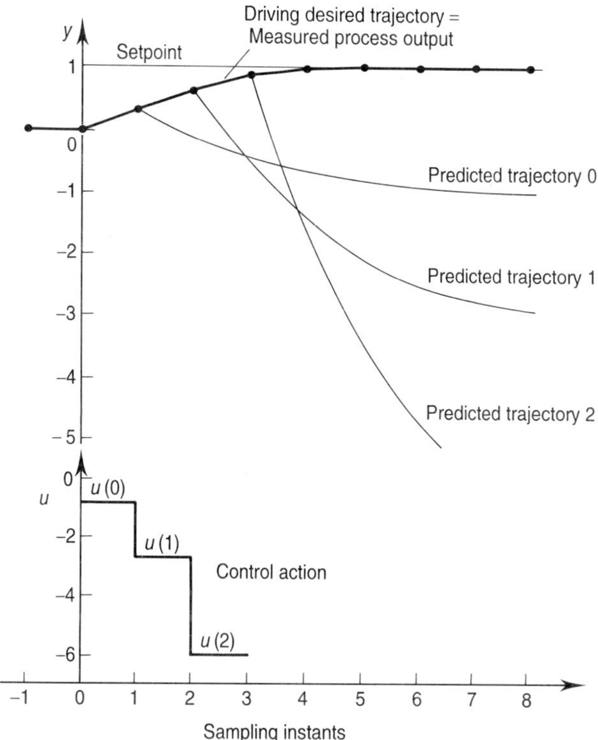

Fig. 2.6. Driving desired trajectory, predicted trajectories and unstable control action.

sampling instant k is able to ensure that both the predicted trajectory and the driving desired trajectory coincide at $k+1$. However, the consecutive predicted trajectories differ increasingly from the setpoint, and the control requires increasingly negative increments, as can be observed in Figure 2.6.

The behaviour described in this example can readily be interpreted using the classical concepts of stability for discrete time systems as described in Appendix A and used in the following.

By applying the Z transform to the control law of (2.7) we have

$$y_d(z) = \sum_{i=1}^{n} a_i z^{-i+1} y(z) + \sum_{i=1}^{m} b_i z^{-i+1} u(z) \tag{2.14}$$

where we have assumed that $\hat{n} = n$, $\hat{m} = m$, $\hat{a}_i = a_i$ and $\hat{b}_i = b_i$ because, as has been assumed in the example, there are no discrepancies between the predictive model and the process model.

By also applying the Z transform to the process model of (2.3) without the disturbance vector, we obtain

$$y(z) = \sum_{i=1}^{n} a_i z^{-i} y(z) + \sum_{i=1}^{m} b_i z^{-i} u(z) \tag{2.15}$$

or, equivalently, the discrete transfer function

$$\frac{y(z)}{u(z)} = \frac{B(z^{-1})}{A(z^{-1})} = H(z^{-1}) \tag{2.16}$$

where A and B are polynomials of the form

$$\begin{aligned} A(z^{-1}) &= 1 - \sum_{i=1}^{n} a_i z^{-i} \\ B(z^{-1}) &= \sum_{i=1}^{m} b_i z^{-i} \end{aligned} \tag{2.17}$$

By comparing (2.14) and (2.15), we readily obtain

$$y(z) = z^{-1} y_d(z) \tag{2.18}$$

Substitution of (2.16) into (2.18) gives

$$u(z) = \frac{A(z^{-1})}{B(z^{-1})} z^{-1} y_d(z) \tag{2.19}$$

Equation (2.18) guarantees that the sequence of outputs $y(\cdot)$ is equal, with a delay of one sampling period, to the sequence of desired outputs $y_d(\cdot)$. On the other hand, (2.19) is the inverse of the process transfer function of (2.16). Therefore, if the polynomial $B(z^{-1})$ has roots with modulus $|z| > 1$, then the process has an unstable inverse and, consequently, an unbounded sequence of values of $u(\cdot)$ will be necessary for the process to follow an arbitrary sequence of values of $y_d(\cdot)$. Consequently, in a case such as this, the control, which in theory is able to drive the process to the setpoint, is physically unrealizable in practice.

In the preceding analysis we have assumed a sequence $y_d(\cdot)$ completely arbitrarily. However, as we have illustrated in the example, it may be generated by an equation such as (2.6). In the following, we will extend the analysis to include this way of generating the desired output.

We may consider increasing the order of the driver block equation (2.6) to obtain the extended expression

$$y_d(k+1|k) = \sum_{i=1}^{p} \alpha_i y(k+1-i) + \sum_{i=1}^{q} \beta_i y_{sp}(k+1-i) \tag{2.20}$$

By applying the Z transform to (2.20) we have

$$y_d(z) = \Phi(z^{-1})y(z) + \Delta(z^{-1})y_{sp}(z) \tag{2.21}$$

where Φ and Δ are polynomials of the form

$$\begin{aligned} \Phi(z^{-1}) &= \sum_{i=1}^{p} \alpha_i z^{-i+1} \\ \Delta(z^{-1}) &= \sum_{i=1}^{q} \beta_i z^{-i+1} \end{aligned} \tag{2.22}$$

By substituting (2.21) into (2.18) and using (2.16) we obtain

$$u(z) = \frac{A(z^{-1})\Delta(z^{-1})}{B(z^{-1})[1 - z^{-1}\Phi(z^{-1})]} z^{-1} y_{sp}(z) \tag{2.23}$$

Equation (2.23) gives the closed loop relationship between the setpoint and the predictive control action and shows clearly the effect of the dynamic terms within the design of the driver block. In particular, we can see that the unstable zeros in $B(z^{-1})$ could be cancelled by designing $\Delta(z^{-1})$ to include the same unstable zeros. In this way, an unbounded control sequence could be prevented. In our example, equation (2.6) has been defined without taking the unstable zero of the process into account, thus resulting in an unbounded control sequence.

However, the requirement of accurate knowledge of the unstable zeros of the process renders the above cancelling procedure a theoretical solution only which cannot be used in a practical context. The following chapter presents a general solution to the problem by extending the way in which predictive control may be applied.

Chapter 3

EXTENDED STRATEGY AND PROPERTIES

3.1 INTRODUCTION

The need to solve the instability problem related to the application of the basic strategy of predictive control to processes with an unstable inverse was the motivating factor in the development of a more general form of application of predictive control, which additionally could exploit all its potential. This form is referred to as the *extended strategy* in a dual sense. First, while the basic strategy considered only a single-step prediction of the process output, the extended strategy considers a multistep prediction horizon. Second, the extended strategy verifies the principle of predictive control in the first prediction step, as does the basic strategy.

This chapter will present the general framework of the extended strategy of predictive control, based on the evaluation of the predicted process input and output sequences within the multistep prediction horizon, which is redefined at each sampling instant. In order to carry out the evaluation, a performance criterion is defined within this horizon. As an example of the evaluation of the process input and output sequences, this chapter proposes a linear quadratic performance index. The minimization of this index is solved technically and the corresponding control law is derived. With the aim of reducing the computational complexity associated with this control law, an alternative one is obtained by imposing specified step-shaped control sequences within the prediction horizon. The basic stability and robustness properties of this last control law will be analyzed in the remainder of this chapter in connection with the length of the multistep prediction horizon. This analysis will be performed using elemental tools that are summarized in Appendix A and illustrated by means of two numerical examples. This particular control law will be used in the practical applications of adaptive predictive control described in subsequent chapters of this book.

3.2 EXTENDED PREDICTIVE CONTROL STRATEGY

3.2.1 The extended strategy

The method of solving the stability problem discussed in Section 2.4 encompasses the application of the basic predictive control strategy and the appropriate selection, at each of the successive control instants k, of the projected desired trajectories (PDT) that guarantee a driving desired trajectory (DDT) that is compatible with a bounded control signal.

The generation of the PDT in the preceding chapter, by means of equation (2.6), took the preceding outputs of the process into account, but did not account for its dynamic nature. It seems logical to use knowledge of this dynamic nature to evaluate the evolution of the process variables in a future interval and to make the appropriate selection of the PDT from this evaluation. The information on the dynamics of the process applies to the predictive model, and the evaluation of the process behaviour may be achieved through the introduction of a performance criterion within the said future interval or prediction horizon. This way of applying predictive control is known as the *extended strategy*.

Figure 3.1 illustrates the fact that, from information on the known inputs and outputs of the process at instant k, we can consider a prediction horizon $[k, k+\lambda]$ defined by a number of control periods λ, where a sequence of outputs $\hat{y}(k+j|k)$ can be predicted by means of the predictive model as a function of a control sequence $\hat{u}(k+j-1|k)$, where $j = 1, \ldots, \lambda$. From all the possible predicted trajectories, we will chose as the PDT, $y_d(k+j|k)$, $j = 1, \ldots, \lambda$, the one that, together with the control sequence which produces it, satisfies a performance criterion.

The first values of the PDTs $y_d(k+1|k)$ at each of the successive instants k define the DDT in the manner described in Chapter 2. The control action applied to the process at each instant k corresponds to the value $y_d(k+1|k)$ of the DDT, in accordance with the basic principle of predictive control. As a result, it is equal to the first value in the control sequence $\hat{u}(k+j-1|k)$ that produces the PDT, that is, $u(k) = \hat{u}(k|k)$.

Given that this procedure for choosing a new PDT is redefined at each instant k, the prediction horizon $[k, k+\lambda]$ does not represent a real time but, rather, a fictitious scenario used solely for the purpose of generating the PDT_k.

As we will discuss in the following sections, the control action applied at each instant k can be derived directly by satisfying the performance criterion used to choose the PDT_k. However, this control action produces a predicted output at $k+1$ equal to the desired output contained in the DDT. As a result, under this extended strategy, the principle of predictive control retains its validity completely.

After the introduction of predictive control [Mar74, Mar76a], the extended strategy was suggested in [Mar77b], formally defined in [Mar80] and analyzed in [Rod82]. Other authors have presented a variety of approaches that can be

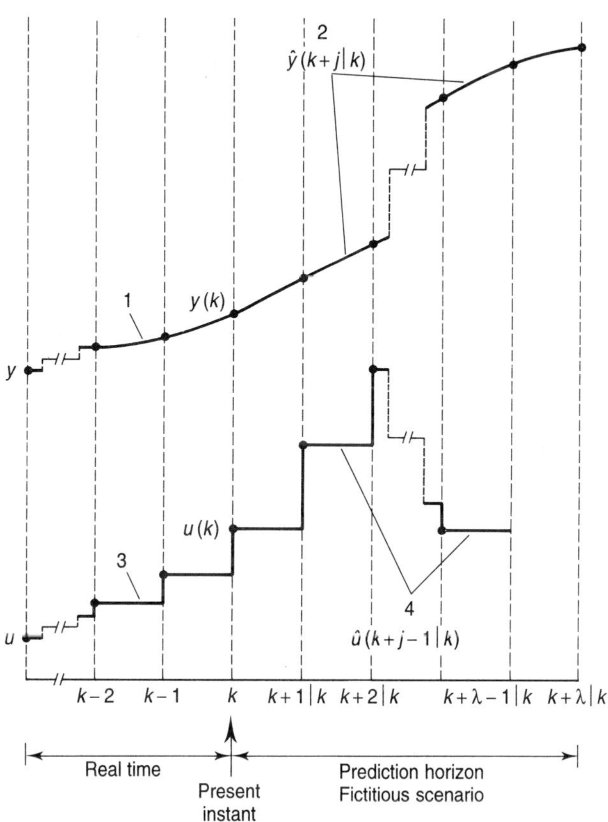

Fig. 3.1. Prediction horizon of the extended strategy.

cast within the framework of the predictive control methodology and its extended strategy. Some of these approaches are identified by names such as Identification and Command (IDCOM) [RRTP78], Dynamic Matrix Control (DMC) [CR80, PG80], Model Algorithmic Control (MAC) [RM82, MRER82], Internal Model Control (IMC) [GM82, MG84], Predictor-Based Self Tuning Control [Pet84], Extended Horizon Adaptive Control (EHAC) [Yds84, YKS85], Extended Predictive Self-Adaptive Control (EPSAC) [DV85, DVD85], Generalized Predictive Control (GPC) [CM87, CMT87, TC88], Multistep Multivariable Adaptive Control (MUSMAR) [GMMZ84, MZM84, MZL89] and others such as [LL83, IFF86, GWK87, MMS88, Gri93]. Surveys on predictive control, including many other references, may be found in [KU88] and [GPM89]. Books related to the subject of predictive control have also been published [BGW90, Soe92, MGPL93, Mos94, CB95].

The problems of the prediction and generation of the desired projected trajectory and the control action at each instant k are considered, respectively, in the following two subsections.

3.2.2 Extended prediction

As shown in Figure 3.1, we define an interval $[k, k+\lambda]$ at the present sampling instant k, within which we can predict a sequence of outputs $\hat{y}(k+j|k)$ as a function of a control sequence $\hat{u}(k+j-1|k)$, where $j = 1, \ldots, \lambda$. In order to carry out this prediction we can use the predictive model of (2.4) at instant k, increasing the number of time steps in the form

$$\hat{y}(k+j|k) = \sum_{i=1}^{\hat{n}} \hat{a}_i \hat{y}(k+j-i|k) + \sum_{i=1}^{\hat{m}} \hat{b}_i \hat{u}(k+j-i|k) \qquad (j = 1, 2, \ldots, \lambda) \tag{3.1}$$

where

$$\begin{aligned} \hat{y}(k+1-i|k) &= y(k+1-i) \qquad & i &= 1, \ldots, \hat{n} \\ \hat{u}(k+1-i|k) &= u(k+1-i) \qquad & i &= 1, \ldots, \hat{m} \end{aligned} \tag{3.2}$$

where $y(k+1-i)$ and $u(k+1-i)$ are the measured outputs and the inputs already applied at instant k. This extended prediction thus includes the specific case of $\lambda = 1$, which was considered previously in Chapter 2.

As in Chapter 2, here we use a particular predictive model based on the difference equation description of (2.4) for single-input, single-output processes. We note that other models, including also the multivariable case, can be used within the extended strategy of predictive control. While all the subsequent formulations in this chapter will be based on the model of (3.1), Appendix B presents formulations based on other predictive models.

Another remark concerns the time delay. We do not explicitly include a time delay in the control sequence of the model of (3.1) other than the discretization delay of one sampling period due to D/A conversion in the control loop. In the case of the existence of an r sampling time delay, we could substitute $\hat{u}(k+j-1-r|k)$ for $\hat{u}(k+j-1|k)$ in the second summation of (3.1). But for the sake of simplicity, we leave the model as it is written in (3.1), assuming that the first r leading $\hat{b}_i$ coefficients are zero, without loss of generality.

3.2.3 Performance criterion, projected desired trajectory and control action

In accordance with the previously described extended strategy, the PDT at each instant k is the trajectory predicted by the predictive model in response to a control sequence such that the trajectory and control sequence satisfy a performance criterion. To clarify this concept, let us consider the following linear quadratic cost function as the performance criterion:

$$J_k = \frac{1}{2}\sum_{j=1}^{\lambda} Q_j[\hat{y}(k+j|k) - y_r(k+j|k)]^2 + \frac{1}{2}\sum_{j=0}^{\lambda-1} R_j\hat{u}(k+j|k)^2 \tag{3.3}$$

where $y_r(k+j|k)$ is a *reference trajectory* which can be generated in a form equivalent to the way that the PDT for the basic strategy was generated in Chapter 2 and irrespective of the dynamic nature of the process to be controlled. Thus, using equation (2.5) within a prediction interval $[k, k+\lambda]$, y_r may be defined in the form

$$y_r(k+j|k) = \sum_{i=1}^{p} \alpha_i y_r(k+j-i|k) + \sum_{i=1}^{q} \beta_i y_{sp}(k+j-i) \qquad (j = 1, 2, \ldots, \lambda) \tag{3.4}$$

where

$$y_r(k+1-i|k) = y(k+1-i) \qquad i = 1, \ldots, p \tag{3.5}$$

and recalling that $y_{sp}(\cdot)$ denotes the setpoint.

The cost function (3.3) imposes a compromise between the resultant PDT being as close to the reference trajectory as possible and the required control action not being excessive. Q_j and R_j are weighting factors which are chosen to weight the tracking of the reference trajectory or the magnitude of the control action. Within this criterion, a predicted trajectory which requires an unlimited control sequence can obviously not be the resulting PDT. By selecting the reference trajectory, the designer can define, in a simple way and independently of the dynamics of the process, the time response and damping that would be desirable for the PDT.

We know that there exists a predicted trajectory and a control sequence, both linked by the predictive model of (3.1), which minimize the value of the performance index of (3.3). The corresponding control sequence $\hat{u}(k+j|k)$ (for $j = 0, 1, \ldots, \lambda - 1$) is optimum in the sense that it minimizes the index J_k.

According to the guidelines of the extended strategy, the only control action that will be applied is the first value in the control sequence, which corresponds to the first value of the PDT. That is, $u(k) = \hat{u}(k|k)$. A new PDT and the corresponding control sequence will be redefined at instant $k+1$, and so on at each successive time instant.

The extended strategy of predictive control is not restrictive in the type of performance criteria used to generate the PDTs [Mar80]. In any case, the final objective is that the resulting driving desired trajectory (DDT), consisting of the initial values of the PDTs, is able to drive the process output to the setpoint in a stable and efficient manner. The cost function defined in (3.3) is nothing more than an example in this context, though it represents a performance criterion with a clear physical meaning. However, the formulation of the performance criteria has to take into account the implementation cost associated with the technical problem of finding the optimal control sequence, which can be laborious. This problem is particularly serious if the intention is to apply the strategy in an adaptive context, combined with an estimation of the model's parameters, since the solution of the performance criteria must be performed in real time at each sampling instant.

The minimization of the cost function (3.3) is a finite horizon optimization problem involving λ unknowns $\hat{u}(k|k), \hat{u}(k+1|k), \ldots, \hat{u}(k+\lambda-1|k)$. Different alternative solutions can be derived depending on the predictive model used to relate the input/output sequences, as well as on the specific way of calculating the extremum condition. In the following section we use the difference equation type of predictive model of (3.1) to derive a minimization procedure and its resultant control law. With the aim of reducing the computational effort and by the way of example an alternative solution will be derived in a further section which is very easy to implement and which upholds the requirement of yielding an efficient and stable control for a wide class of processes.

Other alternative formulations are developed in Appendix B.

3.3 MINIMIZATION OF THE COST FUNCTION AND CONTROL LAW

Let consider the minimization problem stated in the following terms: at each sampling time k, find the control sequence $\hat{u}(k|k), \hat{u}(k+1|k), \ldots, \hat{u}(k+\lambda-1|k)$

such that the cost function J_k in (3.3) is minimum under the following conditions:

1. the process outputs at k and previous instants $y(k), \ldots, y(k+1-\hat{n})$ and the controls at $k-1$ and previous instants $u(k-1), \ldots, u(k+1-\hat{m})$ are known;

2. the predicted output sequence $\hat{y}(k+1|k), \ldots, \hat{y}(k+\lambda|k)$ is related to the control sequence $\hat{u}(k|k), \hat{u}(k+1|k), \ldots, \hat{u}(k+\lambda-1|k)$ by the model of (3.1)–(3.2);

3. the values of the reference trajectory $y_r(k+1|k), \ldots, y_r(k+\lambda|k)$ are calculated independently using the setpoint values over the prediction horizon;

By using the predictive model of (3.1) recursively from the initial conditions (3.2), we may write

$$\begin{aligned}\hat{y}(k+j|k) = &\sum_{i=1}^{\hat{n}} \hat{e}_i^{(j)} y(k+1-i) + \sum_{i=2}^{\hat{m}} \hat{g}_i^{(j)} u(k+1-i) \\ &+ \sum_{i=0}^{j-1} \hat{g}_1^{(j-i)} \hat{u}(k+i|k) \\ &(j = 1, 2, \ldots, \lambda)\end{aligned} \tag{3.6}$$

where $\hat{e}_i^{(j)}$ and $\hat{g}_i^{(j)}$ are coefficients that can be obtained from the parameters $\hat{a}_i$ and $\hat{b}_i$ of the predictive model using the following recursive algorithms:

$$\begin{aligned}\hat{e}_i^{(j)} &= \hat{e}_1^{(j-1)} \hat{a}_i + \hat{e}_{i+1}^{(j-1)} \qquad i = 1, \ldots, \hat{n} \qquad j = 2, \ldots, \lambda \\ \hat{g}_i^{(j)} &= \hat{e}_1^{(j-1)} \hat{b}_i + \hat{g}_{i+1}^{(j-1)} \qquad i = 1, \ldots, \hat{m} \qquad j = 2, \ldots, \lambda\end{aligned} \tag{3.7a}$$

with

$$\begin{aligned}\hat{e}_i^{(1)} &= \hat{a}_i \qquad i = 1, \ldots, \hat{n} \\ \hat{g}_i^{(1)} &= \hat{b}_i \qquad i = 1, \ldots, \hat{m} \\ \hat{e}_{\hat{n}+1}^{(j-1)} &= 0 \qquad j = 2, \ldots, \lambda \\ \hat{g}_{\hat{m}+1}^{(j-1)} &= 0 \qquad j = 2, \ldots, \lambda\end{aligned} \tag{3.7b}$$

Equation (3.6) gives the predicted output for each instant $k+j$ as a function of the information on the process inputs and outputs at instant k, as well as of the unknown control inputs $\hat{u}(k|k), \ldots, \hat{u}(k+j-1|k)$. The set of λ equations (3.6) can be written in matrix-vector form as follows:

$$\hat{Y} = E\,Y_k + G\,U_k + G_0\,\hat{U} \tag{3.8}$$

where $\hat{Y}$ and $\hat{U}$ are the $\lambda \times 1$ vectors

$$\hat{Y} = [\hat{y}(k+1|k), \hat{y}(k+2|k), \ldots, \hat{y}(k+\lambda|k)]^T$$

$$\hat{U} = [\hat{u}(k|k), \hat{u}(k+1|k), \ldots, \hat{u}(k+\lambda-1|k)]^T$$

Y_k and U_k are, respectively, the $\hat{n} \times 1$ and $(\hat{m}-1) \times 1$ vectors

$$Y_k = [y(k), y(k-1), \ldots, y(k-\hat{n}+1)]^T$$

$$U_k = [u(k-1), u(k-2), \ldots, u(k-\hat{m}+1)]^T$$

E, G and G_0 are matrices, with dimensions $\lambda \times \hat{n}$, $\lambda \times (\hat{m}-1)$ and $\lambda \times \lambda$, respectively, defined as

$$E = \begin{pmatrix} \hat{e}_1^{(1)} & \hat{e}_2^{(1)} & \ldots & \hat{e}_{\hat{n}}^{(1)} \\ \hat{e}_1^{(2)} & \hat{e}_2^{(2)} & \ldots & \hat{e}_{\hat{n}}^{(2)} \\ \vdots & \vdots & \ddots & \vdots \\ \hat{e}_1^{(\lambda)} & \hat{e}_2^{(\lambda)} & \ldots & \hat{e}_{\hat{n}}^{(\lambda)} \end{pmatrix} \qquad G = \begin{pmatrix} \hat{g}_2^{(1)} & \hat{g}_3^{(1)} & \ldots & \hat{g}_{\hat{m}}^{(1)} \\ \hat{g}_2^{(2)} & \hat{g}_3^{(2)} & \ldots & \hat{g}_{\hat{m}}^{(2)} \\ \vdots & \vdots & \ddots & \vdots \\ \hat{g}_2^{(\lambda)} & \hat{g}_3^{(\lambda)} & \ldots & \hat{g}_{\hat{m}}^{(\lambda)} \end{pmatrix}$$

$$G_0 = \begin{pmatrix} \hat{g}_1^{(1)} & 0 & 0 & \ldots & 0 \\ \hat{g}_1^{(2)} & \hat{g}_1^{(1)} & 0 & \ldots & 0 \\ \vdots & \vdots & \vdots & \ddots & \vdots \\ \hat{g}_1^{(\lambda)} & \hat{g}_1^{(\lambda-1)} & \hat{g}_1^{(\lambda-2)} & \ldots & \hat{g}_1^{(1)} \end{pmatrix}$$

We include the values of the reference trajectory in the $\lambda \times 1$ vector in the same setting

$$Y_r = [y_r(k+1|k), y_r(k+2|k), \ldots, y_r(k+\lambda|k)]^T$$

Using the above matrix-vector notation, the index of (3.3) can be written in the form

$$J_k = \frac{1}{2}[\hat{Y} - Y_r]^T Q [\hat{Y} - Y_r] + \frac{1}{2}\hat{U}^T R \hat{U} \tag{3.9}$$

where Q and R are the weighting $\lambda \times \lambda$ matrices

$$Q = \text{diag}[Q_1, Q_2, \ldots, Q_\lambda] \qquad R = \text{diag}[R_0, R_1, \ldots, R_{\lambda-1}]$$

Substituting (3.8) into (3.9), we write

$$\begin{aligned} J_k =& \frac{1}{2}[E Y_k + G U_k + G_0 \hat{U} - Y_r]^T Q [E Y_k + G U_k + G_0 \hat{U} - Y_r] \\ &+ \frac{1}{2}\hat{U}^T R \hat{U} \end{aligned} \tag{3.10}$$

Now, noting that the vector $\hat{U}$ is the only unknown in (3.10), we can impose the following necessary condition for J_k to reach an extremum provided that there is no constraint on $\hat{U}$:

$$\frac{\partial J}{\partial \hat{U}} = 0$$

Imposing this condition on (3.10), we write

$$G_0^T Q [E Y_k + G U_k + G_0 \hat{U} - Y_r] + R \hat{U} = 0$$

which gives

$$\hat{U} = [G_0^T Q G_0 + R]^{-1} G_0 Q [E Y_k + G U_k - Y_r] \tag{3.11}$$

Sufficient conditions for $\hat{U}$ rendering J_k a minimum value are satisfied when the matrices Q and R are non-negative definite.

As previously explained, although (3.11) gives the complete control sequence minimizing J_k over the prediction interval $[k, k+\lambda]$, only the first value is actually applied to the process as the control signal $u(k)$ at instant k. Thus the final control law has the form

$$u(k) = \hat{u}(k|k) = g_0^T [E Y_k + G U_k - Y_r] \tag{3.12}$$

g_0^T being the first row of the matrix $[G_0^T Q G_0 + R]^{-1} G_0 Q$.

The parameters involved in the implementation of this control law are the length of the prediction horizon λ, the weighting factors in Q, R and the parameters α_i, β_i defining the reference trajectory. These parameters are associated with the performance criterion. The other parameters define the predictive model: the order given by $\hat{n}, \hat{m}$ and the coefficients $\hat{a}_i, \hat{b}_i$. Matrices E, G, G_0 are computed using these coefficients. This computation is performed only once, prior to the real time implementation of the control law, when the model parameters $\hat{a}_i, \hat{b}_i$ are fixed. However, the computation must be repeated at each sampling instant if they are updated in real time using the adaptive capabilities of the APCS systems.

Another way of solving the minimization of the predictive control problem under index (3.3) would be to use a matrix Riccati equation [Rod82, RBM87]. This type of equation was previously used in a standard way in linear quadratic (LQ) optimal control [KS72, SW77]. Thus, in this case, predictive control may borrow from optimal control theory the procedure for evaluating the process evolution in the prediction horizon. However, it is important to point out that this particular way of implementing predictive control differs essentially from standard LQ optimal control in the sense that the prediction horizon is redefined and the minimization procedure is repeated at each sampling instant k. The idea of redefining the control objective on line, which is inherent to the basic concept of predictive control, was also considered in the context of LQ optimal control, leading to the concept of the so-called *receding horizon* [Tho75, KP77, KP78, CS82, KG88, MM90, MM91]. Some authors have studied analogies between receding horizon LQ optimal control and predictive control [Pet90, BGW90, Mos94]. It is clear that they are two basically different phylosophical concepts with different origins. As a matter of fact, the motivation for the introduction of the redefinition of the horizon in optimal control lay in the requirement of a finite time approximation of the standard LQ optimal control problem with infinite time to simplify the solution of the associated Riccati equation. On the other hand, predictive control was motivated to exploit the potential of the real time prediction by the digital systems, allowing the selection of different control objectives in the prediction horizon and the use of additional concepts which are essential for practical applications, as illustrated in this and the following chapters of this book. It is apparent that LQ optimal control, when redefined at each sampling instant, approaches particular solutions of the extended strategy of predictive control that also involve the use of the Riccati equation [Rod82, RBM87]. Within the predictive control methodology, the role to be played by the Riccati equation is simply that of a tool appearing in a particular method of application. However this tool is computationally disadvantageous and more simple methods of application can be derived in practice, as we will discuss in the following section.

3.4 A PARTICULAR SOLUTION OF THE EXTENDED STRATEGY

The complexity introduced by the cost function (3.3) is basically due to the number of unknowns λ, the number of values in the control sequence $\hat{u}(k+j|k)$. One way of reducing the number of unknowns is to predetermine the form of the control sequence. It has proved useful to impose a step control sequence together with a cost function such as the one given in (3.3), thus reducing the number of unknowns to a single one. This type of solution was first proposed in [Mar80] and analyzed in [Rod82]. Other authors have considered the same kind of step control sequence within the performance index [Yds84, DV85, CMT87, RBM87].

In this section, as a simple example, we consider the following selection of weighting factors in the performance index (3.3):

$$Q_j = 0 \quad (j = 1, \ldots, \lambda - 1), \quad Q_\lambda = 1, \quad R_j = 0 \quad (j = 0, 1, \ldots \lambda - 1)$$

Under this choice, the performance index reduces to

$$J_k = \hat{y}(k+\lambda|k) - y_r(k+\lambda|k) \tag{3.13}$$

with the condition that the control sequence remain constant during the prediction interval; that is to say,

$$\hat{u}(k|k) = \hat{u}(k+1|k) = \ldots = \hat{u}(k+\lambda-1|k) \tag{3.14}$$

In accordance with the previously defined criterion, the projected desired trajectory (PDT) will be the same in this case as a trajectory predicted by the predictive model in response to a step, such that the value of the reference trajectory and the value of the predicted trajectory coincide at the end of the prediction interval. Imposing a step control signal within the prediction horizon is equivalent to exploring the process evolution if no further change in the control action is made. This is intuitively appealing and we may conclude that the evolution within the prediction interval is determined by the process dynamics itself. Therefore, the longer the horizon λ, the less demanding is condition (3.13).

The calculation of the control action is now reduced to obtaining a single unknown $u(k) = \hat{u}(k|k)$ and can easily be solved. Using the prediction equation of (3.6) for instant $k+\lambda$ and imposing condition (3.14) we obtain the following result:

$$\hat{y}(k+\lambda|k) = \sum_{i=1}^{\hat{n}} \hat{e}_i^{(\lambda)} y(k+1-i) + \sum_{i=2}^{\hat{m}} \hat{g}_i^{(\lambda)} u(k+1-i) + \hat{h}^{(\lambda)} \hat{u}(k|k) \tag{3.15}$$

where

$$\hat{h}^{(\lambda)} = \hat{g}_1^{(\lambda)} + \hat{g}_1^{(\lambda-1)} + \ldots + \hat{g}_1^{(1)} \tag{3.16}$$

Equation (3.15) expresses the predicted output at $k+\lambda$ as a function of the known input/output information at instant k and the input $\hat{u}(k|k)$ which has to be calculated.

To obtain the value of the predictive control action $u(k)$, simply substitute (3.15) in (3.13), cancel the index J_k and solve:

$$u(k) = \hat{u}(k|k) = \frac{y_r(k+\lambda|k) - \sum_{i=1}^{\hat{n}} \hat{e}_i^{(\lambda)} y(k+1-i) - \sum_{i=2}^{\hat{m}} \hat{g}_i^{(\lambda)} u(k+1-i)}{\hat{h}^{(\lambda)}} \tag{3.17}$$

The operations to implement this control law are indeed very simple, even where the calculation of the parameters is performed within an adaptive context. This makes it very attractive for real applications. In fact, most of the applications described in further chapters in this book involve the use of this control law. Because of this, in the next section it may be interesting to discuss and illustrate some stability and robustness properties associated with the implementation of (3.17).

3.5 ILLUSTRATIVE STABILITY AND ROBUSTNESS ANALYSIS

In this section we will analyze the stability and robustness of the predictive control law derived in the preceding section using the basic tools given in Appendix A. The analysis involves two steps: (1) the formulation of transfer functions relating the sequence of setpoints to the process output and input sequences respectively; and (2) the evolution of the roots of the corresponding characteristic polynomials when extending the prediction horizon to the limit. Two simple examples will be used to illustrate the stability and robustness analysis respectively. A further study within the framework of the APCS stability theory will be developed later on in Chapter 5.

3.5.1 Theoretical analysis: transfer functions

Let us assume that the process dynamics is described by the following discrete time transfer function :

$$y(z) = \frac{B(z^{-1})}{A(z^{-1})} u(z) \tag{3.18a}$$

$B(z^{-1})$ and $A(z^{-1})$ being the polynomials

$$\begin{aligned} B(z^{-1}) &= b_1 z^{-1} + b_2 z^{-2} + \ldots + b_m z^{-m} \\ A(z^{-1}) &= 1 - a_1 z^{-1} - a_2 z^{-2} - \ldots - a_n z^{-n} \end{aligned} \tag{3.18b}$$

The control law of (3.17) can be written in the form

$$y_r(k+\lambda|k) = \sum_{i=1}^{\hat{n}} \hat{e}_i^{(\lambda)} y(k+1-i) + \sum_{i=2}^{\hat{m}} \hat{g}_i^{(\lambda)} u(k+1-i) + \hat{h}^{(\lambda)} u(k) \tag{3.19}$$

and, applying the Z transform to this expression, we may write

$$y_r(z) = \hat{E}_\lambda(z^{-1}) y(z) + \hat{G}_\lambda(z^{-1}) u(z) \tag{3.20a}$$

with the polynomials

$$\begin{aligned} \hat{E}_\lambda(z^{-1}) &= \hat{e}_1^{(\lambda)} + \hat{e}_2^{(\lambda)} z^{-1} + \ldots + \hat{e}_{\hat{n}}^{(\lambda)} z^{-\hat{n}+1} \\ \hat{G}_\lambda(z^{-1}) &= \hat{h}^{(\lambda)} + \hat{g}_2^{(\lambda)} z^{-1} + \ldots + \hat{g}_{\hat{m}}^{(\lambda)} z^{-\hat{m}+1} \end{aligned} \tag{3.20b}$$

By using equation (3.4) recursively, which generates the reference trajectory, we obtain the value of $y_r(k+j|k)$ in the following form:

$$\begin{aligned} y_r(k+j|k) = &\sum_{i=1}^{p} \varphi_i^{(j)} y(k+1-i) + \sum_{i=2}^{q} \delta_i^{(j)} y_{sp}(k+1-i) \\ &+ \sum_{i=0}^{j-1} \delta_1^{(j-i)} \hat{y}_{sp}(k+i|k) \\ &(j = 1, 2, \ldots, \lambda) \end{aligned} \tag{3.21}$$

We can see that this result is similar to the one obtained in (3.6) and that the parameters φ_i and δ_i may be calculated from the parameters α_i and β_i by recursive expressions similar to those given in (3.7). $\hat{y}_{sp}(k+i|k)$ represents the values of the setpoint within the prediction horizon. Assuming that these values are equal to $y_{sp}(k)$, which is the most usual case, we obtain the value $y_r(k+\lambda|k)$ in the form

$$y_r(k+\lambda|k) = \sum_{i=1}^{p} \varphi_i^{(\lambda)} y(k+1-i) + \sum_{i=2}^{q} \delta_i^{(\lambda)} y_{sp}(k+1-i) + \mu^{(\lambda)} y_{sp}(k) \tag{3.22}$$

where

$$\mu^{(\lambda)} = \delta_1^{(\lambda)} + \delta_1^{(\lambda-1)} + \ldots + \delta_1^{(1)} \tag{3.23}$$

By applying the Z transform to (3.22), we obtain

$$y_r(z) = \Phi_\lambda(z^{-1})y(z) + \Delta_\lambda(z^{-1})y_{sp}(z) \tag{3.24a}$$

with the polynomials:

$$\begin{aligned} \Phi_\lambda(z^{-1}) &= \varphi_1^{(\lambda)} + \varphi_2^{(\lambda)} z^{-1} + \ldots + \varphi_p^{(\lambda)} z^{-p+1} \\ \Delta_\lambda(z^{-1}) &= \mu^{(\lambda)} + \delta_2^{(\lambda)} z^{-1} + \ldots + \delta_q^{(\lambda)} z^{-q+1} \end{aligned} \tag{3.24b}$$

Solving for $u(z)$ in (3.20a) and substituting into (3.18a), the following equation is easily obtained:

$$y(z) = \frac{B(z^{-1})}{\hat{\theta}'_\lambda(z^{-1})} y_r(z) \tag{3.25}$$

where $\hat{\theta}'_\lambda(z^{-1})$ is a characteristic polynomial of the form

$$\hat{\theta}'_\lambda(z^{-1}) = \hat{E}_\lambda(z^{-1})B(z^{-1}) + \hat{G}_\lambda(z^{-1})A(z^{-1}) \tag{3.26}$$

Direct substitution of (3.18a) into (3.25) obtains the following equation:

$$u(z) = \frac{A(z^{-1})}{\hat{\theta}'_\lambda(z^{-1})} y_r(z) \tag{3.27}$$

Equation (3.25) defines a transfer function between the sequence formed by the value of the reference trajectory $y_r(k+\lambda|k)$ at successive real time instants k and the sequence of process output values $y(k)$ produced by successive application of the control law. It can also be interpreted as being the relationship between the values of the projected desired trajectory (PDT) at the end of the prediction horizon and the measured values of the process output, which, under the assumption of no prediction errors, are also the values of the driving desired trajectory (DDT).

Relationship (3.27) is a transfer function between the values of the PDT$_k$ at $k+\lambda$ and the predictive control action generated by the control law to produce it.

Substituting (3.24a) into (3.25) we obtain

$$y(z) = \frac{B(z^{-1})\Delta_\lambda(z^{-1})}{\hat{\theta}_\lambda(z^{-1})} y_{sp}(z) \tag{3.28}$$

where $\hat{\theta}_\lambda(z^{-1})$ is the characteristic polynomial of the closed loop

$$\hat{\theta}_\lambda(z^{-1}) = [\hat{E}_\lambda(z^{-1}) - \Phi_\lambda(z^{-1})]B(z^{-1}) + \hat{G}_\lambda(z^{-1})A(z^{-1}) \tag{3.29}$$

On the other hand, substituting (3.18a) into (3.28) we obtain

$$u(z) = \frac{A(z^{-1})\Delta_\lambda(z^{-1})}{\hat{\theta}_\lambda(z^{-1})} y_{sp}(z) \tag{3.30}$$

Expression (3.28) defines the dynamic relationship between the setpoint and the process output, that is to say, the closed loop transfer function. Since no prediction error is assumed, (3.28) also defines the relationship between the setpoint and the DDT. Equation (3.30) defines the transfer function between the setpoint and the control action. In both transfer functions (3.28) and (3.30) the denominator is the same characteristic polynomial $\hat{\theta}_\lambda(z^{-1})$ which determines the stability. The stability of relationship (3.28) signifies that, given a bounded sequence of setpoints, the controlled output will also be bounded. The stability of transfer function (3.30) implies that the control action generated to obtain a bounded sequence of setpoint values is also bounded. In both cases the stability condition is that the polynomial $\hat{\theta}_\lambda(z^{-1})$ has its roots in the modulus $|z| < 1$.

A similar consideration can be made for transfer functions (3.25) and (3.27) in relation to the sequence of reference trajectory values at $k + \lambda$. The characteristic polynomial that determines stability in this case is $\hat{\theta}'_\lambda(z^{-1})$. Comparing (3.26) and (3.29) we can see that the difference between $\hat{\theta}_\lambda(z^{-1})$ and $\hat{\theta}'_\lambda(z^{-1})$ is determined by the dynamic term $\Phi_\lambda(z^{-1})$, which is selected by the designer to define the reference trajectory. The effect of this choice in the closed loop dynamics is clearly shown in this way.

For the case where $\lambda = 1$, the control law of (3.17) reduces to the basic predictive control law of (2.7). In this case, if the inverse of the process is unstable, the control sequence capable of ensuring that the process follows a given series of reference values will not be bounded, as discussed in Section 2.4 of the preceding chapter.

As set out in the following subsection, we will see that, for stable processes, assuming only that the predictive model is also stable, extending the prediction horizon to values of $\lambda > 1$ overcomes the above instability problem. Additionally, the following analysis also focuses on robustness, since it is important to emphasize that the dynamics of the predictive model may differ from that of the process.

3.5.2 Theoretical analysis: stability and robustness

To examine the effect that extending the prediction horizon has on stability and robustness, we will take the value of λ to the limit of infinity. In this case, since the predictive model is stable and given that $\hat{u}(k|k)$ is constant during the prediction interval according to (3.14), the predicted output will reach a stationary final value which will satisfy

$$\lim_{\lambda \to \infty} \hat{y}(k+\lambda|k) \stackrel{\text{def}}{=} \hat{y}(k+\infty|k) = \hat{G}_s \hat{u}(k|k) \tag{3.31}$$

where $\hat{G}_s$ is the static gain of the predictive model which satisfies the known relation (Appendix A)

$$\hat{G}_s = \frac{\hat{b}_1 + \hat{b}_2 + \ldots + \hat{b}_{\hat{m}}}{1 - \hat{a}_1 - \hat{a}_2 - \ldots - \hat{a}_{\hat{n}}} \tag{3.32}$$

On the other hand, assuming the infinite limit of λ in (3.15), we can write

$$\hat{y}(k+\infty|k) = \sum_{i=1}^{\hat{n}} \hat{e}_i^{(\infty)} y(k+1-i) + \sum_{i=2}^{\hat{m}} \hat{g}_i^{(\infty)} u(k+1-i) + \hat{h}^{(\infty)} \hat{u}(k|k) \tag{3.33}$$

where the superscript (∞) indicates the limit as $\lambda \to \infty$.

Comparing the two forms of expressing the prediction $\hat{y}(k+\infty|k)$, equations (3.31) and (3.33), the following limiting properties for the control law coefficients can be deduced directly:

$$\begin{aligned} \hat{e}_i^{(\infty)} &= 0 & \quad i &= 1, 2, \ldots, \hat{n} \\ \hat{g}_i^{(\infty)} &= 0 & \quad i &= 2, 3, \ldots, \hat{m} \\ \hat{h}^{(\infty)} &= \hat{G}_s & & \end{aligned} \tag{3.34}$$

Consequently, the polynomials defined in (3.20) satisfy the following limiting properties:

$$\begin{aligned} \lim_{\lambda \to \infty} \hat{E}_\lambda(z^{-1}) &= 0 \\ \lim_{\lambda \to \infty} \hat{G}_\lambda(z^{-1}) &= \hat{G}_s \end{aligned} \tag{3.35}$$

Extending the above limit analysis to the reference trajectory equations (3.4) and (3.22), taking the facts that the static gain of (3.4) is usually equal to one and the setpoint is considered constant within the prediction horizon into account, the following similar results to those of equations (3.34) and (3.35) may be obtained:

$$\begin{aligned} \varphi_i^{(\infty)} &= 0 & \quad i &= 1, 2, \ldots, p \\ \delta_i^{(\infty)} &= 0 & \quad i &= 2, 3, \ldots, q \\ \mu^{(\infty)} &= 1 & & \end{aligned} \tag{3.36}$$

$$\lim_{\lambda \to \infty} \Phi_\lambda(z^{-1}) = 0$$
$$\lim_{\lambda \to \infty} \Delta_\lambda(z^{-1}) = 1 \tag{3.37}$$

Therefore, the characteristic polynomials $\hat{\theta}'_\lambda(z^{-1})$ and $\hat{\theta}_\lambda(z^{-1})$, defined in (3.26) and (3.29), respectively, will converge in the limit to the expression

$$\lim_{\lambda \to \infty} \hat{\theta}'_\lambda(z^{-1}) = \lim_{\lambda \to \infty} \hat{\theta}_\lambda(z^{-1}) = \hat{G}_s A(z^{-1}) \tag{3.38}$$

From (3.38) it can be deduced that the roots of the characteristic polynomials tend towards those of the polynomial $A(z^{-1})$, the denominator of the process transfer function. Since the process is stable, these roots have modulus $|z|<1$. Since the roots of $\hat{\theta}'_\lambda(z^{-1})$ and $\hat{\theta}_\lambda(z^{-1})$ approach those of $A(z^{-1})$ as the value of λ increases, there will be values of λ for which both polynomials will have roots with modulus $|z|<1$ and, in particular, a value λ_0 such that the control loop will be stable for any $\lambda>\lambda_0$. With these values of λ and whatever the setpoint, we produce an input $u(k)$ and an output $y(k)$, both of which are bounded, irrespective of whether the inverse of the process is stable or not.

We will complete this illustrative analysis by describing how the process output will approach the setpoint under the predictive control law of (3.17). Assuming that the setpoint has a constant value of $\bar{y}_{sp}$, we can apply the discrete time final value theorem (Appendix A) to (3.28) in order to show that the output has a steady state value $\bar{y}$ that verifies

$$\bar{y} = \lim_{k \to \infty} y(k) = \lim_{z \to 1} \frac{B(z^{-1})\Delta_\lambda(z^{-1})}{\hat{\theta}_\lambda(z^{-1})} \bar{y}_{sp} \tag{3.39}$$

Now we note that the polynomials $\hat{G}_\lambda(z^{-1})$ and $\hat{E}_\lambda(z^{-1})$ from equation (3.20b) verify

$$\frac{\hat{G}_\lambda(1)}{1 - \hat{E}_\lambda(1)} = \hat{G}_s \tag{3.40}$$

where $\hat{G}_s$ is the static gain of the predictive model defined in (3.32). In a similar vein, the static gain of the process, defined in (3.18), is given by

$$G_s = \frac{B(1)}{A(1)} = \frac{b_1 + b_2 + \ldots + b_m}{1 - a_1 - a_2 - \ldots - a_n} \tag{3.41}$$

Since the reference trajectory of equation (3.4) has a static gain equal to one, we may write

$$1 = \frac{\Delta_\lambda(1)}{1 - \Phi_\lambda(1)} \tag{3.42}$$

Using (3.40)–(3.42) in (3.39) we readily obtain

$$\bar{y} = \frac{G_s[1 - \Phi_\lambda(1)]}{[G_s - \hat{G}_s]\hat{E}_\lambda(1) + \hat{G}_s - \Phi_\lambda(1)G_s}\bar{y}_{sp} \tag{3.43}$$

Therefore, we can see that, if the gains of the predictive model and the process are equal ($G_s = \hat{G}_s$), equation (3.43) reduces to $\bar{y} = \bar{y}_{sp}$ which guarantees that the process output tends towards the constant setpoint.

It is important to note the robustness aspect of this result, since, although we assume a knowledge of the process gain, we do not make any other assumption concerning other modelling errors. As is shown in Appendix B, the requirement of knowing the static gain of the process is not necessary to guarantee that the process output will tend towards the setpoint when using an incremental formulation of the extended predictive control strategy. Appendix B also shows that the incremental formulation is able to handle potential steady state deviations from the setpoint due to load disturbances. Although the analysis in this section, whose purpose has been to illustrate the main concepts behind the extended strategy of predictive control, has not considered the transient performance of the process output, it must be expected that this performance will deteriorate as the magnitude of the modelling errors between the predictive model and the process increases. This will be confirmed in the stability analysis of Chapter 5.

Numerical examples are presented in the following two subsections illustrating, respectively, the stability and robustness issues previously discussed.

3.5.3 Illustrative example 1

Let us consider the same stable process with the same unstable inverse as used in Section 2.4, defined by the transfer function (2.12), whose response to a step is shown in Figure 2.5.

For a sampling control period of one second, the process model has the following transfer function in discrete time:

$$\frac{y(z)}{u(z)} = \frac{B(z^{-1})}{A(z^{-1})} = \frac{b_1 z^{-1} + b_2 z^{-2}}{1 - a_1 z^{-1} - a_2 z^{-2}} \tag{3.44}$$

with

$$\begin{aligned} a_1 &= 1.1467 \qquad & a_2 &= -0.2865 \\ b_1 &= -0.4637 \qquad & b_2 &= 0.6035 \end{aligned} \tag{3.45}$$

Let us take the process model as our predictive model, assuming that there are no modelling errors. Therefore, the order of the model is 2 and the coefficients of the control law of (3.17) are

$$\hat{e}_1^{(\lambda)}, \hat{e}_2^{(\lambda)}, \hat{g}_2^{(\lambda)}, \hat{h}^{(\lambda)}$$

which are obtained using the algorithm of (3.7) and equation (3.16) from the process coefficients given in (3.45). In this case the value of the reference trajectory at the end of the prediction interval $y_r(k+\lambda|k)$ is taken to be equal to the setpoint. Therefore, equation (3.22) becomes

$$y_r(k+\lambda|k) = y_{sp}(k) \tag{3.46}$$

and the polynomials in (3.24) reduce to

$$\Phi_\lambda(z^{-1}) = 0 \quad ; \quad \Delta_\lambda(z^{-1}) = 1 \tag{3.47}$$

As a consequence, the characteristic polynomials in (3.26) and (3.29) are

$$\begin{aligned} \hat{\theta}'_\lambda(z^{-1}) = \hat{\theta}_\lambda(z^{-1}) = [\hat{e}_1^{(\lambda)} + \hat{e}_2^{(\lambda)} z^{-1}][b_1 z^{-1} + b_2 z^{-2}] \\ + [\hat{h}^{(\lambda)} + \hat{g}_2^{(\lambda)} z^{-1}][1 - a_1 z^{-1} - a_2 z^{-2}] \end{aligned} \tag{3.48}$$

For the stability analysis let us use the variable z instead of z^{-1}, with the polynomial $\bar{\theta}(z)$:

$$\bar{\theta}(z) = z^3 \hat{\theta}(z^{-1}) \tag{3.49}$$

The stability condition is that all the roots of $\bar{\theta}(z)$ have a modulus less than unity. Figure 3.2 shows the value of the two most significant roots for different values of λ. It can be seen that it is necessary to increase the prediction horizon to a value of $\lambda \geq 5$ so that the roots meet the stability condition.

Figure 3.3 shows the output of the process and the control applied within the range of stability for different values of λ when two successive changes of setpoint are requested.

It is interesting to observe that the setpoint is reached whatever the value of λ, but that the way of achieving it differs. As the value of λ is increased, the typical response of the process with an unstable inverse is less pronounced and the evolution towards the setpoint is slower and smoother. This has a clear intuitive interpretation from condition (3.13) for the performance criterion used to define the projected desired trajectory (PDT). Indeed, (3.13) requires that the predicted output be equal to the reference value (in this example this is the setpoint) at the end of the prediction interval. Therefore, a short interval implies

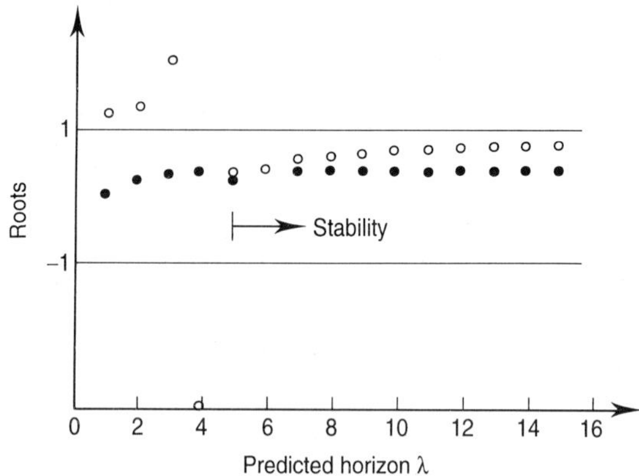

Fig. 3.2. Roots of the characteristic polynomial of the predictive control law.

a more demanding condition and requires a more drastic control action, which reduces as the prediction horizon is increased.

The behaviour of the predictive control for different values of λ described previously can also be explained by the limit properties of the control law's coefficients (3.34) and the characteristic polynomials (3.38). For a value of λ tending towards infinity, we can substitute (3.38) into the transfer function (3.30) and write

$$u(z) = \frac{y_{SP}(z)}{\hat{G}_S} \tag{3.50}$$

Since $\hat{G}_S = G_S = 1$ in this example, relationship (3.50) indicates that the control $u(k)$, when λ is infinity, becomes equal to the setpoint. In this case, when there is a step setpoint change, the control action will also be a step of the same magnitude. Therefore, the response of the process to the control action will be the natural response to a step and, consequently, it will be the slowest and smoothest response which can be obtained by applying predictive control for different values of λ.

It can be seen in Figure 3.3 that a value of $\lambda = 12$ already produces a control signal that is very close to the stepped control action limit. As can be seen, there is a wide margin within which to choose a value for λ which guarantees stability, and which rapidly reaches the setpoint without excessive control action.

To analyze in greater depth why the extended predictive control law of (3.17) generates a stable action, it would be helpful to consider Figure 3.4 and compare the desired trajectories generated in this case with those obtained using the basic

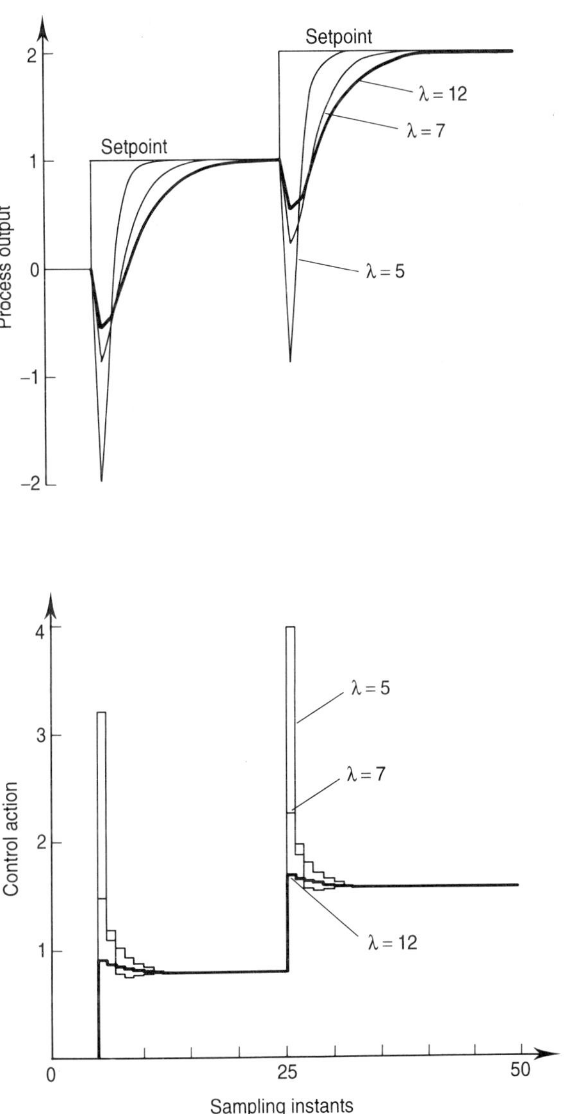

Fig. 3.3. Process output and control action for different values of λ.

control law (see the example presented in Section 2.4, and especially Figure 2.6). The results shown in Figure 3.4 are obtained by applying the predictive control law of (3.17) with $\lambda = 7$.

At instant $k = 0$ a projected desired trajectory PDT_0 is generated which satisfies the performance index defined by (3.13)–(3.14); in this case the reference value at $k + \lambda$ is the setpoint. Consequently, as can be seen in Figure 3.4, the PDT_0 reaches the setpoint at instant 7, that is, $y_d(7|0) = 1$. As we know, this trajectory corresponds to the response of the predictive model to a step equal to $u(0)$ and, therefore, it deviates from the setpoint, reaching a steady state equal to $u(0)$ since the gain of the process and the predictive model is equal to one. This deviation from the setpoint does not matter since only the first value of the PDT_0 is used to define the driving desired output value $y_d(1|0)$. Likewise, only the value $u(0)$ from the constant control sequence is applied as the control at instant 0.

The procedure described above is redefined at instant 1, generating a projected desired trajectory PDT_1 which, in response to a step, passes through the setpoint at instant 8. This trajectory defines a driving desired output $y_d(2|1)$ and a control $u(1)$. It can be seen that, although the PDT_1 also does not reach the setpoint in its steady state, this is closer to the setpoint than is the steady state of the preceding PDT_0. The same occurs with the trajectory defined at instant 2. In general, if we represented the successive projected desired trajectories for instants $k = 3, 4, \ldots$, we would see that in their steady states these tend towards the setpoint.

This tendency of the projected desired trajectories is the opposite of that observed in the example presented in Section 2.4, where the successive predicted trajectories generated by the basic control law became ever more removed from the setpoint, generating a sequence of unbounded controls, as shown in Figure 2.6. The control sequence in this case tends towards a bounded value equal to one.

The driving desired trajectory is, by construction, the one made up of the successive desired outputs $y_d(1|0)$, $y_d(2|1)$, $y_d(3|2), \ldots$. These values coincide with the measured outputs $y(1)$, $y(2)$, $y(3), \ldots$ in Figure 3.4 since there are no prediction errors in this example. In general, the existence of prediction errors is foreseeable due to control limits (as was considered in Section 2.4) or, more generally, to modelling errors, with the result that the measured output and the driving desired output may differ. In such a case, as explained in Section 2.2.4 and illustrated in the example of Figure 2.4, the successive projected desired trajectories are defined starting from the successive measured values $y(1), y(2), y(3), \ldots$. A practical case which illustrates this point clearly for this extended strategy may be found in [RGM89].

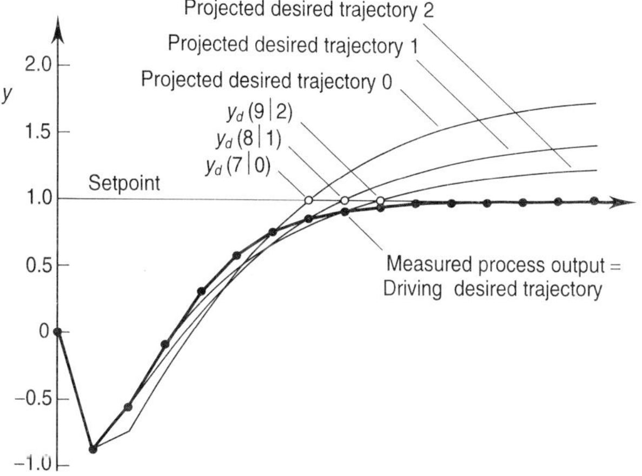

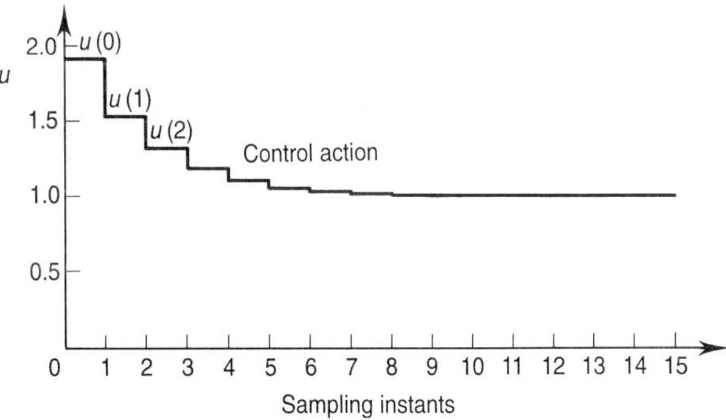

Fig. 3.4. Desired trajectories and stable predictive control action.

3.5.4 Illustrative example 2

This example considers a flexible mechanical model with one degree of freedom, as represented in Figure 3.5. It has been selected as being a typical prototype in system theory that can represent various mechanical systems or structures vibrating in response to exciting forces [CP75, Mei90].

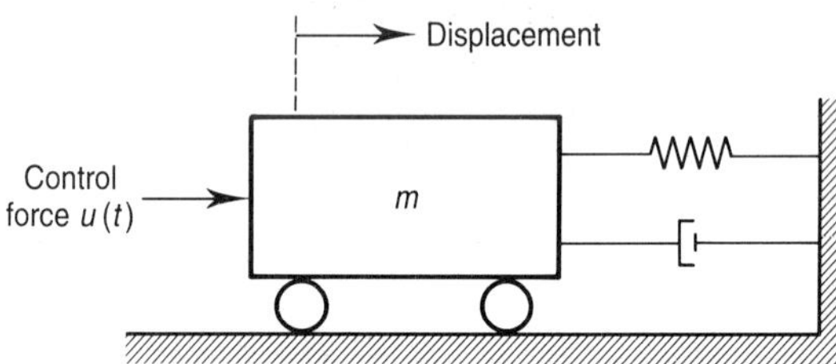

Fig. 3.5. Mechanical system with one degree of freedom.

If we consider the movement in the horizontal direction, the displacement $y(t)$ with respect to the equilibrium position is given by the differential equation

$$\frac{d^2y}{dt^2} + 2\nu\omega\frac{dy}{dt} + \omega^2 y = \frac{u(t-\tau)}{m} + \frac{v(t)}{m} \tag{3.51}$$

where m is the mass, ν is the damping and ω is the natural frequency. $v(t)$ is an exciting force that produces an undesirable vibration in the system, whilst $u(t-\tau)$ is the active control force produced by an actuator. It is assumed that the actuator has a significant inertia, which results in a time delay represented by τ.

The actual values of the parameters are $m = 2922.7$ kg, $\nu = 0.0124$ and $\omega = 21.79$ rad/s.

The problem of active control essentially consists of generating the control force u automatically as a feedback function of the measured position y in order to mitigate the vibration induced by the excitation v, which is considered to be a disturbance. Practical examples of active control in civil engineering structures, and specifically the application of predictive control for reduction of the response in experimental buildings under seismic loads, will be dealt with later on, in Chapter 9.

The purpose of considering this case here is to illustrate the robustness of the predictive control law of (3.17) when it is applied in the presence of discrepancies between the predictive model and the model describing the process. In this example we use the following discrete time model to describe the relationship between the control u and the displacement y at each sampling instant k:

$$y(k) = \sum_{i=1}^{2} a_i y(k-i) + \sum_{i=1}^{2} b_i u(k-i-r) \tag{3.52}$$

where, for a sampling period T,

$$\begin{aligned} b_1 &= \frac{m}{\omega^2} - \frac{m}{\omega^2} e^{-\nu\omega T}[\cos\omega\sqrt{1-\nu^2}T + \frac{\nu\sin\omega\sqrt{1-\nu^2}T}{\sqrt{1-\nu^2}}] \\ b_2 &= \frac{m}{\omega^2} e^{-2\nu\omega T} + \frac{m}{\omega^2} e^{-\nu\omega T}[\frac{\nu\sin\omega\sqrt{1-\nu^2}T}{\sqrt{1-\nu^2}} - \cos\omega\sqrt{1-\nu^2}T] \\ a_1 &= 2e^{-\nu\omega T}\cos\omega\sqrt{1-\nu^2}T \\ a_2 &= -e^{-2\nu\omega T} \\ r &= \frac{\tau}{T} \qquad \text{expressed as an integer number.} \end{aligned} \tag{3.53}$$

These expressions have been obtained by discretizing the equation of motion (3.51) using the procedure outlined in Appendix A (Section A.1.4).

A sampling period $T = 1$ second and the second-order predictive model

$$\hat{y}(k+j|k) = \sum_{i=1}^{2} \hat{a}_i \hat{y}(k+j-i|k) + \sum_{i=1}^{2} \hat{b}_i \hat{u}(k+j-i|k) \tag{3.54}$$

is used for the application of the predictive control law of (3.17).

The parameters of the predictive model of (3.54) are obtained using expressions (3.53), with erroneous values for the mass, the frequency and the time delay. These erroneous values for the mass and the frequency are taken as $m + \Delta m$ and $\omega + \Delta\omega$ respectively. The time delay is always made equal to zero in the predictive model, while in the process equation it is represented by the integer number r.

As in the preceding example, the value of the reference trajectory at the end of the prediction horizon is assumed equal to the setpoint, which in this case is zero (equilibrium position).

In order to assess the robustness of the predictive control law of (3.17) when these erroneous parameters are used, the roots of the characteristic polynomial (3.29) are calculated for different values of λ. Table 3.1 shows the modulus of the most significant root for different values of $\Delta m, \Delta\omega$ (in percentages of the actual values m, ω) and that of the time delay r.

It can be seen in the first row in Table 3.1, that a value of $\lambda = 1$ is sufficient to guarantee stability. The remaining cases correspond to the existence of errors, and it can be seen that the prediction horizon must be extended to achieve a stable control. The value of λ required in each case depends on the magnitude of the errors. The value of λ must be greater in the case of discrepancies in the time delay than in the case of errors in the mass and the frequency.

Table 3.1

Modulus of the main root of the characteristic polynomial

r	$\Delta\omega$	Δm	λ: 1	2	3	4	5	6	7	8
0	0	0	**0.99**	0.50	0.67	0.76	0.81	0.85	0.88	0.90
1	0	0	1.86	1.21	**0.96**	0.82	0.76	0.79	0.84	0.87
2	0	0	1.76	1.34	1.15	1.04	**0.96**	0.90	0.86	0.84
3	0	0	1.65	1.34	1.20	1.11	1.05	**0.99**	0.95	0.92
0	−20%	20%	1.71	**0.32**	0.58	0.70	0.77	0.81	0.84	0.87
1	−20%	20%	2.01	1.31	1.03	**0.88**	0.79	0.74	0.77	0.82
3	−20%	20%	1.71	1.39	1.24	1.15	1.09	1.04	1.00	**0.97**

Now let us consider the application of the control law of (3.17) to case number 4 in Table 3.1, in which the time delay in the process is $r = 3$. The exciting force is assumed to be an initial impulse which places the mass out of equilibrium. The objective of the control is to damp out the vibration and to return the mass to the equilibrium position as soon as possible.

Figure 3.6 shows the unstable displacement of the system for various values of $\lambda \leq 5$. Figure 3.7 shows the already stable control for $\lambda = 6$. The effectiveness of the control in damping the response after the initial impulse is remarkable as compared to the case shown in Figure 3.8 with no control. Note that, under predictive control, the vibration disappears within less than one second, while it still continues after five seconds in the uncontrolled case.

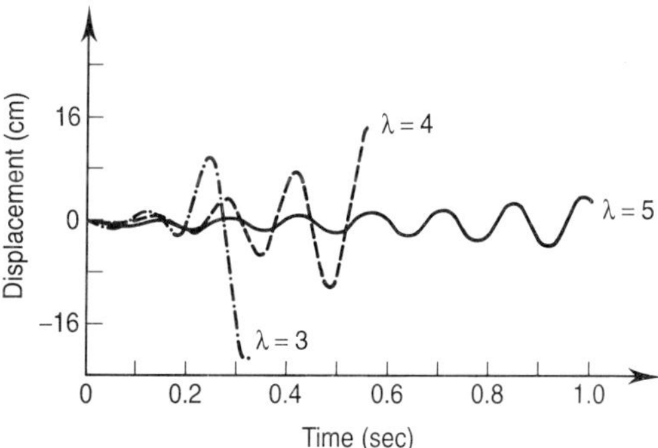

Fig. 3.6. Unstable responses for $\lambda \leq 5$.

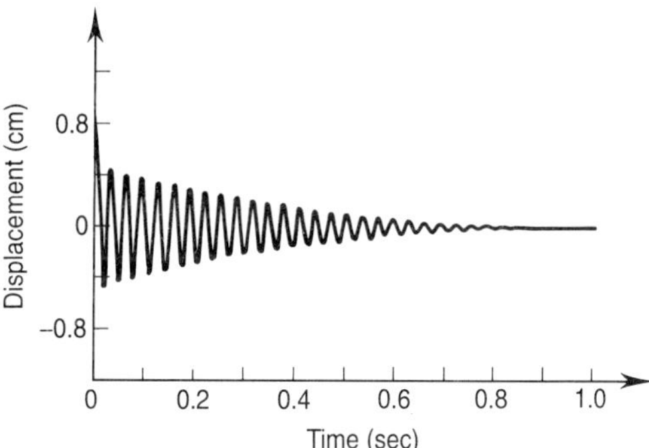

Fig. 3.7. Stable response for $\lambda = 6$.

3.6 CONCLUSIONS

The need to solve the instability problem related to the application of predictive control to processes with an unstable inverse led to the development of the extended strategy presented in this chapter. The new strategy is essentially based on an evaluation of the predicted process input/output sequences within a prediction horizon. In order to carry out this evaluation, a performance criterion is redefined at each sampling instant within this horizon. The length of the prediction horizon (λ) is an important design parameter in this strategy. Its intuitive and physical meaning makes its selection a simple matter. Through an appropriate choice of λ the designer may ensure a stable and robust predictive control

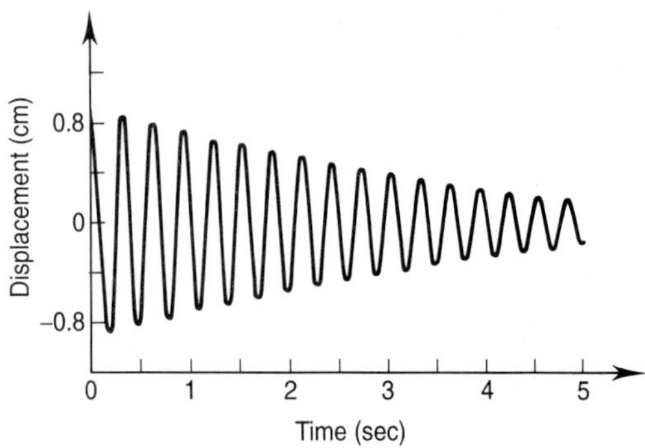

Fig. 3.8. Response without control.

in the presence of unstable inverse dynamics, time delay and modelling errors. Other relevant parameters in the design of the predictive control scheme are those defining the reference trajectory and also the weighting factors entering in the performance criterion. Using this set of tools, the designer can accommodate the control system for a more or less demanding control effort or, in other words, he or she can generate the driving desired trajectory (DDT) that would require a more abrupt or smoother control action. As a matter of fact, the physical realizability of the DDT is an essential concept in the framework of APCS stability, as will be pointed out in Chapter 5.

This chapter has presented an illustrative theoretical analysis with the purpose of showing the significance of the concepts behind the extended strategy. This analysis has been supported by numerical examples. Further chapters will illustrate a variety of industrial applications of these concepts. Nevertheless, the ultimate guarantee for the efficient performance of predictive control is based on the assumption of a good prediction. In practice, in a real environment, this may only be fulfilled by the adaptive formulation outlined in Chapter 1. The following two chapters focus on the design of adaptive systems in the context of predictive control and on the stability theory of APCS.

Chapter 4

ADAPTIVE SYSTEMS WITHIN THE CONTEXT OF PREDICTIVE CONTROL

4.1 INTRODUCTION

4.1.1 The need for adaptive control and first developments

Industrial processes are typically of a non-linear, time varying type. Therefore, modelling is very difficult in most cases, especially when the process operating conditions frequently vary. This is why many modern control theory methods requiring exact knowledge of the process model cannot be applied satisfactorily in the industrial context. In most cases control strategies with constant parameters cannot adapt to changes in the operating conditions. The non-linear and time varying nature of industrial processes is the main motivating factor for the development of *adaptive control* techniques.

The area of adaptive control has been the subject of intense research over the last two decades. A well known paper [Lan74] surveyed the developments of the first generation of adaptive systems, called *model reference adaptive systems* (MRAS). The introduction of adaptive control did not signify a change in the control strategy, but only an adaptation of the controller parameters to prevent a decrease in the global performance of the control system in the presence of changes in the process dynamics. Thus model reference adaptive systems used traditional negative feedback methodology as the basic control strategy and adapted the controller parameters using different algorithms.

Later, in an attempt to overcome the problems inherent in the use of the negative feedback strategy, research into adaptive control systems considered the control strategy of the so-called *linear quadratic optimal control* [PA73]. However, the complexity required for the calculation of the control signal in accordance with the formulation of this strategy rendered this alternative barely practicable in an adaptive context.

However, as a specific case within the preceding formulation, an adaptation mechanism based on identification by least squares was combined with the simplest form of quadratic optimal control law, which is that of *minimum variance*

[Ast70], leading to the so-called *self-tuning regulators* (STR) [AW73]. An extension to the minimum variance strategy led to the formulation of the so-called *self-tuning controllers* (STC) [CG75]. However, the simplification achieved in the implementations of the STRs and STCs with respect to the application of the general quadratic optimal control law did not compensate for the performance limitations of these solutions when applied in industry.

From all this it could be inferred that the problems with the first developments in adaptive control lay not only in the adaptation mechanism but, fundamentally, in the control strategy adopted.

4.1.2 Predictive control and the approach of stability

Predictive control, as described in the preceding chapters, originated [Mar74, Mar76a, Mar76b, Mar80] in an attempt to solve the theoretical and practical problems that adaptive control systems were encountering. Using predictive control, if the prediction made by the model is good, then, by definition, the evolution of the process variables can be driven perfectly. Consequently, under predictive control, the objective of the adaptation mechanism was clearly defined: it consisted of achieving the convergence of the model prediction error to zero as efficiently as possible, or, in a more realistic case, in rendering the prediction error bounded within a close margin around zero.

The aforementioned objective can be expressed in terms of stability similar to the classical concepts outlined in Appendix A. Therefore, in the initial work on adaptive predictive control [Mar74, Mar76a], the solution given to the adaptation problem was based on the stability approach, rather than on an approach based on optimization. Also, a stability approach was previously used in the design of MRAS [Lan74, NK74].

While the perspective of stability implies a tendency or a boundedness, it is well known that the perspective of optimization generally implies the minimization of a performance criterion. In the area of modelling, where the purpose is the identification of the model parameters, the perspective of optimization was perfectly defined in an academic context by the *least squares method* for linear processes with constant parameters. However, where the process parameters vary unpredictably, as occurs in the industrial context, the choice of optimization index is, at best, unclear.

Within the context of predictive control, the role of the adaptive system is to solve the problem of predicting the evolution of the process variables and not necessarily that of performing the parametric identification of the equations that define the complete dynamics of the process. That is to say, rather than identifying the process, it is necessary to predict efficiently. If the prediction error is associated with the state of the adaptive system and this system is designed in such a way that the prediction error approaches zero as time approaches infinity or, at least, remains bounded, then we may assume that the adaptive system behaves in a stable manner.

From all this, within the context of predictive control, it seems almost natural to consider the problem of synthesis of an adaptive system from the perspective of stability. However, within this perspective the existence of an optimality criterion is not excluded.

The results discussed in this and the following chapter are based on, and extend, those previously published on the analysis and synthesis of adaptive predictive control systems from the perspective of stability [Mar76a, Mar84, MSF84, Mar86, CMSF88].

4.1.3 Outline of this chapter

In this chapter we will present the basic concepts of adaptive systems within the context of predictive control. Section 4.2 introduces notations and concepts that will be used in the rest of the book, and also defines different scenarios, characterized by different hypotheses, progressively approaching the real industrial environment. Sections 4.3–4.5 outline the principles of the design of adaptive systems. First, Section 4.3 describes, from an intuitive basis, the desired performance objectives that a global adaptive predictive system should achieve. Second, in Section 4.4 these objectives are recalled and translated into stability concepts, defining what will be referred to as *APCS stability*. Third, Section 4.5 states, by means of conjecture, the specific conditions for the adaptive systems which, if satisfied, would lead to achievement of the desired objectives. Sections 4.6 – 4.9 solve the problem of synthesis of the adaptive system for the different scenarios considered. The properties derived for the adaptive system in these sections match the conditions of the conjecture under a certain assumption. Later, in the following chapter, these properties are used to prove that, when the desired objective is physically realizable and predictive control is combined with the adaptive system, the said assumption is satisfied. In this way, the present chapter and Chapter 5 prove that the overall adaptive predictive control system is *globally stable*, which is to say, it achieves the desired performance objectives.

All the theory presented in this chapter has been developed for single-input, single-output systems because of the simplicity of the notations. Its extension to the multivariable case, which has been considered in [Mar84, MSF84, Mar86], is direct and easy.

4.2 SCENARIOS, NOTATION AND BASIC CONCEPTS

In this section we introduce some notation and concepts that will be useful in subsequent sections to deal in a unified framework with different cases that can be considered, depending on the severity of the hypothesis used to describe the process approaching the real environment in an industrial context. First, we describe the cases for which the adaptive system will be formulated. Second, we present the models describing the process. Third, we introduce the basic definitions and notation for adaptive predictive control systems.

4.2.1 Scenarios

The problem of the synthesis of an adaptive system can initially be approached in a theoretical manner for an *ideal case*, based on the following hypotheses:

(a) The process is described by linear equations with constant parameters.

(b) The equations of the process and the model have the same order.

(c) There exist no measurement noises, or unmeasurable disturbances acting on the process.

However, if we want to guarantee that the adaptive system works satisfactorily in an industrial environment, the synthesis problem must be approached using hypotheses that accord with such an environment. These hypotheses may be the following:

(a_1) The process is described by linear equations, but with time varying parameters.

(b_1) The process and model equations may have different orders.

(c_1) There exist measurement noises and unmeasurable disturbances randomly acting on the process.

The hypotheses give above are useful in defining the different scenarios or *real cases* that we will consider in this chapter, and are as follows:

1. **The real case with no difference in structure**: In this case hypotheses (a) and (b) of the ideal case will be maintained, but hypothesis (c) will be substituted by hypothesis (c_1).

2. **The real case with difference in structure**: In this case only hypothesis (a) of the ideal case will be maintained, while (b) and (c) will be substituted by hypotheses (b_1) and (c_1).

3. **The real case with time varying parameters**: This case takes into account hypotheses (a_1), (b_1) and (c_1). The first hypothesis accounts

for the basically non-linear and variable nature of the industrial process. For this, when describing it through linear equations, parametric changes will occur due to any kind of variation in the conditions of the operation environment.

4.2.2 Description of the process

For the real case with time varying parameters, let us consider a single-input, single-output process where the relation between the inputs and outputs, using notation similar to that used in preceding chapters, may be given by

$$\begin{aligned} y_a(k) = \sum_{i=1}^{n} a_i(k) y_a(k-i) + \sum_{i=1}^{m} b_i(k) u_a(k-r-i) \\ + \sum_{i=1}^{p} c_i(k) w_a(k-r_1-i) + \xi(k) \end{aligned} \tag{4.1}$$

where y_a, u_a and w_a are the actual, present or previous values of the process output, input and measurable disturbance respectively; r and r_1 represent the pure time delays related to the process input and measurable disturbances respectively; $\xi(k)$ represents the effect of the unmeasured disturbances on the process output at instant k; and a_i, b_i and c_i are the process parameters which, in the context of adaptive systems, may be unknown and time variant.

Using equation (4.1) to express the outputs y_a at instants $k-1, \ldots, k-r$ and substituting the results recursively into the right-hand side of (4.1), the above process equation may be written in the following form

$$\begin{aligned} y_a(k) = \sum_{i=1}^{n} a'_i(k) y_a(k-r-i) + \sum_{i=1}^{m+r} b'_i(k) u_a(k-r-i) \\ + \sum_{i=1}^{p+r} c'_i(k) w_a(k-r_1-i) + \xi(k) \end{aligned} \tag{4.2}$$

where parameters a'_i, b'_i and c'_i are easily obtained from a_i, b_i and c_i.

In order to use a more manageable notation, this equation will be expressed in the form

$$y_a(k) = \theta(k)^T \phi_a(k-d) + \xi(k) \tag{4.3}$$

where

$$\begin{aligned}\phi_a(k-d)^T = [&y_a(k-d), y_a(k-d-1), \ldots, y_a(k-d-n+1),\\ &u_a(k-d), u_a(k-d-1), \ldots, u_a(k-2d-m+2),\\ &w_a(k-d_1), w_a(k-d_1-1), \ldots, w_a(k-p-d-d_1+2)]\end{aligned}$$

and

$$\begin{aligned}\theta(k)^T = [&a_1'(k), a_2'(k), \ldots, a_n'(k), b_1'(k), b_2'(k), \ldots, b_{m+r}'(k),\\ &c_1'(k), c_2'(k), \ldots, c_{p+r}'(k)]\end{aligned}$$

where d represents the time delay related to the process input and includes the discretization delay plus the pure delay, that is, $d = r + 1$; d_1 is the equivalent for the measurable disturbances, $d_1 = r_1 + 1$; $\theta(k)$ is the vector of the parameters of the process.

The measured values of the process variables differ from their actual values due to measurement errors, noise, etc., as expressed by the following equations:

$$\begin{aligned}y(k) &= y_a(k) + n_y(k)\\ u(k) &= u_a(k) + n_u(k)\\ w(k) &= w_a(k) + n_w(k)\end{aligned} \tag{4.4}$$

where, for each variable, subscript a denotes its actual value and n denotes the additional corrupting signal.

As a result, the corresponding measured vector ϕ becomes

$$\phi(k) = \phi_a(k) + n_\phi(k) \tag{4.5}$$

This vector is known as the *input/output (I/O) vector* or *regression vector*. Substituting (4.4) and (4.5) into process equation (4.3) we obtain

$$y(k) = \theta(k)^T \phi(k-d) + \Delta(k) \tag{4.6}$$

where

$$\Delta(k) = n_y(k) - \theta(k)^T n_\phi(k-d) + \xi(k)$$

which can be referred to as the *perturbation signal* and represents the effect of the unmeasured perturbations and measurement noise acting on the process. The previous assumptions about the relation between the inputs and outputs of the process come from hypotheses (a_1) and (c_1) of the real case, in which the existence of measurement noise and unmeasured perturbations is considered and where the process parameters may vary with time.

The process equation (4.6) may also be written in the form:

$$y(k) = \theta_o(k)^T \phi_o(k-d) + \theta_1(k)u(k-d) + \Delta(k) \tag{4.7}$$

where $\theta_1(k)$ is the single parameter included in vector $\theta(k)$ in the process equation (4.6) that multiplies the control signal in the inner product at instant $k-d$, $u(k-d)$. $\theta_o(k)$ and $\phi_o(k-d)$ result after the exclusion of the parameter $\theta_1(k)$ and the control signal $u(k-d)$ from the parameter vector $\theta(k)$ and the I/O vector $\phi(k-d)$ respectively.

The parameter vector $\theta_o(k)$ and the parameter $\theta_1(k)$ are always assumed to be bounded, while the absolute value of $\theta_1(k)$ is assumed to be greater than a certain positive constant, that is, $|\theta_1(k)| > \nu > 0$.

4.2.3 Description of the adaptive predictive control system

An adaptive predictive control system (APCS) can be represented by the block diagram shown in Figure 4.1, which, in essence, results from the combination of a predictive control system and an adaptive system. Looking at Figure 4.1, we may observe the presence of the predictive model in both the adaptive system and the predictive controller according to the double role it plays in the global system at each sampling instant k.

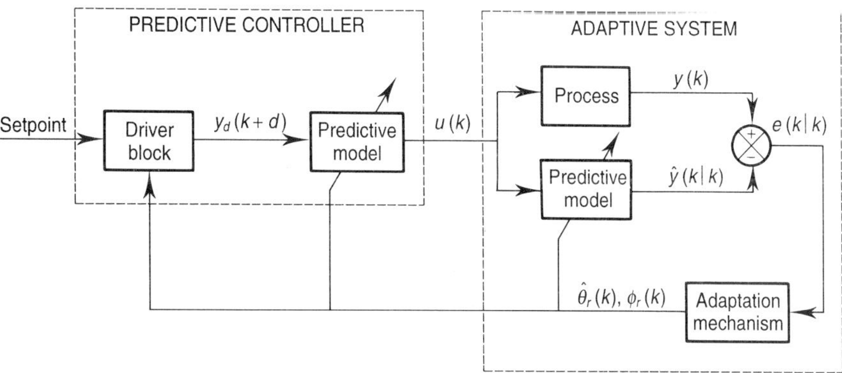

Fig. 4.1. Overall block diagram of an adaptive predictive control system.

In the adaptive system, the predictive model gives an estimation of the process output at instant k using the model parameters also estimated at instant k, which will be denoted by $\hat{\theta}_r(k)$, and the control signals and process outputs already applied or measured at previous instants, which are included in the I/O vector. This estimation is expressed in the form

$$\hat{y}(k|k) = \hat{\theta}_r(k)^T \phi_r(k-d) \tag{4.8}$$

where

$$\begin{aligned}\phi_r(k-d)^T = [&y(k-d), y(k-d-1), \ldots, y(k-d-n_r+1,\\ &u(k-d), u(k-d-1), \ldots, u(k-d-m_r+1),\\ &w(k-d_1), w(k-d_1-1), \ldots, w(k-d_1-p_r+1)]\end{aligned}$$

and

$$\begin{aligned}\hat{\theta}_r(k) = [&\hat{a}_1(k), \hat{a}_2(k), \ldots, \hat{a}_{n_r}(k), \hat{b}_1(k), \hat{b}_2(k), \ldots, \hat{b}_{m_r}(k),\\ &\hat{c}_1(k), \hat{c}_2(k), \ldots,, \hat{c}_{p_r}(k)]\end{aligned}$$

The dimensions of ϕ_r and θ_r are usually less than or equal to the dimensions of ϕ and θ. $\phi_r(k-d)$ contains a subset of the most recent process inputs and outputs included in $\phi(k-d)$. These assumptions account for hypothesis (b_1) of the real case, in which it was considered that the process and model equations had different orders.

The estimated parameter vector $\hat{\theta}_r(k)$ is generated by the adaptation mechanism using the information available on the process inputs and outputs up to instant k.

The following notation is adopted for the estimation error, the difference between the process output and that of the model:

$$e(k|k) = y(k) - \hat{y}(k|k) = y(k) - \hat{\theta}_r(k)^T \phi_r(k-d) \tag{4.9}$$

$e(k|k)$ is called the *a posteriori estimation error* because it is calculated after the model parameters have been adjusted at instant k.

Another estimation error which, as we will see in this chapter, is extremely important in the analysis and synthesis of adaptive systems, is the *a priori estimation error*, which is defined by the equation

$$e(k|k-1) = y(k) - \hat{y}(k|k-1) = y(k) - \hat{\theta}_r(k-1)^T \phi_r(k-d) \tag{4.10}$$

where $\hat{y}(k|k-1)$ is the *a priori* estimation for the process output $y(k)$ calculated from the model parameter vector not yet adjusted for instant k but adjusted at instant $k-1$, that is, $\hat{\theta}_r(k-1)$.

According to the principle of predictive control, already stated in Chapter 2 and now invoked by predicting the output at instant $k+d$ instead of $k+1$, the predictive model calculates the control action $u(k)$ in order to make the predicted output at instant $k+d$ equal to the driving desired output at the same instant. Using equation (4.8) we may express this predicted output in the form

$$\hat{y}(k+d|k) = \hat{\theta}_r(k)^T \phi_r(k)$$

When $r > r_1$, some terms within $\phi_r(k)$ related to the disturbance w will not have been measured at instant k yet and, since they will be generally unknown, their value will be taken as equal to zero. For the sake of simplicity we will assume in the following that $r \leq r_1$.

Denoting now the driving desired trajectory as $y_d(k+d)$, the principle of predictive control can be translated into the equation

$$y_d(k+d) = \hat{\theta}_r(k)^T \phi_r(k) \tag{4.11a}$$

This equation may also be written in the form

$$y_d(k+d) = \hat{\theta}_{ro}(k)^T \phi_{ro}(k) + \hat{\theta}_1(k) u(k) \tag{4.11b}$$

where $\hat{\theta}_1(k)$ is the parameter included in vector $\hat{\theta}_r(k)$ in equation (4.11a) that corresponds to the control signal $u(k)$ in the inner product. $\hat{\theta}_{ro}(k)$ and $\hat{\phi}_{ro}(k)$ result from excluding the parameter $\hat{\theta}_1(k)$ and the control signal $u(k)$, respectively, from $\hat{\theta}_r(k)$ and $\hat{\phi}_r(k)$.

The predictive control law can be written from (4.11b) in the form

$$u(k) = \frac{y_d(k+d) - \hat{\theta}_{ro}(k)^T \phi_{ro}(k)}{\hat{\theta}_1(k)} \tag{4.12}$$

Clearly, the adaptation mechanism must always guarantee that the parameter $\hat{\theta}_1(k)$ is not zero for any instant k.

The difference between the process output and the driving desired output is defined as the *control* or *tracking error*

$$\epsilon(k) = y(k) - y_d(k) \tag{4.13}$$

which will play an important role in characterizing the performance of adaptive predictive control systems, as is considered in the following section.

4.3 CONTROL OBJECTIVES FOR ADAPTIVE PREDICTIVE CONTROL SYSTEMS

Recalling the analysis performed in Chapter 3, we may extract two essential features associated with the application of predictive control. One is that the process output follows the driving desired trajectory (DDT) to reach the setpoint. The other is that the DDT is bounded and *physically realizable*, which means that the control signal able to produce it is also bounded.

The above features of predictive control are based on the implicit assumption that the process dynamics is known and that this knowledge is contained in the predictive model. In practice, if a 'good' model is available *a priori* to describe the dynamic relationship between the process inputs and outputs, we may use it as the predictive model. For a certain level of discrepancy between the process dynamics and the predictive model, the extended control strategy may even be adequate to maintain a satisfactory control performance. However, in most cases it is difficult to obtain precise information about the process *a priori.* Moreover, even if it were available, the process would frequently vary its dynamics in its evolution over time, as usually happens in an industrial context. The purpose of adding an adaptive system to the predictive control system is to reach, in this time varying environment, the satisfactory results that would be obtained by the predictive control system if the process dynamics were known.

Therefore, the objectives that we would expect to obtain from an adaptive predictive control system (APCS) can be summarized by the following two points:

1. After a certain time for adaptation, the process output should follow the driving desired trajectory (DDT) with a tracking error that is always bounded in the real case or is zero at the limit in the ideal case.

2. The DDT should be maintained bounded, and the control action able to achieve the above objective 1 should also be bounded. In other words, the DDT to be followed should be physically realizable and bounded.

In fact, the boundedness of the control action considered in objective 2 above is physically imposed in practical applications by the actuator limits. Likewise, the boundedness condition on the DDT is naturally associated with the limited variation rank of the sensors that allow the measurement of the process variables.

The boundedness conditions considered above do not represent limitations of the control capabilities in practice, as any variable that, due to its nature, would evolve in an unbounded way could be controlled by means of a variable associated with it according to an incremental or derivative relation, whose values would evolve over a certain bounded variation range.

In short, the above two objectives state that the goal of the APCS is to make the process output follow a physically realizable and bounded DDT.

4.4 DESIGN FROM A STABILITY PERSPECTIVE

In this section we transfer the above intuitive control objectives to a mathematical setting in terms of stability. Thus we will have a framework with which to work out all the subsequent formulation involved in the design of APCS in this and the following chapter.

The first aspect to consider concerns the adaptive system, which includes the adaptation mechanism to adjust the parameters of the predictive model, with the purpose of making this model produce an output that is as close to the process output as possible when both receive the same input. Thus it is natural to characterize the performance of the adaptive system by the difference between the process output and the predictive model output. In our case, this difference is represented by the estimation error $e(k|k)$ defined in (4.9).

In the ideal case considered in Section 4.2.1, under hypotheses (a)–(c), the result that we should expect from a good solution to the problem of synthesis of the adaptation mechanism is that the error $e(k|k) \to 0$ as $k \to \infty$ from any initial condition. If we obtain this result, associating the estimation error $e(k|k)$ with the state of the adaptive system with an equilibrium state at zero, we may say that the adaptive system is globally asymptotically stable, as defined classically in Appendix A.

In the real cases considered in Section 4.2.1 it is not realistic to require the estimation error to go asymptotically to zero since, for example, simply the measurement noise could cause deviation of this error from zero, even in the hypothetical case that this value had been reached. Therefore, the result that we should expect from a good solution to the problem of designing the adaptation mechanism is that, from any initial condition, the error should become bounded after a certain sampling instant k_f and the corresponding bound should be the smallest possible taking the level of noise, disturbances of all types and parametric changes acting on the process into account. This result can be expressed mathematically in the form

$$|e(k|k)| < \bar{M} \text{ for all } k \geq k_f$$

In the real cases, we may also associate the error $e(k|k)$ with the state of the adaptive system, but now considering the disturbances and noise as exogenous inputs. Then we may interpret the above boundedness condition of this error in terms of the stability concepts outlined in Appendix A by saying that the adaptive system is externally stable.

When considering the overall adaptive predictive control system (APCS), we would like to relate the desired performance with the corresponding stability concepts considered above for the adaptive system. Thus we may assume that the tracking error $\epsilon(k)$, which represents the difference between the driving desired output and the process output, is related to the state of the overall system and, if the adaptive system has been designed stable, the tracking error should satisfy the stability properties previously considered for the estimation error $e(k|k)$.

However, this result is not sufficient to cover the desired performance objectives for APCS as stated in Section 4.3. In fact, as a control system, APCS has to generate a control signal within conditions imposed by practice, being mainly that it must be bounded. We may recall the analysis in Chapters 2 and 3 showing how, in order to make the process output follow certain trajectories, it may be necessary to apply unbounded control signals. In such cases, even with stable estimation and tracking errors, we would obtain an undesirable APCS performance. In the same way, the process output driven by APCS will have to evolve bounded and within the range limits of the sensors. The same consideration may also be extended to the different kinds of perturbation. All these requirements on the input/output signals may be included in a single condition by stating that the input/output vector $\phi(k)$ must be bounded. This condition has necessarily to be included for satisfactory performance of APCS.

Taking the preceding considerations into account, we may now state the following definitions of *global stability* for APCS that correspond to the desired performance.

Definition 4.1: An adaptive predictive control system is said to be *globally stable* if the following conditions are satisfied:

1. $|\epsilon(k)| \leq M < +\infty \qquad \forall k \geq k_f > 0.$

2. $\|\phi(k)\| \leq \Omega < +\infty \qquad \forall k \geq k_f > 0.$

$\|\cdot\|$ denotes the euclidean norm. □

The above definition corresponds to the stability result that may be expected in the real cases. For the ideal case, the expected result will correspond to the following definition.

Definition 4.2: An adaptive predictive control system is said to be *globally asymptotically stable* if the following conditions are satisfied:

1. $\epsilon(k) \to 0$ as $k \to \infty$.

2 $\|\phi(k)\| \leq \Omega < +\infty \qquad \forall k \geq k_f > 0.$

□

The purpose of all the theoretical developments presented in this chapter and in Chapter 5 is to prove the above stability results corresponding to the ideal and the real cases. While this chapter deals with the analysis and synthesis of stable adaptive systems included in APCS, Chapter 5 will present the formal results of stability for the overall system. Both chapters form the body of what we refer to as *APCS stability theory.*

4.5 CONTRIBUTION OF THE ADAPTIVE SYSTEM TO APCS STABILITY

The remainder of this chapter is devoted to the synthesis of an adaptation mechanism within APCS. The design principle will be stated by means of Conjecture 1 in terms of specific conditions that, if satisfied by the adaptation mechanism, would make APCS globally stable in the sense of Definitions 4.1 (for the real cases) and 4.2 (for the ideal case) and, as a result, the desired control objectives would be achieved.

Conjecture 1: Assume that the DDT is physically realizable and bounded and that, for certain values M and k_f, the adaptive system of Figure 4.1 satisfies the following conditions:

1. $\hat{\theta}_r(k) = \hat{\theta}_r(k-d) \qquad \forall k \geq k_f > 0.$

2. $|e(k|k)| \leq M < +\infty \qquad \forall k \geq k_f > 0.$

Then the adaptive predictive control system represented in Figure 4.1 will have the following properties:

(a) $|\epsilon(k)| = |y(k) - y_d(k)| \leq M < +\infty \qquad \forall k \geq k_f > 0.$

(b) $\|\phi(k)\| \leq \Omega < +\infty \qquad \forall k \geq k_f > 0.$

□

Although in the following we are going to prove the above proposition, we call it a 'conjecture' since assumptions 1–2 need to be satisfied through an appropriate synthesis of the adaptation mechanism and this has not yet been performed, it being the content of Sections 4.6–4.9.

Equation (4.11a) can be written in the delayed form

$$y_d(k) = \hat{\theta}_r(k-d)^T \phi_r(k-d) \tag{4.14}$$

By comparing equations (4.8) and (4.14) it is clear that, if condition 1 holds, then we may write

$$y_d(k) = \hat{y}(k|k) \qquad \forall k \geq k_f > 0$$

From this property and condition 2, along with (4.9), property (a) of Conjecture 1 immediately follows. Property (b) is derived solely from the boundedness and physical realizability of the DDT.

Note that condition 1 defines a form of convergence of the AP model parameters, while condition 2 states the stability of the adaptive system as considered in the preceding section.

Once we have verified that the control objectives are achieved if the stability and convergence conditions of the conjecture hold, the problem to solve seems to be that of the synthesis of the adaptation mechanism that is able to satisfy these conditions. The following sections of this chapter will present a particular and illustrative solution of this synthesis problem for the different situations of the ideal case, the real case with no difference in structure and the real case with difference in structure.

It will be proved that the proposed adaptive systems only verify both conditions of Conjecture 1 if the sequence $\{\|\phi_r(k)\|\}$ is bounded. Clearly, the boundedness of the I/O vector cannot be guaranteed by the adaptive system alone, since the control signal is produced by the whole adaptive predictive control system. Only this global system can guarantee the bounding condition. However, it will be proved that the proposed adaptive systems verify properties that are close to the conditions of Conjecture 1 and are inherent in the adaptive system on its own, irrespective of the boundedness or otherwise of the sequence $\{\|\phi_r(k)\|\}$. In Chapter 5, these properties will be combined with the predictive control principle to ensure the boundedness of the sequence $\{\|\phi_r(k)\|\}$ and, as a result, the attainment of the desired control objective provided the DDT is physically realizable and bounded.

In the real case with time varying parameters, condition 1 of Conjecture 1 may be incompatible with the desired control objective when the process parameters exhibit a permanent variation. However, in this chapter, when dealing with the synthesis of the adaptation mechanism for that case, we will derive a form of convergence of the AP model parameters that will be sufficient to meet the desired control objectives in practice within constraints that are reasonable in an industrial environment, as shown in the application chapters of this book.

4.6 THE SYNTHESIS PROBLEM IN THE IDEAL CASE

4.6.1 Solution strategy

In this section we present a strategy for the design of an asymptotically stable adaptive system in the ideal case defined in Section 4.2. Later we will analyze how this adaptive system complies with conditions 1–2 of Conjecture 1. The desired stability result is defined in the form

$$\lim_{k\to\infty} e(k|k) = 0 \tag{4.15}$$

The following proposition defines the design strategy alone.

Proposition 4.1: Property (4.15) holds if the following condition is satisfied:

$$s(k_t) = \sum_{k=1}^{k_t} e(k|k)^2 \leq \delta^2 < +\infty \qquad \forall k_t > 0 \tag{4.16}$$

□

Proof: Clearly, since it sums the squares of the error, $s(k_t)$ is a non-decreasing sequence that may begin to grow from the instant at which the adaptation mechanism starts to operate, as is illustrated in Figure 4.2. If condition (4.16) holds, this sequence is bounded by δ^2 and thus its increments must tend to zero, that is, $e(k|k) \to 0$ as $k \to \infty$.

■

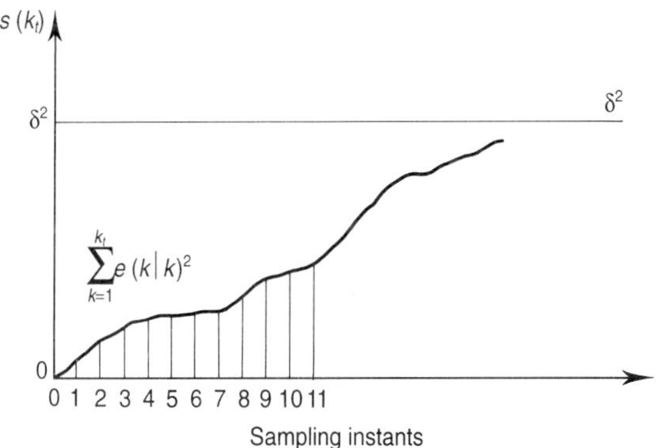

Fig. 4.2. Graphical illustration of condition (4.16).

Therefore, an adaptive system satisfying (4.16) will be asymptotically stable.

4.6.2 Example of synthesis of the adaptive system in the ideal case

In the ideal case the process will be described by the following equation:

$$y(k) = \theta^T \phi(k-d) \tag{4.17}$$

This equation is a particular case of equation (4.3) in which the perturbation signal $\Delta(k)$ is null and the parameter vector θ is constant.

The AP model is now described by

$$\hat{y}(k|k) = \hat{\theta}(k)^T \phi(k-d) \tag{4.18}$$

where the dimensions of $\hat{\theta}(k)$ and θ are equal.

A frequently used measure of the difference between the process and the model parameters is given by the *parameter identification error vector*, which, for this ideal case, is defined as follows:

$$\tilde{\theta}(k) = \theta - \hat{\theta}(k) \tag{4.19}$$

However, our synthesis strategy, as described in the preceding section, is based on the *a posteriori* estimation error $e(k|k)$. From the definition of such an error and from equations (4.17) and (4.18), we obtain the following equation:

$$e(k|k) = [\theta - \hat{\theta}(k)]^T \phi(k-d) \tag{4.20}$$

Using (4.20), the summation in (4.16) can be written in the form of

$$-s(k_t) = \sum_{k=1}^{k_t} e(k|k)[\hat{\theta}(k) - \theta]^T \phi(k-d) \tag{4.21}$$

Now we define the adaptation mechanism.

Proposition 4.2: Condition (4.16) is satisfied if the AP model parameter vector is generated at instant k by means of the following algorithm:

$$\hat{\theta}(k) = \sum_{h=1}^{k} e(h|h)B\phi(h-d) + \hat{\theta}(0) \tag{4.22}$$

where B is a positive definite matrix.

□

Proof: Let us consider the following expression:

$$\begin{aligned}\sum_{k=1}^{k_t}\Big[\sum_{h=1}^{k} z(h) - c\Big]^T B^{-1} z(k) = \frac{1}{2}\Big[\sum_{k=1}^{k_t} z(k) - c\Big]^T + B^{-1}\Big[\sum_{k=1}^{k_t} z(k) - c\Big] \\ + \frac{1}{2}\sum_{k=1}^{k_t} z(k)^T B^{-1} z(k) - \frac{1}{2} c^T B^{-1} c \geq -\frac{1}{2} c^T B^{-1} c\end{aligned} \tag{4.23}$$

which is valid for any time variant vector z and constant vector c. By substituting (4.22) into (4.21) we may write

$$-s(k_t) = \sum_{k=1}^{k_t}\Big[\sum_{h=1}^{k} e(h|h)B\phi(h-d) + \hat{\theta}(0) - \theta\Big]^T e(k|k)\phi(k-d) \tag{4.24}$$

Now, identifying

$$\begin{aligned} z(k) &= e(k|k)B\phi(k-d) \\ c &= \theta - \hat{\theta}(0) \end{aligned} \tag{4.25}$$

in (4.24), we may apply (4.23) to obtain $-s(k_t) \geq \delta^2$ with

$$\delta^2 = \frac{1}{2}[\theta - \hat{\theta}(0)]^T B^{-1}[\theta - \hat{\theta}(0)] = \frac{1}{2}\tilde{\theta}(0)^T B^{-1}\tilde{\theta}(0) \tag{4.26}$$

This proves condition (4.16).

■

If B is equal to the identity matrix I, the bound becomes

$$\delta^2 = \frac{1}{2}||\tilde{\theta}(0)||^2$$

This result shows that the bound δ^2 depends on the size of the parameter identification error at the initial instant. If this is zero, the estimation error $e(k|k)$ will be zero for all k. However, if the parameter identification error is not zero, the sequence $s(k_t)$ approaches the bound and thus the distance between $s(k_t)$ and δ^2 decreases, as illustrated in Figure 4.2. Note that in this analysis we may consider any time other than 0 as the initial instant in (4.16). Then we may intuitively associate the distance from $s(k_t)$ to the bound δ^2 with the square of the euclidean norm of the parameter identification error. Since this distance decreases, we may conclude that the square of the norm of the parameter identification error will also decrease. Since this norm is positive, its increments will tend to zero, which implies that the increments of the AP model parameters will also tend to zero. This intuitive fact will be rigorously proved in the following section.

4.6.3 Convergence of the AP model parameters

For the case where $B = I$, the algorithm of (4.22) can be written in the recursive form:

$$\hat{\theta}(k) = e(k|k)\phi(k-d) + \hat{\theta}(k-1) \tag{4.27a}$$

and the parametric increment at each instant k:

$$\Delta\hat{\theta} = \hat{\theta}(k) - \hat{\theta}(k-1) = e(k|k)\phi(k-d) \tag{4.27b}$$

From (4.19) and (4.27a) we may write

$$\tilde{\theta}(k) + e(k|k)\phi(k-d) = \tilde{\theta}(k-1) \tag{4.28}$$

$$\|\tilde{\theta}(k)\|^2 + 2e(k|k)\tilde{\theta}(k)^T \phi(k-d) + e(k|k)^2\|\phi(k-d)\|^2 = \|\tilde{\theta}(k-1)\|^2 \quad (4.29)$$

From (4.20) and (4.29) we obtain

$$\|\tilde{\theta}(k)\|^2 - \|\tilde{\theta}(k-1)\|^2 = -2e(k|k)^2 - e(k|k)^2\|\phi(k-d)\|^2 \quad (4.30)$$

Given that $\{\|\tilde{\theta}(k)\|\}$ is a non-negative and non-increasing sequence as observed in (4.30), it converges to zero; thus:

$$\lim_{k\to\infty} [-2e(k|k)^2 - e(k|k)^2\|\phi(k-d)\|^2] = 0 \quad (4.31)$$

In the preceding section we saw that $e(k|k) \to 0$ as $k \to \infty$. Using this result and (4.27b) we may write

$$\lim_{k\to\infty} [-e(k|k)^2\|\phi(k-d)\|^2] = -\lim_{k\to\infty} \|\Delta\hat{\theta}(k)\|^2 = 0 \quad (4.32)$$

which implies

$$\lim_{k\to\infty} \Delta\hat{\theta}(k) = 0 \quad (4.33)$$

This proves the asymptotic convergence of the AP model parameters. Note that this result is independent of the fact that the sequence $\|\phi(k)\|$ may be unbounded.

4.6.4 Analysis of the results

In the preceding sections we illustrated the synthesis of an adaptation mechanism in the ideal case that guarantees the following asymptotic properties for the adaptive system:

1. $\lim_{k\to\infty} \Delta\hat{\theta}(k) = 0$
2. $\lim_{k\to\infty} e(k|k) = 0$

These properties are independent of the fact that the input/output vector sequence $\{\|\phi(k)\|\}$ may be bounded or not. However, it is important to remark that the first of these properties does not take the form of the first condition of Conjecture 1. Thus the fulfillment of the control objectives associated with that conjecture must be analyzed carefully. We will now discuss the influence of the boundedness of the sequence $\{\|\phi(k)\|\}$ in achieving these objectives.

If $\{\|\phi(k)\|\}$ is bounded, using the above property 1 in (4.8) and (4.14), it can readily be seen that $y_d(k) \to \hat{y}(k|k)$ as $k \to \infty$. Using this result and the above property 2 in (4.9), we obtain

$$\lim_{k\to\infty}[y(k) - y_d(k)] = 0$$

which, since we are in the ideal case, is clearly a result of asymptotic stability even stronger than that considered in property (a) of Conjecture 1.

However, if the sequence $\{\|\phi(k)\|\}$ is unbounded, it cannot be ensured that $y_d(k) \to \hat{y}(k|k)$ as $k \to \infty$; not even that the difference $y_d(k) - \hat{y}(k|k)$ is bounded.

Nevertheless, the above properties 1 and 2 of the adaptive system, which are also independent of any control law, will be combined in the next chapter with a predictive control law and the boundedness of $\{\|\phi(k)\|\}$ will be proved by a stability analysis, thus achieving the desired control objectives, provided the DDT is physically realizable and bounded.

4.6.5 A priori estimation error and general adaptation expression

The algorithm of (4.22) can be rewritten in the following recursive form, which is better suited for a practical implementation:

$$\hat{\theta}(k) = e(k|k)B\phi(k-d) + \hat{\theta}(k-1) \tag{4.34}$$

However, although we know that the adaptation mechanism will meet the desired stability criteria if (4.34) is satisfied, such an adaptation mechanism is still incomplete in its definition, as we explain below.

It can be seen that the algorithm of (4.34) adjusts the model parameters at instant k using the error $e(k|k)$. However, this error depends on the model output $\hat{y}(k|k)$, which is calculated by means of the AP model of equation (4.18) from the value of the parameters already adjusted for instant k. To break this circle, we will derive the relation between the *a posteriori* estimation error and the *a priori* estimation error, which in this ideal case takes the form

$$e(k|k-1) = y(k) - \hat{y}(k|k-1) = y(k) - \hat{\theta}(k-1)^T\phi(k-d) \tag{4.35}$$

Subtracting (4.35) from (4.9) and taking into account the fact that $\theta_r(k) = \theta(k)$ and $\phi_r(k) = \phi(k)$ in this case, we obtain

$$e(k|k) - e(k|k-1) = [\hat{\theta}(k-1) - \hat{\theta}(k)]^T\phi(k-d) \tag{4.36}$$

From (4.34) and (4.36) we may write

$$e(k|k) - e(k|k-1) = -e(k|k)\phi(k-d)^T B\phi(k-d) \tag{4.37}$$

$$e(k|k) = \frac{e(k|k-1)}{1 + \phi(k-d)^T B\phi(k-d)} \tag{4.38}$$

This relation between the *a posteriori* and *a priori* estimation errors allows us to conclude the definition of the adaptation mechanism by writing (4.34) in the form

$$\hat{\theta}(k) = \frac{e(k|k-1)B\phi(k-d)}{1+\phi(k-d)^T B\phi(k-d)} + \hat{\theta}(k-1) \tag{4.39a}$$

which can also be written

$$\hat{\theta}(k) = G(k)[y(k) - \hat{\theta}(k-1)^T \phi(k-d)] + \hat{\theta}(k-1) \tag{4.39b}$$

where

$$G(k) = \frac{B\phi(k-d)}{1+\phi(k-d)^T B\phi(k-d)}$$

With this formulation, (4.39b) takes the form of a linear recursive filter with a variable gain: the new vector of the estimated parameters is obtained by adding to the previous one an increment equal to the estimation error obtained using the preceding vector multiplied by the gain vector $G(k)$.

Other techniques of parametric estimation, such as those of the gradient [Men73], which minimize a function of the square of the prediction error in the direction of the gradient (from which the name is derived), or the estimation techniques based on optimization criteria [Lju87], converge with small differences towards the general expression that we have derived in this case from a perspective of stability.

4.7 REAL CASE WITH NO DIFFERENCE IN STRUCTURE

4.7.1 Solution strategy

In the preceding section we performed the synthesis of an asymptotically stable adaptive system for the ideal case, illustrating in this way the application of stability concepts in the solution of this type of problem.

The main difference in the real case considered here is that we must now deal with perturbations and noise acting on the process. Therefore, as already discussed in Section 4.4, the purpose of this section will be to design a stable adaptive system and to analyze its compliance with conditions 1 and 2 of Conjecture 1.

The stability and convergence properties of the adaptive system in the ideal case were based on two issues: (i) the boundedness of the non-decreasing function $s(k_t)$ in (4.16) by a constant δ^2 which results in dependence on the parametric identification error at the initial time $||\tilde{\theta}(0)||$; and (ii) the non-increasing behaviour of $\|\tilde{\theta}(k)\|$ as shown in (4.30). The boundedness (4.16) leads to $e(k|k) \to 0$ as $k \to \infty$, while (4.30) leads to $\Delta\hat{\theta}(k) \to 0$ as $k \to \infty$.

An important feature of the adaptation mechanism (4.39) is that it operates continuously, that is, it adjusts the model parameters at each instant k. In the ideal case the only source contributing to the estimation error is the parameter identification error. Thus it seems logical to have a continuous adaptation in order to make the convergence as fast as possible. The situation is different when noise and disturbances are present, since they also contribute to the estimation error. The problem arises from the fact that their contribution is unpredictable, since noise and disturbances are not measurable.

In [Mar86] a general methodology is presented for the design of stable adaptive systems, which can be applied in the real case with no difference in structure and with variations in the process parameters. This methodology can be interpreted as an extension of the basic stability approach presented in the preceding section and it also considers a continuous adaptation of the AP models. The key point is the design of adaptation mechanisms such that, in a vein similar to that described for the ideal case, the non-decreasing function of the error $s(k_t)$ is always bounded by a non-decreasing function $\mu(k_t)$, whose increase depends on the disturbances and noises acting on the process as well as on the parameter variations. In this way it is guaranteed that the *a posteriori* estimation error $e(k|k)$ is bounded with bounds that depend on the initial parameter identification error, the size of the non-measurable noise and disturbances and on the variations of the process parameters. If the conditions of the real case disappear, that is, if the disturbances, noise and parameter variations cease, the error $e(k|k) \to 0$ as $k \to \infty$, and the adaptive system will then be asymptotically stable.

The main problem with this kind of result is that the non-decreasing function of the estimation error $s(k_t)$ and the boundedness function $\mu(k_t)$ may separate or may approach each other. This depends on whether the estimation error is mainly due to the effect of noise and disturbances or to that of the parameter identification error respectively. When the estimation error is due mainly to noise and disturbances, the information on which the parameter adaptation is based will be deceiving, and, consequently, the corresponding adaptation will be negative in the sense that the norm of the parameter identification error will increase and the boundedness function μ will grow more than the function of the estimation error s. On the other hand, if the estimation error is due mainly to the parameter identification error, the adaptation will be positive in the sense that the norm of the parameter identification error will decrease and functions s and μ will approach each other. This fact suggests the convenience of performing the adaptation only when the second case occurs, instead of doing it continuously.

This idea, first presented for the real case with no difference in structure in [Mar84, MSF84], is the core of the approach that we follow here. It is essentially based on the introduction of a criterion that allows the adaptation of the model parameters only when it is certain that this adaptation will lead to a reduction in the norm of the parameter identification error. This criterion is based on some estimated knowledge about the size of the noise and disturbances that can influence the process output unpredictably, and can be described in the following simple terms:

1. If the *a priori* estimation error is of the same (or a lesser) order than a function of the maximum level that the perturbation signal can reach, it may be possible that such an error is due more to the perturbation signal than to the parameter identification error. In such a case, the adaptation will be stopped.

2. If the *a priori* estimation error is greater than such a function, such an estimation error is mainly due to the parameter identification error and, consequently, the adaptation will be performed.

Clearly, the problem of synthesis will basically consist of determining the aforementioned function of the maximum level of the perturbation signal that is able to guarantee that, when adaptation is performed, the norm of the parameter identification error will be reduced, and also that the highest number of possible adaptations will be allowed.

At this point it is important to note that the effect of the non-measurable disturbances in the process output is only totally unpredictable at an initial stage. However, since the disturbances contribute to the evolution of the process output, their effect can subsequently be predicted in part by the AP model itself. Therefore, the contribution of the non-measurable disturbances to the maximum level of the perturbation signal is not usually as large as it might seem *a priori*.

4.7.2 Example of solution to the synthesis problem

Since in this case non-measurable disturbances and noise are present, which are represented by the perturbation signal $\Delta(k)$, the process equation is

$$y(k) = \theta^T \phi(k-d) + \Delta(k) \tag{4.40}$$

Here we use the same AP model as for the ideal case defined in (4.18), since we are again assuming no structure difference between the process and the AP model.

In the following we use the simplest form of adaptation mechanism (4.39), defined by (4.27a):

$$\hat{\theta}(k) = e(k|k)\phi(k-d) + \hat{\theta}(k-1)$$

and derive a criterion for the execution, or not, of the adaptation algorithm that always leads to a decrease in $\|\tilde{\theta}(k)\|$.

By subtracting (4.18) from (4.40) and using the parameter identification error defined in (4.19), we may write

$$e(k|k) - \Delta(k) = \tilde{\theta}(k)^T \phi(k-d) \tag{4.41}$$

From (4.27a), equation (4.29) is obtained as shown in Section 4.6.3. From (4.29) and (4.41) we can derive the following expressions

$$\begin{aligned} \|\tilde{\theta}(k)\|^2 - \|\tilde{\theta}(k-1)\|^2 = & -2e(k|k)[e(k|k) - \Delta(k)] \\ & - e(k|k)^2 \phi(k-d)^T \phi(k-d) \end{aligned} \tag{4.42}$$

$$\|\tilde{\theta}(k)\|^2 - \|\tilde{\theta}(k-1)\|^2 = -e(k|k)^2[2 + \phi(k-d)^T \phi(k-d)] + 2e(k|k)\Delta(k) \tag{4.43}$$

In (4.43) it can be seen that $\|\tilde{\theta}(k)\|^2 - \|\tilde{\theta}(k-1)\|^2$ will be negative if the following condition is met:

$$|e(k|k)| > \frac{2|\Delta(k)|}{2 + \phi(k-d)^T \phi(k-d)} \tag{4.44}$$

For $B = I$ the relation between the *a priori* and *a posteriori* identification errors (4.38) may be written in the form

$$e(k|k) = \frac{e(k|k-1)}{1 + \phi(k-d)^T \phi(k-d)} \tag{4.45}$$

Using (4.45), condition (4.44) may be written in the form

$$|e(k|k-1)| > \frac{2[1 + \phi(k-d)^T \phi(k-d)]}{2 + \phi(k-d)^T \phi(k-d)} |\Delta(k)| \tag{4.46}$$

In terms of the *a priori* estimation error, the adaptation mechanism of (4.27a) can be written in the following form:

$$\hat{\theta}(k) = \frac{e(k|k-1)\phi(k-d)}{1 + \phi(k-d)^T \phi(k-d)} + \hat{\theta}(k-1) \tag{4.47}$$

If condition (4.46) is verified at instant k, then it is convenient to perform the adaptation of (4.47) since this ensures that $\|\tilde{\theta}(k)\|^2 - \|\tilde{\theta}(k-1)\|^2$ will be negative. Thus, this condition may give us a criterion for executing the adaptation. We may note that all variables involved in (4.46) are known or can be calculated at instant k, except the perturbation signal $|\Delta(k)|$. One way of coping with this uncertainty is to use an estimate of an upper bound of this signal. To do this, we may consider this bound to be defined as follows:

$$\Delta_b \geq \max_{0<k<\infty} |\Delta(k)| + \rho \qquad \rho > 0 \tag{4.48}$$

We now propose modification of (4.47) to the following adaptation mechanism:

$$\hat{\theta}(k) = \frac{\psi(k)e(k|k-1)\phi(k-d)}{1 + \psi(k)\phi(k-d)^T\phi(k-d)} + \hat{\theta}(k-1) \tag{4.49}$$

where the value of the variable $\psi(k)$ is defined in the following way:

$$\psi(k) = 0 \;\text{ if }\; |e(k|k-1)| < \frac{2[1+\phi(k-d)^T\phi(k-d)]}{2+\phi(k-d)^T\phi(k-d)}\Delta_b < 2\Delta_b \tag{4.50a}$$

$$\psi(k) = 1 \;\text{ if }\; |e(k|k-1)| \geq \frac{2[1+\phi(k-d)^T\phi(k-d)]}{2+\phi(k-d)^T\phi(k-d)}\Delta_b \geq \Delta_b \tag{4.50b}$$

Conditions (4.50) define a criterion for turning on/off the execution of the adaptation algorithm (4.49), which guarantees that when the variation of the AP model parameters occurs, condition (4.46) will be verified and, consequently, the norm of the identification error will decrease as desired.

The above result can be summarized in the following lemma, whose proof has been presented along with the construction of the adaptation mechanism.

Lemma 4.1: Along with the operation of the adaptive system described by the process equation (4.40), the model equation (4.18) and the adaptation mechanism defined in (4.48)–(4.50), the parameter identification error satisfies the following property:

$$\|\tilde{\theta}(k)\|^2 - \|\tilde{\theta}(k-1)\|^2 \leq 0 \qquad \forall k > 0 \tag{4.51}$$

□

It is interesting to note that the smaller the value of ρ in equation (4.48), that is, the more accurate the estimation of the upper limit Δ_b, the greater will be the number of adaptations allowed and, logically, the better will be the results of adaptation, which will be analyzed in the following section.

4.7.3 Properties of the adaptive system

The adaptation mechanism derived above guarantees the stability and convergence properties of the adaptive system that we state in the following lemmas.

Lemma 4.2: Along with the operation of the adaptive system considered in Lemma 4.1, if the norm of the I/O vector remains bounded, that is, $\|\phi(k-d)\| < \Omega < +\infty \; \forall k > 0$, there exists a time instant $k_f < +\infty$ such that

$$\begin{aligned} &\text{(a)} \quad \hat{\theta}(k) \; = \; \hat{\theta}(k-1) && \forall k \geq k_f > 0 \\ &\text{(b)} \quad |e(k|k)| \; < \; 2\Delta_b && \forall k \geq k_f > 0 \end{aligned}$$

□

Proof: Let us consider an instant h at which adaptation is performed so that condition (4.50b) is satisfied. Thus,

$$|e(h|h-1)| \; \geq \; \Delta_b \tag{4.52}$$

From (4.52), (4.45) and the fact that $\|\phi(k-d)\|$ is bounded as expressed in the lemma statement, we obtain

$$|e(h|h)| > \omega > 0 \qquad \text{with} \;\; \omega = \frac{\Delta_b}{1+\Omega^2} \tag{4.53}$$

From (4.45) and (4.50b), the following can also be derived:

$$|e(h|h)|[2+\phi(h-d)^T \phi(h-d)] \geq 2\Delta_b \tag{4.54}$$

From (4.54) and (4.43):

$$\|\tilde{\theta}(h)\|^2 - \|\tilde{\theta}(h-1)\|^2 \; \leq \; -|e(h|h)|2\Delta_b + 2e(h|h)\Delta(h) \tag{4.55}$$

Therefore,

$$\|\tilde{\theta}(h)\|^2 - \|\tilde{\theta}(h-1)\|^2 \; \leq \; -2|e(h|h)|[\Delta_b - |\Delta(h)|] \tag{4.56}$$

From (4.56), (4.48) and (4.53) we finally obtain

$$\|\tilde{\theta}(h)\|^2 - \|\tilde{\theta}(h-1)\|^2 \; \leq \; -\gamma^2 < 0 \qquad \text{with} \;\; \gamma^2 = 2\omega\rho \tag{4.57}$$

From (4.57), it can be seen that, at each instant h at which adaptation is performed, $\|\tilde{\theta}(h)\|^2$ decreases, at least in γ^2. Since $\{\|\tilde{\theta}(h)\|^2\}$ is a non-negative sequence and its initial value $\|\tilde{\theta}(0)\|^2$ is bounded, the number of adaptation instants must be finite. Thus it may be concluded that there exists an instant k_f

from which there will be no more parameter adaptation, which proves property (a) of the lemma, and also, from (4.50a), property (b).

■

Lemma 4.3: Along with the operation of the adaptive system considered in Lemma 4.1, if the norm of the I/O vector is unbounded and there exists a subsequence $\{h\}$ characterized by $\psi(h) = 1 \;\; \forall h$, then

$$(a) \quad \lim_{h\to\infty} e(h|h) = 0$$
$$(b) \quad \lim_{h\to\infty} [\hat{\theta}(h) - \hat{\theta}(h-1)] = 0$$

□

Proof: Since $\{\|\tilde{\theta}(h)\|\}$ is a non-increasing sequence with a lower bound, it converges. Thus

$$\lim_{h\to\infty} [\|\tilde{\theta}(h)\|^2 - \|\tilde{\theta}(h-1)\|^2] = 0 \tag{4.58}$$

From (4.43) and (4.58) we obtain

$$\lim_{h\to\infty} [-e(h|h)^2[2 + \phi(h-d)^T \phi(h-d)] + 2e(h|h)\Delta(k)] = 0 \tag{4.59}$$

and therefore

$$e(h|h)^2[2 + \phi(h-d)^T \phi(h-d)] \to 2e(h|h)\Delta(h) \quad \text{as} \quad h \to \infty \tag{4.60}$$

From (4.48) and (4.54) we obtain

$$|e(h|h)|[2 + \phi(h-d)^T \phi(h-d)] \;\geq\; 2\rho + 2|\Delta(h)| \quad \forall h \tag{4.61}$$

Given (4.61), property (4.60) is only possible if $e(h|h) \to 0$ as $h \to \infty$, which proves property (a) of this lemma.

From (4.42) and (4.58) we may write

$$\lim_{h\to\infty} [2e(h|h)[e(h|h) - \Delta(h)] + e(h|h)^2\phi(h-d)^T \phi(h-d)] = 0 \tag{4.62}$$

From the chosen adaptation algorithm (4.27a) we write

$$\|\hat{\theta}(h) - \hat{\theta}(h-1)\|^2 = e(h|h)^2\phi(h-d)^T \phi(h-d) \tag{4.63}$$

and from (4.62) and (4.63) we derive

$$\lim_{h\to\infty} \{2e(h|h)[e(h|h) - \Delta(h)] + \|\hat{\theta}(h) - \hat{\theta}(h-1)\|^2\} = 0 \qquad (4.64)$$

Assuming property (a) of this lemma (already proved) and the fact that $|\Delta(h)|$ is bounded, (4.64) implies

$$\lim_{h\to\infty} \|\hat{\theta}(h) - \hat{\theta}(h-1)\| = 0$$

which proves property (b) and concludes the proof of this lemma.

■

From Lemma 4.3 and the definition of the adaptation mechanism, the following corollary, which is stated without proof, is directly derived.

Corollary 4.3.1: During the operation of the adaptive system considered in Lemma 4.1, the following properties hold:

(a) $\exists k_f$ such that $|e(k|k)| < 2\Delta_b \qquad \forall k \geq k_f > 0$

(b) $\lim_{k\to\infty} [\hat{\theta}(k) - \hat{\theta}(k-1)] = 0$

□

4.7.4 Analysis of the results

The adaptive system designed in this case verifies the conditions of Conjecture 1 only when the norm of the I/O vector remains bounded. However, we have derived convergence properties for the *a posteriori* estimation error and the AP model parameters that will be recalled in Chapter 5, allowing the APCS stability theory to prove the boundedness of $\{\|\phi(k)\|\}$ and, as a result, to attain the desired global stability result.

Note that the ideal case considered in Section 4.6 can be approached in the real case considered here, simply by assuming Δ_b arbitrarily small.

It must also be emphasized that, following the didactic objective of this book, we have chosen the simplest algorithm for the synthesis performed. The reader can find an extension of this algorithm in [Mar84] which allows us, for example, to guarantee the non-singularity of the parameter $\hat{\theta}_1(k)$ for all time and to minimize the boundedness function determining the execution criterion for the adaptation algorithm.

4.8 REAL CASE WITH DIFFERENCE IN STRUCTURE

4.8.1 Solution strategy

In preceding sections we have synthesized asymptotically stable and stable adaptive systems under the hypotheses of the ideal case and the real case respectively, with no difference in structure. In the latter case, the existence of noise and non-measurable disturbances acting on the process has been considered, but it has been assumed that the order of the equations that govern its behaviour is equal to the order of the adaptive model chosen.

However, the equations governing the process behaviour are generally of a high order and, consequently, in practice the order of the adaptive model chosen will be lower. This leads us to consider analysis and synthesis of adaptive systems able to handle not only the problem of noise and non-measurable disturbances, but also the problem of difference in structure.

When the solution for the real case with no difference in structure was published [Mar84, MSF84], it had already been suggested that some terms of the process dynamics could be included in the perturbation signal, thus reducing the AP model order while still preserving the validity of the method, provided that such terms were bounded. Clearly, a general solution to the problem of difference in structure cannot admit this restriction, since, for instance, the boundedness of terms within the process dynamics related to the process output cannot be assumed *a priori*.

Therefore, in the real case with difference in structure, the stability results obtained in the design of the adaptive system must be as those previously considered, irrespective of the boundedness or unboundedness of all the terms of the process I/O vector. As in the preceding cases, a desirable design objective for the adaptive system would be that, under the hypothesis of boundedness of the process I/O vector, the conditions of Conjecture 1 are satisfied, and that, in the absence of such a hypothesis, conditions able to allow APCS stability theory to demonstrate the boundedness of the I/O vector are derived.

Various authors considered this problem in the adaptive control area [Pra83, Pra84, KJ84, OPL85, KF85, KA86], leaning in some cases towards stability solutions that were only local, and in all cases assuming *a priori* the existence of a set of control parameters able to ensure stability. This hypothesis is not required in the solution provided by APCS stability theory, which was first presented in [CMSF88].

In the same context as [CMSF88], in this section we will design an adaptive system that is able to support the APCS stability results and that accords with the strategy outlined in the following:

1. Definition of a *normalized* adaptive system, so that all the process input/output signals become bounded, allowing dynamic terms to be included in an extended perturbation signal and, therefore, assuming a

reduced order AP model.

2. Similar to the case with no difference in structure, an adaptation mechanism is defined which guarantees that the norm of a reduced parameter identification error vector is a non-increasing function.

3. From the result obtained in point 2, the properties of convergence of the *a posteriori* estimation error and the AP model parameters are derived for the normalized adaptive system.

4. Finally, by the inverse process to normalization, the corresponding convergence properties are derived for the adaptive system without normalization.

4.8.2 Definition of the normalized system

Since the hypotheses for the process are the same as in the case given in Section 4.7, here we will use the same equation (4.40) to describe it:

$$y(k) = \theta^T \phi(k-d) + \Delta(k)$$

The AP model is now defined by equation (4.8)

$$\hat{y}(k|k) = \hat{\theta}_r(k)^T \phi_r(k-d)$$

In this model, the input/output vector $\phi_r(k-d)$ and the parameter vector $\hat{\theta}_r(k)$ have dimension n_r, which is less than the dimension (denoted n_p) of the process input/output vector $\phi(k-d)$ and parameter vector θ.

Let us introduce a normalization factor defined as

$$n(k) = \max\{\max_{1\le i\le n_p} |\phi_i(k-d)|, c\} \tag{4.65}$$

where ϕ_i is the ith component of ϕ and c is a positive constant that can be chosen arbitrarily.

The input/output variables in the normalized system are defined as follows:

$$y^n(k) = \frac{y(k)}{n(k)} \tag{4.66}$$

$$x(k-d) = \frac{1}{n(k)}\phi(k-d) \tag{4.67}$$

$$x_r(k-d) = \frac{1}{n(k)}\phi_r(k-d) \tag{4.68}$$

Using these variables in (4.40), the normalized process equation can be written in the form:

$$y^n(k) = \theta^T x(k-d) + \frac{\Delta(k)}{n(k)} \tag{4.69}$$

Taking the reduced dimension of the AP model into account, (4.69) may be written in the form

$$y^n(k) = {\theta_r}^T x_r(k-d) + {\theta_u}^T x_u(k-d) + \frac{\Delta(k)}{n(k)} \tag{4.70}$$

where

$$\begin{aligned} \theta^T &= [{\theta_r}^T, {\theta_u}^T] \\ x(k-d)^T &= [x_r(k-d)^T, x_u(k-d)^T] \end{aligned}$$

The dimension of θ_r is n_r, as for x_r; $x_u(k-d)$ contains the terms of $x(k-d)$ not included in $x_r(k-d)$ and thus its dimension is $n_p - n_r$.

According to the solution strategy outlined previously, the last two terms in (4.70) can be jointly interpreted as a normalized perturbation signal acting on the normalized process output $y^n(k)$. Thus we may write

$$y^n(k) = {\theta_r}^T x_r(k-d) + \Delta^n(k) \tag{4.71}$$

with

$$\Delta^n(k) = {\theta_u}^T x_u(k-d) + \frac{\Delta(k)}{n(k)} \tag{4.72}$$

The following lemma proves the boundedness of the sequence $\{\Delta^n(k)\}$ defined in (4.72).

Lemma 4.4: The sequence of normalized perturbation signals $\{\Delta^n(k)\}$, defined by equation (4.72), will be bounded if the sequence of perturbation signals $\{\Delta(k)\}$ is bounded.

□

Proof: From the definitions of $n(k)$ given in (4.65) and $x(k-d)$ given in (4.67) we can see that $|x_i(k-d)| \leq 1$ for all the components $i = 1, \ldots, n_p$ of vector $x(k-d)$. Since, as defined in (4.70), $x_u(k-d)$ has $n_p - n_r$ of these components, we may write

$$\|x_u(k-d)\| \leq (n_p - n_r)^{1/2} \tag{4.73}$$

Using the Cauchy–Schwarz inequality, we write

$$|\theta_u^T x_u(k-d)| \leq (n_p - n_r)^{1/2}\|\theta_u\| \tag{4.74}$$

From the boundedness of $\|\theta_u\|$ and $\{\Delta(k)\}$, and given that $n(k) \geq c > 0$, the boundedness of $\{\Delta^n(k)\}$ is immediately concluded from (4.72).

■

Note that the boundedness of $\{\Delta^n(k)\}$ only requires the boundedness of the non-measurable disturbances and noise included in $\Delta(k)$, which is not a restrictive assumption and which renders the result of Lemma 4.4 practicable and general.

By comparing the normalized process output with the estimations given by the AP model, we can define normalized *a priori* and *a posteriori* estimation errors as follows:

$$e^n(k|k-1) = y^n(k) - \hat{\theta}_r(k-1)^T x_r(k-d) \tag{4.75}$$

$$e^n(k|k) = y^n(k) - \hat{\theta}_r(k)^T x_r(k-d) \tag{4.76}$$

Next, we consider the problem of the synthesis of the adaptation mechanism for the normalized system.

4.8.3 Synthesis of the adaptation mechanism

Since Lemma 4.4 states the boundedness of the signal $\Delta^n(k)$, the problem of the synthesis of the adaptation mechanism for the normalized system can be approached in terms that are completely equivalent to those considered in Section 4.7 in the problem of synthesis for the real case with no difference in structure. Thus we propose an adaptation mechanism with an on/off criterion similar to the one defined in (4.49)–(4.50). This mechanism is now defined in the following form:

$$\hat{\theta}_r(k) = \frac{\psi(k)e^n(k|k-1)x_r(k-d)}{1+\psi(k)x_r(k-d)^T x_r(k-d)} + \hat{\theta}_r(k-1) \tag{4.77}$$

where the value of the variable $\psi(k)$ is now defined as follows:

$$\psi(k) = 0 \;\text{ if }\; |e^n(k|k-1)| < \frac{2[1+x_r(k-d)^T x_r(k-d)]}{2+x_r(k-d)^T x_r(k-d)}\,\Delta_b^n < 2\Delta_b^n \tag{4.78}$$

$$\psi(k) = 1 \text{ if } |e^n(k|k-1)| \geq \frac{2[1 + x_r(k-d)^T x_r(k-d)]}{2 + x_r(k-d)^T x_r(k-d)} \Delta_b^n \geq \Delta_b^n \quad (4.79)$$

where in this case Δ_b^n is an upper bound for the sequence $\{\Delta^n(k)\}$ which we define in the form

$$\Delta_b^n \geq \max_{0<k<\infty} |\Delta^n(k)| + \delta \qquad \delta > 0 \qquad (4.80)$$

The estimation of an upper bound for the norm of the unmodelled parameter vector $\|\theta_u\|$ and another one for the absolute value of the perturbation signal $|\Delta(k)|$ will allow estimation of the bound Δ_b^n. Note that the influence of the perturbation signal $\Delta(k)$ on the estimation of this bound can be reduced as desired by choosing the positive constant c in (4.65) appropriately large. When the norm of the I/O vector is larger than c, the first term on the right-hand side of equation (4.72), which accounts for the unmodelled dynamics, will not be affected by the value of c, but will remain bounded according to expression (4.74).

Conditions (4.78)–(4.80) define a criterion for turning on/off the adaptation algorithm (4.77) which leads to the result stated in the following lemma.

Lemma 4.5: Along with the operation of the normalized system described by equations (4.71), (4.72) and (4.75)–(4.80), the parameter identification error $\tilde{\theta}_r(k) = \theta_r - \hat{\theta}_r(k)$, satisfies the following property:

$$\|\tilde{\theta}_r(k)\|^2 - \|\tilde{\theta}_r(k-1)\|^2 \leq 0 \quad \forall k > 0$$

□

Proof: Subtracting (4.76) from (4.75) we obtain

$$e^n(k|k-1) = [\hat{\theta}_r(k) - \hat{\theta}_r(k-1)]^T x_r(k-d) + e^n(k|k) \qquad (4.81)$$

Substituting in (4.81) the value of $[\hat{\theta}_r(k) - \hat{\theta}_r(k-1)]$ by that obtained from (4.77), we obtain the relation between the normalized *a priori* and *a posteriori* estimation errors:

$$e^n(k|k) = \frac{e^n(k|k-1)}{1 + \psi(k) x_r(k-d)^T x_r(k-d)} \qquad (4.82)$$

From (4.82), the adaptation algorithm (4.77) can be written in the form

$$\hat{\theta}_r(k) = \hat{\theta}_r(k-1) + \psi(k) e^n(k|k) x_r(k-d) \qquad (4.83)$$

Subtracting θ_r from both sides of (4.83), we obtain

$$\tilde{\theta}_r(k) + \psi(k)e^n(k|k)x_r(k-d) = \tilde{\theta}_r(k-1) \tag{4.84}$$

$$\begin{aligned}\|\tilde{\theta}_r(k)\|^2 &+ 2\psi(k)e^n(k|k)\tilde{\theta}_r(k)^T x_r(k-d) \\ &+ \psi(k)^2 e^n(k|k)^2 x_r(k-d)^T x_r(k-d) = \|\tilde{\theta}_r(k-1)\|^2\end{aligned} \tag{4.85}$$

From (4.71) and (4.76), $e^n(k|k)$ can be expressed in the form

$$e^n(k|k) = \tilde{\theta}_r(k)^T x_r(k-d) + \Delta^n(k) \tag{4.86}$$

By combining (4.85) and (4.86) we obtain the expressions

$$\begin{aligned}\|\tilde{\theta}_r(k)\|^2 - \|\tilde{\theta}_r(k-1)\|^2 &= -2\psi(k)e^n(k|k)[e^n(k|k) - \Delta^n(k)] \\ &\quad - \psi(k)^2 e^n(k|k)^2 x_r(k-d)^T x_r(k-d)\end{aligned} \tag{4.87}$$

$$\begin{aligned}\|\tilde{\theta}_r(k)\|^2 - \|\tilde{\theta}_r(k-1)\|^2 &= -\psi(k)e^n(k|k)^2[2 + \psi(k)x_r(k-d)^T x_r(k-d)] \\ &\quad + 2\psi(k)e^n(k|k)\Delta^n(k)\end{aligned} \tag{4.88}$$

The left-hand side of (4.88) will be equal to zero if $\psi(k) = 0$, and will be less than zero if the following inequality holds for $\psi(k) = 1$:

$$|e^n(k|k)[2 + x_r(k-d)^T x_r(k-d)]| > 2|\Delta^n(k)| \tag{4.89}$$

which can be written in the form

$$|e^n(k|k)| > \frac{2|\Delta^n(k)|}{2 + x_r(k-d)^T x_r(k-d)} \tag{4.90}$$

From (4.82), inequality (4.90) can be expressed in the form

$$|e^n(k|k-1)| > \frac{2[1 + x_r(k-d)^T x_r(k-d)]}{2 + x_r(k-d)^T x_r(k-d)} |\Delta^n(k)| \tag{4.91}$$

By definition, (4.79) holds when $\psi(k) = 1$, which implies, assuming (4.80), that inequality (4.91) will be satisfied in this case. This concludes the proof of Lemma 4.5.

∎

4.8.4 Properties of the adaptive system

From the result of Lemma 4.5 we will first derive the stability and convergence properties for the normalized system, and later for the system without normalization. These properties are stated in Lemmas 4.6 and 4.7 respectively.

Lemma 4.6: Along with the operation of the normalized system considered in Lemma 4.5, there is a time instant k_f such that

$$\begin{array}{lll} \text{(a)} \ \hat{\theta}_r(k) & = \hat{\theta}_r(k-1) & \forall k \geq k_f > 0 \\ \text{(b)} \ |e^n(k|k)| & < 2\Delta_b^n & \forall k \geq k_f > 0 \end{array}$$

□

Proof: Using the same argument as given in (4.73), we may write

$$\|x_r(k-d)\|^2 \leq n_r \tag{4.92}$$

Let h be any instant within the sequence $\{k\}$ in which $\psi(h) = 1$. Then, according to (4.79), we have

$$|e^n(h|h-1)| \geq \Delta_b^n \tag{4.93}$$

Thus, from (4.82), (4.92) and (4.93), we obtain

$$|e^n(h|h)| \geq \omega_n > 0 \qquad \text{with } \omega_n = \frac{\Delta_b^n}{1+n_r} \tag{4.94}$$

Also, from (4.79) and (4.82), the following may be derived:

$$|e^n(h|h)| \, [2 + x_r(h-d)^T x_r(h-d)] \geq 2\Delta_b^n \tag{4.95}$$

Using (4.95) in (4.88) we readily obtain

$$\|\tilde{\theta}_r(h)\|^2 - \|\tilde{\theta}_r(h-1)\|^2 \leq -2|e^n(h|h)| \, [\Delta_b^n - |\Delta^n(k)|] \tag{4.96}$$

and from (4.80), (4.94) and (4.96) we obtain

$$\|\tilde{\theta}_r(h)\|^2 - \|\tilde{\theta}_r(h-1)\|^2 \leq -\gamma_n^2 < 0 \qquad \text{with } \gamma_n^2 = 2\omega_n\delta \tag{4.97}$$

Consequently, at each adaptation instant h, in which $\psi(h) = 1$, $\|\tilde{\theta}_r(h)\|^2$ decreases by at least γ_n^2. Therefore, following the same reasoning as in Lemma 4.2, the number of adaptation instants will be finite, which implies that properties (a) and (b) of Lemma 4.6 are satisfied.

■

Lemma 4.7: Along with the operation of the normalized system considered in Lemma 4.5, the *a posteriori* estimation error $e(k|k)$ for the system without normalization satisfies the following property:

$$\exists k_f < +\infty \text{ such that } |e(k|k)| < 2\Delta_b^n \max\{\|\phi(k-d)\|, c\} \qquad \forall k \geq k_f > 0$$

□

Proof: From the definition of $e(k|k)$ given in (4.9) and $e^n(k|k)$ given in (4.76), we observe that

$$e^n(k|k) = \frac{e(k|k)}{n(k)} \tag{4.98}$$

From (4.98) and property (b) of Lemma 4.6 we obtain

$$|e(k|k)| \; < \; 2\Delta_b^n \, n(k) \qquad \forall k \geq k_f > 0 \tag{4.99}$$

and using the definition of $n(k)$ in (4.65):

$$\begin{aligned} |e(k|k)| \; &< \; 2\Delta_b^n \max\{\max_{1\leq i\leq n_p} |\phi_i(k-d)|, c\} \\ &\leq 2\Delta_b^n \max\{\|\phi(k-d)\|, c) \qquad \forall k \geq k_f > 0 \end{aligned} \tag{4.100}$$

which completes the proof of Lemma 4.7.

■

4.8.5 Analysis of the results

From the above synthesis, a convergence property has been derived for the AP model parameters which fully guarantees condition 1 of Conjecture 1: the value of these parameters does not vary after a certain instant k_f. A convergence property has also been derived for the *a posteriori* estimation error $e(k|k)$, which will remain bounded provided that the norm of the process I/O vector $\phi(k-d)$ does not exceed an arbitrarily chosen positive constant c. Otherwise, this error will be bounded by a function proportional to such a norm, where the proportionality factor depends on the norm of the unmodelled parameter vector $\|\theta_u\|$.

Clearly, if the norm of the I/O vector remains bounded, the second condition of Conjecture 1 will also be met, that is, the stability of the adaptive system. The conditions under which the norm will be bounded will be defined within the framework of APCS stability theory in Chapter 5.

The real case with difference in structure considers hypotheses that are more similar to the real industrial context. In this sense, the results obtained for this case approach the desired control objectives of a more realistic environment. Furthermore, the cases previously studied in this chapter, ideal and real with no difference in structure, can clearly be included in the real case with difference in structure considered here. An important advantage in this case is that, while precise knowledge of the structure of the process equations was required before in order to define an AP model with the same structure, in this case a reduced-order AP model can be chosen without requiring such knowledge.

In this case the only knowledge of the process structure is that required to calculate the normalization factor $n(k)$, in which the dimension of the process I/O vector ϕ is involved, and to consider a bound for the norm of the unmodelled parameter vector θ_u. Nevertheless, both the dimension of vector ϕ and the norm of vector θ_u can be over-estimated to some degree without affecting the stability result obtained, and thus this does not represent a practical restriction.

In the study of the real case with difference in structure we have been guided by the didactic purpose of this book. As in the preceding cases, this is why we have chosen the simplest algorithms, which, while simplifying the analysis, are useful for illustrating the basic concepts. The reader can find an extension of the algorithm presented here in reference [CMSF88] which guarantees the non-singularity of $\hat{\theta}_1(k)$ and minimizes the boundedness function that determines the criterion for the execution of the adaptation.

It is important to note that the results presented in [CMSF88] have been corrected and simplified here: while in [CMSF88] it was assumed that the number of adaptations within the normalized system could tend to infinity, we have proved here that this number must be finite. The reader can compare both results and can also find an example in [CMSF88] illustrating how the use of an adaptive system designed for the real case with difference in structure is able to stabilize the operation of an APCS system in the desired form.

4.9 TIME VARYING PARAMETERS

4.9.1 Nature of the problem and solution strategy

The analysis of the real case with time varying parameters represents another step towards the real conditions of industrial processes. Within the scope of this book, time varying parameter processes must be considered within an engineering perspective, taking advantage of the available experience on the dynamic behaviour of such processes.

The dynamic nature of industrial processes is basically non-linear, but in general their behaviour can be described approximately by linear equations such as those considered in preceding sections for the real cases, with their parameters within certain values, provided that the process operates around a certain equilibrium state. These values will usually undergo variations due to the effect of external perturbations acting on the process. When the action of these perturbations is slow, the parametric variation will occur slowly, while if the action is rapid, the value of the parameters of the equations describing the process behaviour can vary rapidly. The parametric variations can also be originated by a change in the point or conditions in which the process is operating. The value of the process parameters will evolve dynamically, with more or less velocity depending on the characteristics of the change, from the initial to the final state around which the process operates.

In both cases the stabilization of the process operation depends to a considerable extent on the actuation of the control system, and this stabilization of the operation implies the stabilization of the value of the process parameters. In practice, an interactive effect can be recognized between the efficiency of the control system and the convergence of the process dynamics towards linear equations with constant parameters such as those considered in preceding sections.

In the context of adaptive predictive control systems, the control efficiency will be determined, on one hand, by the form of the application of the predictive control strategy, which will depend on each specific case; and, on the other hand, by the ability of the AP model to track the parametric change produced in the process equations. If this parametric change is tracked with enough efficiency and the control strategy is adequate, it seems reasonable to assume that the control system will be able to stabilize the operation and the dynamic behaviour of the process simultaneously. This fact is confirmed almost without exception by industrial practice.

In a previous study of analysis and synthesis of adaptive systems for a real case with no difference in structure and with time varying parameters [Mar86], results were presented on the boundedness of the *a posteriori* estimation error, with the only restriction being that the value of the process parameters had to be bounded. However, as has been discussed in Section 4.7.1, the continuous adaptation considered in [Mar86] can decrease or increase the norm of the parameter identification error due to the effect of the measurement noise and non-measurable disturbances acting on the process. For this reason, we have introduced a criterion that guarantees the decrease of the norm in the case of execution of the adaptation in the real cases considered previously.

Now, in the context in which the process parameters may vary at each instant, this criterion must be reformulated in the sense that it cannot be required that the norm of the parametric identification error decreases between two consecutive adaptation instants, since the variation in itself of the process parameters may prevent this decrease. However, the new criterion must ensure that the decrease of the norm occurs during the adaptation instant itself when comparing the parametric error before and after the adaptation has been performed.

The remainder of Section 4.9 develops the mathematical analysis that will allow us to verify that the adaptation algorithms considered previously for the real case with difference in structure does indeed include the new criterion described above in the presence of variations in the value of the process parameters.

4.9.2 Definition of the adaptive system

The adaptive system that we are now going to consider basically differs from that defined for the real case with differences in structure in the fact that the process is described by the general equation (4.6), which assumes time varying parameters $\theta(k)$:

$$y(k) = \theta(k)^T \phi(k-d) + \Delta(k)$$

In a similar way as in Section 4.8.2, the normalized process equation and perturbation signal are now

$$y^n(k) = \theta_r(k)^T x_r(k-d) + \Delta^n(k) \tag{4.101}$$

$$\Delta^n(k) = \theta_u(k)^T x_u(k-d) + \frac{\Delta(k)}{n(k)} \tag{4.102}$$

where

$$\theta(k)^T = [\theta_r(k)^T, \theta_u(k)^T]$$

and the normalization factor $n(k)$ is the one defined in (4.65).

Assuming that the sequence of the norms of the unmodelled parameter vector $\{\|\theta_u(k)\|\}$ is bounded and that we can estimate the bound, it is possible to use similar reasoning to that given in Lemma 4.4 in order to prove that the sequence of absolute values of the normalized perturbation signals $\{|\Delta^n(k)|\}$ is bounded. Also, it allows the estimation of an upper bound Δ_b^n for this sequence that verifies condition (4.80). Using this upper bound, the adaptation algorithms (4.75)–(4.79), formulated in Section 4.8, complete the definition of the adaptive system considered now.

4.9.3 Verification of the convergence condition

In order to prove that the objectives desired for the parameter adaptation are fulfilled, at this point we need to define the following two *reduced* parameter identification error vectors:

$$\tilde{\theta}_r(k|k) = \theta_r(k) - \hat{\theta}_r(k) \tag{4.103}$$

$$\tilde{\theta}_r(k|k-1) = \theta_r(k) - \hat{\theta}_r(k-1) \tag{4.104}$$

where $\tilde{\theta}_r(k|k)$ represents the reduced parameter identification error at instant k once the AP model parameters have been updated at the same instant k. $\tilde{\theta}_r(k|k-1)$ represents such an error before the AP model parameters have been updated. Clearly, if the adaptation algorithm is not executed at instant k, both vectors will be equal. Otherwise, the objective desired in the adaptation is defined in the following lemma.

Lemma 4.8: Along with the operation of the adaptive system described by equations (4.101)–(4.104) and (4.75)–(4.80), the following property is satisfied at each sampling instant h in which the criterion (4.78)–(4.80) allows the adaptation of the AP model parameters ($\psi(h) = 1$):

$$\|\tilde{\theta}_r(h|h)\|^2 - \|\tilde{\theta}_r(h|h-1)\|^2 < 0 \tag{4.105}$$

□

Proof: By subtracting (4.103) and (4.104) and using expression (4.83) for h, taking $\psi(h) = 1$, we obtain

$$\tilde{\theta}_r(h|h) + e^n(h|h)x_r(h-d) = \tilde{\theta}_r(h|h-1) \tag{4.106}$$

$$\begin{aligned} \|\tilde{\theta}_r(h|h)\|^2 &+ 2e^n(h|h)\tilde{\theta}_r(h|h)^T x_r(h-d) \\ &+ e^n(h|h)^2 x_r(h-d)^T x_r(h-d) = \|\tilde{\theta}_r(h|h-1)\|^2 \end{aligned} \tag{4.107}$$

By substituting (4.101) into (4.76) we obtain

$$e^n(h|h) = [\theta_r(h) - \hat{\theta}_r(h)]^T x_r(h-d) + \Delta^n(h) \tag{4.108}$$

and from (4.103) and (4.108) we obtain

$$e^n(h|h) - \Delta^n(h) = \tilde{\theta}_r(h|h)^T x_r(h-d) \tag{4.109}$$

By using (4.109), the following expressions can be derived from (4.107):

$$\begin{aligned} \|\tilde{\theta}_r(h|h)\|^2 - \|\tilde{\theta}_r(h|h-1)\|^2 =& - 2e^n(h|h)[e^n(h|h) - \Delta^n(h)] \\ &- e^n(h|h)^2 x_r(h-d)^T x_r(h-d) \end{aligned} \tag{4.110}$$

$$\begin{aligned} \|\tilde{\theta}_r(h|h)\|^2 - \|\tilde{\theta}_r(h|h-1)\|^2 =& - e^n(h|h)^2[2 + x_r(h-d)^T x_r(h-d)] \\ &+ 2e^n(h|h)\Delta^n(h) \end{aligned} \tag{4.111}$$

From (4.111) it can be deduced that $\|\tilde{\theta}_r(h|h)\|^2 - \|\tilde{\theta}_r(h|h-1)\|^2$ will be negative if the following condition is satisfied:

$$|e^n(h|h)|\,[2 + x_r(h-d)^T x_r(h-d)] \; > \; 2|\Delta^n(h)| \qquad (4.112)$$

As was proved in Lemma 4.5, this condition is satisfied for all h such that $\psi(h) = 1$, which concludes the proof of Lemma 4.8.

■

4.9.4 Analysis of the results

We have analyzed the real case with time varying parameters based on heuristic considerations drawn from the real experience of the dynamic behaviour of industrial processes, rather than pursuing a purely theoretical approach. We have used the same adaptive system derived for the real case with differences in structure whose hypotheses are closer to the conditions of the industrial environment and include, as particular cases, the previous ones studied in this chapter. Lemma 4.8 has proved that this adaptive system is still suitable for the case with time varying parameters, since property (4.105) ensures that the adaptation always causes the norm of the difference between the reduced process parameter vector and the estimated parameter vector to decrease. Thus, when process parameter changes occur, the adaptive system will work in the appropriate direction.

This property of the adaptation will strengthen the predictive control strategy and will continue until the control system stabilizes the operation of the process again. In this case, the convergence results previously obtained in the case with differences in structure will again be valid.

4.10 CONCLUSIONS

This chapter has focused on the problem of the analysis and synthesis of adaptive systems in the context of predictive control. In this context the role to be played by the adaptive system has motivated us to approach its design within a stability perspective.

In a simple way, through a conjecture, we have considered convergence properties for the *a posteriori* estimation error and for the AP model parameters, which, if satisfied by the adaptive system, guarantee the desired control objectives for the overall adaptive predictive control system (APCS).

Having the fulfillment of these properties as a design principle, we have approached the problem of the synthesis of the adaptive system under different hypotheses. First, the ideal case has been used to illustrate the basic concepts, and later we have discussed the real cases without and with differences in structure, the latter including those previously considered as particular cases. In this way we have progressively approached the hypotheses characterizing a real industrial environment.

The results obtained narrowly approach the desired convergence properties and even verify them as long as the process input/output (I/O) vector remains bounded. However, this condition does not depend on the adaptive system in itself, but, rather, on the APCS as a global system. The following chapter, based on the results obtained in this one, will complete the development of the APCS stability, proving the boundedness of the I/O vector and thus guaranteeing achievement of the desired performance objectives.

Finally, we have considered the real case with time varying parameters and we have demonstrated that the algorithms derived for the real case with differences in structure are also adequate in this context since they satisfy convergence properties that are desirable when the process parameters change.

Chapter 5

APCS STABILITY THEORY

5.1 INTRODUCTION

In the preceding chapter it was proved that adaptive systems based on a stability design can guarantee the conditions of Conjecture 1 provided that the input/output (I/O) vector ϕ is bounded. However, the boundedness of this vector cannot be guaranteed by the adaptive system in itself but, rather, it must be guaranteed by the overall control system.

In this chapter we will analyze the conditions under which predictive control and adaptive predictive control can guarantee the boundedness of the I/O vector, and the consequences that this boundedness implies for APCS stability.

Our stability analysis will be based on the following premises:

1. the results derived from the stable design of the adaptive systems considered in Chapter 4;

2. the stability nature of the process;

3. the physical realizability and the boundedness of the desired output.

The second premise given above corresponds to the fact that the stability results that we derive in this chapter are based on properties that are inherent in the stability nature of the process. In particular, within this perspective we will consider three classes of process which will be referred to as follows:

(a) processes of a linear and stable nature;

(b) processes of a linear and stable inverse nature;

(c) unstable processes with an unstable inverse.

The third premise allows us to derive general stability results for predictive and adaptive predictive control irrespective of the specific strategy used for their application, and therefore to unify the body of stability theory. For each class of processes considered, this premise takes a specific form, whose theoretical and practical consequences will be analyzed. The verification of this third premise by the different possible control strategies within the context of predictive and adaptive predictive control is simple in some cases and may seem obvious and intuitive in others, but a rigorous analysis of all possible cases is an extensive and complex open problem. In this book, such a rigorous analysis to prove the verification of the third premise is presented in Appendix D for a particular, but representative, case. However, it is important to emphasize that the theory presented here, according to the objectives of this book, provides the background required for the practical use of the methodology of predictive and adaptive predictive control in the industrial context, and in this sense its scope is barely diminished by the theory to be developed.

In the next section we will derive the input/output properties associated with the stability nature of the processes to be considered, while in Section 5.3 the concept of the physical realizability of the desired output will be defined in mathematical terms. Section 5.4 will translate the above-mentioned third premise into a general condition for processes of a linear and stable inverse nature, and will analyze its practical consequences from the design perspective and its role in the proof of APCS stability in Sections 5.5 and 5.6. Section 5.5 will consider the effect of modelling errors on the stability of the control system when predictive control is applied, as well as the limits on these modelling errors that determine such stability. Section 5.6 will analyze the role of adaptation to overcome the problem of the modelling errors found in predictive control, and will thus establish the general conditions that guarantee stability for the ideal and the real cases. The section will then consider for each case how the operation of an adaptive system can fulfil these general conditions and, therefore, guarantee global stability and the desired control objectives. Section 5.7 performs the parallel stability analysis for processes of a linear and stable nature. Last, Section 5.8 considers the corresponding stability analysis for unstable processes with an unstable inverse, and Section 5.9 presents the conclusions to the chapter.

5.2 INPUT/OUTPUT PROPERTIES ASSOCIATED WITH THE STABILITY NATURE OF THE PROCESS

5.2.1 Definitions

It is a standard result of systems theory that stable linear processes verify the so-called *bounded input/bounded output* property, which means that the output is bounded for any applied bounded input.

As we have already indicated, we will consider in our analysis three classes of process, all of them described in Chapter 4 (Section 4.2.2) by the same equations that we recall here:

$$y(k) = \theta(k)^T \phi(k-d) + \Delta(k) \tag{5.1a}$$

$$y(k) = \theta_o(k)^T \phi_o(k-d) + \theta_1(k)u(k-d) + \Delta(k) \tag{5.1b}$$

These equations, which are obviously equivalent, will be used extensively throughout this chapter. As defined in Section 4.2.2, d is the time delay; ϕ is the input/output (I/O) vector, which includes a number of control inputs u, outputs y and measurable disturbances w at present and previous instants, and we may rewrite it in the form

$$\begin{aligned}\phi(k)^T = [&y(k), y(k-1), \ldots, y(k-p), u(k), u(k-1), \ldots, u(k-q),\\ &w(k), w(k-1), \ldots, w(k-s)]\end{aligned} \tag{5.2}$$

Measurable disturbances are assumed bounded according to the industrial practice. $\theta(k)$ is the parameter vector. $\Delta(k)$ is the perturbation signal. $\theta_1(k)$ is the single parameter included in vector $\theta(k)$ in (5.1a) which multiplies in the inner product to the input $u(k-d)$. $\theta_o(k)$ and $\phi_o(k-d)$ result from the exclusion of $\theta_1(k)$ and $u(k-d)$ from $\theta(k)$ and $\phi(k-d)$ respectively.

The first class of processes to be considered will be referred to as those *of a linear and stable nature*, and is defined in the following.

Definition 5.1: A process is *of a linear and stable nature* when it can be described by an equation such as (5.1) and when, for any finite input sequence, it verifies the following property:

$$\max_{0 \le j \le k} |u(j)| > \tau_1 \max_{0 \le j \le k} |y(j+d)| - \tau_2 \qquad \forall k \ge 0 \tag{5.3}$$

where $0 < \tau_1 < +\infty$ and $0 \le \tau_2 < +\infty$.

□

Appendix C proves that discrete time stable linear processes with constant parameters (i.e. with all poles inside the unit circle) verify property (5.3). Also, the constants τ_1 and τ_2 are calculated and it is shown that, while the value of τ_1 depends solely on the process dynamics, the value of τ_2 depends on the initial conditions, that is, on the state of the process at $k = 0$.

However, property (5.3) is verified by a class of processes which is much wider than that considered in Appendix C, which is the class we have termed 'of a linear and stable nature'. For instance, within this class there will be linear processes with time varying parameters, whose dynamics will be able to describe the behaviour of a wide variety of industrial processes.

Given a process, its inverse process results from considering the original inputs as if they were outputs, and the original outputs as if they were inputs. The concept of an inverse process leads us to define the following second class of processes.

Definition 5.2: A process is *of a linear and stable inverse nature* when it can be described by an equation such as (5.1) and when, for any finite input sequence, it verifies the following property:

$$\max_{0\leq j\leq k} |y(j+d)| > \rho_1 \max_{0\leq j\leq k} |u(j)| - \rho_2 \qquad \forall k \geq 0 \qquad (5.4)$$

where $0 < \rho_1 < +\infty$ and $0 \leq \rho_2 < +\infty$.

□

As can be seen, this definition is derived directly from Definition 5.1, simply by considering the input/output exchange that the concept of inverse process involves.

By inverting the transfer function of discrete time stable linear processes with constant parameters, and by a proof such as that given in Appendix C, it can be shown that linear processes with constant parameters and with a stable inverse (i.e. with all zeros inside the unit circle) verify property (5.4). However, the class of processes with property (5.4) is much wider and closer to industrial reality.

Finally, *unstable processes with an unstable inverse*, which can be described by an equation such as (5.1) form the third class of processes that we will consider in our analysis, which clearly do not verify either Definition 5.1 or Definition 5.2.

5.2.2 Properties

The following lemmas state the input/output properties of the above-mentioned classes of process, which will be used later in the stability analysis.

Lemma 5.1: In a process of a linear and stable nature, where $|u(k)| < +\infty \;\; \forall k \geq 0$, the sequence $\{\|\phi(k)\|\}$ will only be unbounded if there exists a subsequence $\{k_s\}$ of $\{k\}$ such that

1. $|u(k_s)| > \gamma_1 \|\phi(k_s)\| - \gamma_2 \qquad \forall k_s \geq 0$

   with $0 < \gamma_1 < +\infty$ and $0 \leq \gamma_2 < +\infty$

2. $\lim_{k_s \to \infty} \|\phi(k_s)\| = +\infty$

□

Proof: From the definition of ϕ given in (5.2), it is clear that $\{\|\phi(k)\|\}$ will only be unbounded if either the output sequence $\{\|y(k)\|\}$ or the input sequence $\{\|u(k)\|\}$ are unbounded. Property (5.3) states that $\{\|y(k)\|\}$ is only unbounded if $\{\|u(k)\|\}$ is unbounded. Thus, for the class of systems of a linear and stable nature, $\{\|\phi(k)\|\}$ will only be unbounded in the case where $\{\|u(k)\|\}$ is unbounded. This implies that there must exist a subsequence $\{k_s\}$ of $\{k\}$ such that:

$$|u(k_s)| = \max_{0\leq j\leq k_s} |u(j)| \tag{5.5}$$

$$\lim_{k_s\to\infty} |u(k_s)| = +\infty \tag{5.6}$$

By applying the triangular inequality in (5.2) for $k = k_s$, we write

$$|\phi(k_s)| \leq \sum_{i=0}^{p} |y(k_s - i)| + \sum_{i=0}^{q} |u(k_s - i)| + \sum_{i=0}^{s} |w(k_s - i)| \tag{5.7}$$

According to (5.5),

$$\sum_{i=0}^{q} |u(k_s - i)| \leq (q+1)|u(k_s)| \tag{5.8}$$

Using (5.5) in (5.3), we may write

$$\max_{0\leq j\leq k_s} |y(j+d)| < \frac{|u(k_s)| + \tau_2}{\tau_1} \tag{5.9}$$

and thus

$$\sum_{i=0}^{p} |y(k_s - i)| \leq (p+1)\frac{|u(k_s)| + \tau_2}{\tau_1} \tag{5.10}$$

Since measurable disturbances are assumed bounded by a constant w_b:

$$\sum_{i=0}^{s} |w(k_s - i)| < (s+1)w_b \tag{5.11}$$

The substitution of (5.8), (5.10) and (5.11) into (5.7) allows us to write

$$|u(k_s)| > \gamma_1\|\phi(k_s)\| - \gamma_2$$

with

$$\gamma_1 = \frac{\tau_1}{(p+1) + \tau_1(q+1)} \qquad \gamma_2 = \frac{(p+1)\tau_2 + (s+1)w_b\tau_1}{(p+1) + \tau_1(q+1)}$$

This proves property 1 of this lemma. Since $u(k)$ is included in $\phi(k)$, property 2 is a direct consequence of (5.6).

■

Lemma 5.2: In a process of a linear and stable inverse nature, where $|u(k)| < +\infty \ \forall k \geq 0$, the sequence $\{\|\phi(k)\|\}$ will only be unbounded if there exists a subsequence $\{k_s\}$ of $\{k\}$ such that

1. $|y(k_s)| > \alpha_1 \|\phi(k_s - d)\| - \alpha_2 \qquad \forall k_s \geq 0$

   with $0 < \alpha_1 < +\infty$ and $0 \leq \alpha_2 < +\infty$

2. $\lim_{k_s \to \infty} \|\phi(k_s - d)\| = +\infty$

□

Proof: In a similar way as for Lemma 5.1, starting from property (5.4), which holds for all systems of a linear and stable inverse nature, we can deduce that the sequence $\{\|\phi(k)\|\}$ will only be unbounded if $\{|y(k)|\}$ is also unbounded. This implies that there must exist a subsequence $\{k_s\}$ of $\{k\}$ such that

$$|y(k_s)| = \max_{0 \leq j \leq k_s} |y(j)| \tag{5.12}$$

$$\lim_{k_s \to \infty} |y(k_s)| = +\infty \tag{5.13}$$

From (5.12) and property (5.4), following similar arguments to those given in Lemma 5.1, we derive property 1 of this lemma. From the process equation (5.1) and by using triangular and Cauchy–Schwarz inequalities along the subsequence $\{k_s\}$, we may write

$$|y(k_s)| \leq \|\theta(k_s)\| \, \|\phi(k_s - d)\| + |\Delta(k_s)|$$

From this expression and (5.13), given that $\|\theta(k_s)\|$ and $|\Delta(k_s)|$ are bounded, property 2 of this lemma is concluded.

■

Lemma 5.3: In an unstable process with an unstable inverse described by an equation such as (5.1), where $|u(k)| < +\infty \ \forall k \geq 0$, the sequence $\{\|\phi(k)\|\}$ will only be unbounded if one of the following conditions is met:

1. There exists a subsequence $\{k_{s1}\}$ of $\{k\}$ similar to the subsequence $\{k_s\}$ considered in Lemma 5.1.

2. There exists a subsequence $\{k_{s2}\}$ of $\{k\}$ similar to the subsequence $\{k_s\}$ considered in Lemma 5.2.

3. The two subsequences $\{k_{s1}\}$ and $\{k_{s2}\}$ given above exist.

□

Proof: For $\{\|\phi(k)\|\}$ to be unbounded, there must exist a subsequence $\{k_{so}\}$ of $\{k\}$, for which one of the components of $\phi(k_{so})$ reaches an absolute value greater than the absolute value of all the components of $\phi(j)$ in the interval $0 \leq j \leq k_{so}$. This component, which we will refer to as $c(k_{so})$, must verify for $\{k_{so}\}$:

$$\lim_{k_{so}\to\infty} |c(k_{so})| = +\infty \tag{5.14}$$

$$|c(k_{so})| \geq \frac{1}{\sqrt{m}} \max_{0\leq j\leq k_{so}} \|\phi(j)\| \tag{5.15}$$

m being the dimension of ϕ.

Equation (5.14) is needed for $\{\|\phi(k)\|\}$ to be unbounded, and (5.15) is derived from $c(k_{so})$ having the greatest absolute value of all the components of $\{\|\phi(k)\|\}$ in the interval under consideration.

The component $c(k_{so})$ can be $u(k_{so})$ or $y(k_{so})$. It may happen that, for the sequence $\{k_{so}\}$, $c(k_{so})$ is $u(k_{so})$ for an infinite number of time instants. In such a case, we can extract these instants as a subsequence $\{k_{s1}\}$ of $\{k_{so}\}$. Also, if $c(k_{so})$ is $y(k_{so})$ for an infinite number of instants, we can extract a subsequence $\{k_{s2}\}$ of $\{k_{so}\}$. For each of these subsequences, we can apply the arguments of Lemmas 5.1 and 5.2, respectively, which proves Lemma 5.3.

■

It is interesting to note that the properties derived here consist of conditions that must be verified in each of the cases for the sequence $\{\|\phi(k)\|\}$ to be unbounded. In fact, as we will see in the following sections, the stability analysis will consist basically of proving that the aforementioned conditions cannot be verified under predictive control or adaptive predictive control and, consequently, the sequence $\{\|\phi(k)\|\}$ will remain bounded and the control objectives will be reached.

5.3 PHYSICAL REALIZABILITY OF THE DESIRED TRAJECTORY

Although the *physical realizability* of the desired trajectory has already been considered in preceding chapters, in this section we are going to define this concept in mathematical terms that will allow us to use it in the stability analysis that follows. We refer to the driving desired trajectory (DDT), which is the one that finally determines the calculation of the control signal in predictive control.

Definition 5.3: It is said that the DDT is *physically realizable* if, for any instant k, its value $y_d(k+d)$ can be realized through the application at instant k of a control signal $u_d(k)$ that always remains bounded, that is,

$$|u_d(k)| < U < +\infty \qquad \forall k > 0$$

□

According to this definition, for the processes described by (5.1) the relation between $y_d(k+d)$ and $u_d(k)$, or the equivalent relation between $y_d(k)$ and $u_d(k-d)$, is given by the equation

$$y_d(k) = \theta_o(k)^T \phi_o(k-d) + \theta_1(k)u_d(k-d) + \Delta(k) \qquad \forall k \geq d \tag{5.16}$$

Clearly, the random factors involved in equation (5.16) make it impossible to calculate at instant $k-d$ the control signal $u_d(k-d)$ that will later produce a process output at instant k that is equal to $y_d(k)$. Equivalently, it is not possible to calculate at instant k the control signal $u_d(k)$ that is able to produce the output $y_d(k+d)$ at $k+d$. However, for the stability analysis that we are going to develop, it is sufficient to know that the DDT is chosen in such a way that $u_d(k)$ exists, verifies equation (5.16) and is bounded as given in Definition 5.3.

As has already been mentioned, the physical meaning of this mathematical condition of boundedness corresponds to the existence of limits in the control actions that can be applied to real processes.

5.4 A GENERAL CONDITION FOR APCS STABILITY OF PROCESSES OF A LINEAR AND STABLE INVERSE NATURE

Let us consider the class of processes of a linear and stable inverse nature characterized by Definition 5.2. The application of predictive control or adaptive predictive control to this kind of process can be carried out, as we know from previous chapters, through the basic strategy considered in Chapter 2 or through the extended strategy considered in Chapter 3.

The application of the basic strategy according to the principle of predictive control defined by equations (4.11) or its equivalent (4.12) implies knowledge of the process delay d and, in the case of adaptive predictive control, the non-singularity of the parameter $\hat{\theta}_1(k)$. Although in the adaptive systems derived in Chapter 4 the problem of ensuring the non-singularity of $\hat{\theta}_1(k)$ has not been considered in order to simplify the presentation, an extension of these systems considered in [MSF84] allows to guarantee this formally. In any case, this guarantee is only of a theoretical interest because knowledge of the delay d implies the non-singularity of the corresponding process parameter $\theta_1(k)$, and the singularity of $\hat{\theta}_1(k)$ calculated by the adaptive mechanism almost never occurs in practice.

The application of the extended strategy does not require a precise knowledge of the delay d because, as described in Chapter 3, the calculation of the control action at instant k is performed by considering its effect over a prediction interval $[k, k+\lambda]$ and is associated in this interval with the choice of the projected desired trajectory (PDT). The relation between such a control action and the corresponding PDT is determined by the predictive or by the adaptive predictive (AP) model. In the latter, the adaptive mechanism can be either set to zero or can assign non-zero values to the parameters related to the control signal, in this way varying and adjusting the effective delay identified. To allow this adjustment of the effective delay when required, it is clear that the AP model will have to be provided with an adequate parametric structure.

For all this, in those processes without pure delays, or in those where the delay is well defined, the APCS can be applied by using the basic or the extended strategy. However, for those processes where the delay is not well defined or can vary over time, it is advisable to use the extended strategy. Nevertheless, there is a wide variety of forms of application for the extended strategy, and the rigorous stability analysis for each case can have an overwhelming extension and may be extremely laborious and complex.

Therefore, and to make our analysis didactic, intuitive and also unified for all possible methods of application of predictive or adaptive predictive control while still being rigorous we will base our analysis on the fact that the principle of predictive control as defined in (4.11) or (4.12) is always executed implicitly when the basic or the extended strategy is applied. Furthermore, our analysis will be based on a general condition defined in the following assumption.

Assumption 5.1: The driving desired output (DDT) always remains bounded, that is, $|y_d(k+d)| < \omega^2 < +\infty, \quad \forall k \geq 0$, and the delay d is known.

□

Sections 5.5 and 5.6 will derive stability results based on this assumption which will be valid for any method of application of the predictive control strategies which satisfy it. In the simplest case, this verification can be performed by direct selection of the DDT. In other cases the verification may be intuitively clear and confirmed in reality, but it may require a proof that is beyond the objectives of this book.

The stability results will be derived for both the ideal case and the real cases considered in Chapter 4 and will be based on proof of the contradiction between the existence of a subsequence $\{k_s\}$, such as the one defined in Lemma 5.2 determining the unboundedness of the sequence $\{\|\phi(k)\|\}$, and Assumption 5.1 where predictive control and adaptive predictive control respectively are applied.

The stability analysis that we are going to present in the next two sections is an extension of the one which has already been published in [Mar84, MSF84, Mar86, CMSF88] and which provides theoretical support for the application of APCS to any stable or unstable process belonging to the class of processes of a linear and stable inverse nature.

5.5 PREDICTIVE CONTROL OF PROCESSES OF A LINEAR AND STABLE INVERSE NATURE

Where the process is of a linear and stable inverse nature, if the model describing the dynamic behaviour of the process is assumed to be known, the application of predictive control as has been described in Chapters 2 and 3 leads to the achievement of the desired control objective. However, in practice, we will not know that model exactly *a priori* and, moreover, the process may be variable over time. For this reason, our analysis in this section will focus on the effect that dynamic or parametric differences between the predictive model and the process, that is, modelling errors, have on the stability and on the control objective.

5.5.1 Stability results

Consider the predictive control law defined by

$$y_d(k+d) = \hat{\theta}^T \phi(k) \qquad \forall k \geq 0 \tag{5.17}$$

where $\hat{\theta}$ is the parameter vector of the predictive model. This is a particular case of the general principle of predictive control (4.11a) formulated in Chapter 4, in which we are using fixed parameters in the AP model.

The following theorem gives a stability condition for the above predictive control law.

Theorem 5.1: Let a process be of a linear and stable inverse nature described by equation (5.1) and controlled by the predictive control law (5.17). Under Assumption 5.1, the sequence $\{\|\phi(k)\|\}$ is bounded if there exists a time instant $k_f > 0$ for which the following condition is met:

$$\alpha > \|\theta(k) - \hat{\theta}\| \qquad \forall k \geq k_f \tag{5.18}$$

where α is the greatest of all possible α_1 verifying condition 1 of Lemma 5.2.

□

Proof: From the process equations (5.1) and (5.17), the control (or tracking) error $\epsilon(k)$ can be written in the form

$$\epsilon(k) = y(k) - y_d(k) = [\theta(k) - \hat{\theta}]^T \phi(k-d) + \Delta(k) \tag{5.19}$$

Then, using triangular and Cauchy–Schwarz inequalities,

$$|\epsilon(k)| \leq \|\theta(k) - \hat{\theta}\| \, \|\phi(k-d)\| + |\Delta(k)| \tag{5.20}$$

According to Assumption 5.1:

$$|\epsilon(k)| = |y(k) - y_d(k)| \geq |y(k)| - \omega^2 \tag{5.21}$$

From (5.20) and (5.21) we obtain

$$|y(k)| \leq \|\theta(k) - \hat{\theta}\| \, \|\phi(k-d)\| + \omega_1^2 \tag{5.22}$$

with

$$\omega_1^2 = \omega^2 + \Delta_b$$

where Δ_b is an upper bound of $|\Delta(k)|$ that was defined in (4.48).

Let us suppose that the sequence $\{\|\phi(k)\|\}$ is unbounded. In this case there will be a subsequence $\{k_s\}$ of $\{k\}$ that will verify the conditions stated in Lemma 5.2. From condition 1 of that lemma and (5.22) we may write

$$\alpha\|\phi(k_s - d)\| < \|\theta(k_s) - \hat{\theta}\| \, \|\phi(k_s - d)\| + \omega_2^2 \qquad \forall k_s \geq 0 \tag{5.23}$$

where

$$\omega_2^2 = \omega_1^2 + \alpha_2$$

Dividing both sides of (5.23) by $\|\phi(k_s - d)\|$, we have

$$\alpha < \|\theta(k_s) - \hat{\theta}\| + \frac{\omega_2^2}{\|\phi(k_s - d)\|} \qquad \forall k_s \geq 0 \tag{5.24}$$

Condition 2 of Lemma 5.2 states that $\|\phi(k_s - d)\| \to +\infty$ as $k_s \to +\infty$, which implies that the second term on the right-hand side of (5.24) tends to 0 as $k_s \to +\infty$. This clearly contradicts the condition given in (5.18), and Theorem 5.1 is thus proved.

■

Corollary 5.1.1: If condition (5.18) of Theorem 5.1 is satisfied, the sequence $\{|\epsilon(k)|\}$ is also bounded in the form

$$|\epsilon(k)| < \alpha\|\phi(k-d)\| + \Delta_b \qquad \forall k \geq d$$

□

Proof: The proof is derived directly from (5.18), (5.20) and (4.48).

■

Corollary 5.1.2: If the process parameter vector is constant and equal to the parameter vector of the predictive model, then

1. $\|\phi(k)\| < \Omega^2 \qquad \forall k \geq 0$

2. $\epsilon(k) = \Delta(k) \qquad \forall k \geq d$

□

Proof: Clearly, condition (5.18) of Theorem 5.1 is fulfilled for constant $\theta(k) = \hat{\theta}$, which proves property 1 of this corollary. Property 2 is derived directly from equation (5.19), thereby concluding the proof of Corollary 5.1.2.

■

Corollary 5.1.3: The DDT is physically realizable for the conditions considered in Theorem 5.1.

□

Proof: By subtracting equation (5.16) from the process equation (5.1b), we obtain

$$y(k) - y_d(k) = \theta_1(k)[u(k-d) - u_d(k-d)] \qquad \forall k \geq d \qquad (5.25)$$

According to Theorem 5.1, the sequence $\{\|\phi(k)\|\}$is bounded and, therefore, so are the sequences $\{|y(k)|\}$ and $\{|u(k)|\}$. The parameter $\theta_1(k)$ is, for all k, greater in absolute value than a positive constant. Consequently, it can be readily derived from (5.25) that $\{|u_d(k-d)|\}$ must also be bounded, which proves Corollary 5.1.3.

■

5.5.2 Analysis of the results

Theorem 5.1 stated that if the norm of the modelling error $\|\theta(k) - \hat{\theta}\|$ is less than a certain value α, which depends on the process dynamics, the predictive control system will be stable and it will keep both the I/O vector and the control error bounded. However, if the modelling error is beyond α, the stability cannot be guaranteed.

If the predictive model is defined by a parameter vector $\hat{\theta}_r$ of lower dimension than that of the process $\theta(k)$, the results of the theorem remain completely valid because it is sufficient to consider $\hat{\theta} = [\hat{\theta}_r, 0, \ldots, 0]$. In this way Theorem 5.1 takes into account all the different real cases, with and without differences in structure and with time varying parameters.

The boundedness on the control error will decrease with the modelling error according to (5.20). Also, as proved in Corollary 5.1.2, if the process parameters were known and used in the predictive model, the control error would be equal to the perturbation signal in the real cases and zero in the ideal case.

Finally, Corollary 5.1.3 has proved that, in the stability conditions of Theorem 5.1, it is sufficient to choose a bounded DDT in order to render it physically realizable.

5.6 ADAPTIVE PREDICTIVE CONTROL OF PROCESSES OF A LINEAR AND STABLE INVERSE NATURE

5.6.1 The role of adaptation in the stability

The results derived in the preceding section are intuitive and realistic. In fact, intuitively, the stability result depends on how well the predictive model estimates the process dynamics, and this stability cannot be guaranteed if the modelling error is beyond a certain bound. The basic motivation for adaptive predictive control arises from the need to avoid the possibility that the stability guaranteed by Theorem 5.1 may fail due to modelling errors that may occur either initially or throughout the process operation.

As a natural extension of the predictive control law of (5.17), we now consider the adaptive predictive control law

$$y_d(k+d) = \hat{\theta}(k)^T \phi(k) \qquad \forall k \geq 0 \tag{5.26}$$

where $\hat{\theta}(k)$ is the parameter vector of the AP model.

The stability result for predictive control given in Theorem 5.1 may be stated in the following theorem in the context of adaptive predictive control.

Theorem 5.2: If the process described by equation (5.1) is of a linear and stable inverse nature, the application of the adaptive predictive control law of (5.26) guarantees, under Assumption 5.1, that the sequence $\{\|\phi(k)\|\}$ will be bounded if the following condition holds:

$$\alpha > \|\theta(k) - \hat{\theta}(k)\| \qquad \forall k \geq k_f > 0 \tag{5.27}$$

□

Proof: The proof of this theorem is equivalent to that of Theorem 5.1 and similar corollaries may be derived from it.

■

This theorem illustrates the way in which adaptation can help to guarantee stability and so reach the desired control objectives for an adaptive predictive control system. In fact, as long as the adaptation mechanism of the adaptive system is able to adjust the AP model parameters $\hat{\theta}(k)$ in real time, always maintaining condition (5.27), the I/O vector will remain bounded. If we now recall that the adaptive systems designed and analyzed in Chapter 4 were proved to satisfy convergence properties that led to the desired control objectives when the I/O vector was bounded, it would be natural to devote our effort to assessing whether those adaptive systems are able to retain condition (5.27) throughout their operation.

However, condition (5.27) has an important drawback: it is formulated in terms of the modelling error, that is, the difference between the process parameters, which are *a priori* unknown, and the AP model parameters. Fortunately, a general condition for the stability result of Theorem 5.2 may be formulated in terms of the increments of the AP model parameters and the *a posteriori* estimation error. This will allow us to combine this general condition with the convergence properties of the adaptive systems designed in Chapter 4 in order to complete the analysis of global stability for adaptive predictive control.

Two essential advantages of formulating global stability in terms of the parametric increment and the *a posteriori* estimation error are as follows:

1. This formulation avoids the problem of characterizing stability in terms of the process parameters, which are initially unknown. We would encounter this problem if we based the stability solely on the modelling error, as we have done in the case of predictive control in the preceding section. In this case, the analysis has driven us to prove that there exists a stability region defined by inequality (5.18). However, in the context of practical applications, we can't know this stability region initially because we don't know the process parameters.

2. This formulation relaxes the stability condition as the convergence of the AP model parameters toward the process parameters is not required. This means that the desired control objective could be reached in spite of more or less significant modelling errors. In fact, it is sufficient for the adaptation mechanism to guarantee the minimization of the prediction error, which is usually easier than to achieve a proper identification of the process parameters. When this latter problem has a solution, it might require excitation signals to obtain information on the process dynamics. In many cases such signals are unacceptable in the industrial arena, and their results almost always have limited validity over time and under the operation conditions.

The next section states the previously considered general condition for stability. The subsequent sections will use this general condition to prove the global stability of adaptive predictive control for the different ideal and real cases considered in Chapter 4.

5.6.2. General stability condition

Let us recall the general principle of adaptive predictive control as defined in Chapter 4 in the form

$$y_d(k+d) = \hat{\theta}_r(k)^T \phi_r(k) \qquad \forall k \geq 0 \tag{5.28}$$

where $\hat{\theta}_r(k)$ is the parameter vector of the AP model. Subscript r denotes that this parameter vector may have a reduced dimension as compared with the process parameter vector $\theta(k)$, as occurs when dealing with the real case with differences in structure. In the same vein, $\phi_r(k)$ contains a subset of the most recent process inputs and outputs included in $\phi(k)$.

The general condition to guarantee the boundedness of the input/output vector is stated in the following theorem.

Theorem 5.3: Let a process be of a linear and stable inverse nature described by equation (5.1) and controlled by the adaptive predictive control law of (5.28). Under Assumption 5.1, the sequence $\{\|\phi(k)\|\}$ is bounded if there exists a time instant $k_f > 0$ for which the following condition is met:

$$\alpha > \|\hat{\theta}_r(k) - \hat{\theta}_r(k-d)\| + \frac{|e(k|k)|}{\max\{\|\phi(k-d)\|, c\}} \qquad \forall k \geq k_f \tag{5.29}$$

where α is the greatest of all the possible α_1 considered in Lemma 5.2 and c is any positive constant.

□

Proof: The control (or tracking) error can be written from the process equations (5.1) and (5.28) in the form:

$$\epsilon(k) = y(k) - y_d(k) = y(k) - \hat{\theta}_r(k-d)^T \phi_r(k-d) \tag{5.30}$$

where $\hat{\theta}_r(k)$ is the estimated parameter vector of the AP model generated by the adaptation mechanism as considered in Chapter 4.

As we have seen in Chapter 4, equation (4.9), the *a posteriori* estimation error may be expressed by

$$e(k|k) = y(k) - \hat{y}(k|k) = y(k) - \hat{\theta}_r(k)^T \phi_r(k-d)$$

Subtracting this equation from (5.30) we obtain

$$\epsilon(k) = [\hat{\theta}_r(k) - \hat{\theta}_r(k-d)]^T \phi_r(k-d) + e(k|k) \tag{5.31}$$

Using triangular and Cauchy–Schwarz inequalities:

$$|\epsilon(k)| \leq \|\hat{\theta}_r(k) - \hat{\theta}_r(k-d)\| \, \|\phi_r(k-d)\| + |e(k|k)| \tag{5.32}$$

and considering that, by definition, $\|\phi(k-d)\| \geq \|\phi_r(k-d)\|$, we write

$$|\epsilon(k)| \leq \|\hat{\theta}_r(k) - \hat{\theta}_r(k-d)\| \, \|\phi(k-d)\| + |e(k|k)| \tag{5.33}$$

Using (5.21), which is also valid in this case, in (5.33), we obtain

$$|y(k)| \leq \|\hat{\theta}_r(k) - \hat{\theta}_r(k-d)\| \, \|\phi(k-d)\| + |e(k|k)| + \omega^2 \tag{5.34}$$

Let us suppose that the sequence $\{\|\phi(k)\|\}$ is unbounded. In this case there will exist a subsequence $\{k_s\}$ of $\{k\}$ verifying the conditions stated in Lemma 5.2. From condition 1 of that lemma and (5.34) we may write

$$\alpha\|\phi(k_s - d)\| < \|\hat{\theta}_r(k_s) - \hat{\theta}_r(k_s - d)\| \, \|\phi(k_s - d)\| + |e(k_s|k_s)| + \omega_1^2 \tag{5.35}$$

where

$$\omega_1^2 = \omega^2 + \alpha_2 < +\infty$$

Dividing both members of inequality (5.35) by $\|\phi(k_s - d)\|$ we obtain

$$\alpha < \|\hat{\theta}_r(k_s) - \hat{\theta}_r(k_s - d)\| + \frac{|e(k_s|k_s)|}{\|\phi(k_s - d)\|} + \frac{\omega_1^2}{\|\phi(k_s - d)\|} \tag{5.36}$$

Given that $\|\phi(k_s - d)\| \to +\infty$ as $k_s \to +\infty$ (condition 2 of Lemma 5.2), the last term on the right-hand side of (5.36) tends to zero, and consequently inequality (5.36) contradicts condition (5.29) of this theorem, which is thus proved.

■

We note that this stability result is general in the sense that it can be applied to any of the real cases considered in Section 4.2.1 of the preceding chapter. In fact, Theorem 5.3 allows any kind of disturbance and noise that act on the process. It also allows the AP model to be of reduced structure with respect to the process dynamics, and it is not restricted to a process with constant parameters.

As already indicated, in the following four sections, we revisit the adaptive systems designed in Chapter 4, emphasizing the form in which they meet the above stability condition (5.29) for the ideal and the different real cases under consideration.

5.6.3 Ideal case

In the ideal case we deal with a particular case of the general description of the process considered in Theorem 5.3. This is the one defined by hypotheses (a) and (c) of Section 4.2.1, according to which the process parameters are constant and the perturbation signal is zero. Consequently, the process is now described by the equation:

$$y(k) = \theta^T \phi(k - d) \tag{5.37}$$

Likewise, according to hypothesis (b) for this case, the process equations and the AP model equations have the same dimensions, that is, $\hat{\theta}_r(k) = \hat{\theta}(k)$ and, therefore, the application of adaptive predictive control verifies the following equation

$$y_d(k + d) = \hat{\theta}(k)^T \phi(k - d) \tag{5.38}$$

In Chapter 4 (Section 4.6) the existence of an adaptive system has been proved verifying equation (4.27a) and satisfying the following properties in the ideal case:

$$\lim_{k \to \infty} e(k|k) = 0 \tag{5.39}$$

$$\lim_{k \to \infty} [\hat{\theta}(k) - \hat{\theta}(k-1)] = 0 \tag{5.40}$$

From these two properties, the result of global asymptotic stability for the ideal case is stated in the following theorem.

Theorem 5.4: If the process described by equation (5.37) is of a linear and stable inverse nature, the application of the adaptive predictive control law of (5.38) and of the adaptive system satisfying equation (4.27a) guarantees the following properties under Assumption 5.1:

1. $\|\phi(k)\| < \Omega < +\infty \qquad \forall k \geq 0$

2. $\lim_{k \to \infty} \epsilon(k) = 0$

□

Proof: Clearly, the adaptive system satisfying (5.38) and (5.40) verifies condition (5.29) of Theorem 5.3. Consequently, the sequence $\{\phi(k)\|\}$ is bounded, which proves property 1 of Theorem 5.4. Property 2 is readily derived using the above property 1 and equations (5.39)–(5.40) in equation (5.31). The proof of property 2 of the theorem has already been considered in Section 4.6.4.

■

5.6.4 Real case with no difference in structure

According to hypothesis (a) and (c_1) defined in Section 4.2.1, the process dynamics is described in this case by the equation

$$y(k) = \theta^T \phi(k-d) + \Delta(k) \tag{5.41}$$

On the other hand, according to hypothesis (b) for this case, the dimension of the AP model will be identical to that of the process and, consequently, the calculation of adaptive predictive control can also be executed through equation (5.38).

In the analysis performed for this case in Chapter 4 (Section 4.7), the existence of an adaptive system with an adaptation mechanism defined by equations (4.48)–(4.50) has been proved. This system satisfies the following convergence properties according to Corollary 4.3.1:

$$\exists k_f \text{ such that } |e(k|k)| < 2\Delta_b \qquad \forall k \geq k_f > 0 \tag{5.42}$$

$$\lim_{k \to \infty} [\hat{\theta}(k) - \hat{\theta}(k-1)] = 0 \tag{5.43}$$

From these properties, the global stability result is stated in the following theorem.

Theorem 5.5: If the process described by equation (5.41) is of a linear and stable inverse nature, the application of the adaptive predictive control law (5.38) and of the adaptation mechanism defined by (4.48)–(4.50) guarantees the following properties under Assumption 5.1:

1. $\|\phi(k)\| < \Omega < +\infty \qquad \forall k \geq 0$
2. $\exists k_f < +\infty$ such that $\hat{\theta}(k) = \hat{\theta}(k-1) \qquad \forall k \geq k_f > 0$
3. $|\epsilon(k)| = |e(k|k)| < 2\Delta_b \qquad \forall k \geq k_f + d$

□

Proof: The convergence properties satisfied by the adaptive system verify the stability condition (5.29) of Theorem 5.3, and therefore the sequence $\{\|\phi(k)\|\}$ is bounded, which proves property 1 of the theorem. Property 2 is derived from Lemma 4.2, which proves that, if the norm of the I/O vector remains bounded, the number of adaptations of the AP model parameters is finite. Finally, property 3 is derived from property 2 and from (5.31), thus concluding the proof of Theorem 5.5.

■

The result presented in Theorem 5.5 is clearly satisfactory in the sense that it proves the global stability of APCS in the case considered here. However, one may wonder if the bound resulting for the control error is small enough. We know that Δ_b is the bound for the control error that results when the process parameters are known and used in the predictive model, as proved in Corollary 4.1.2. Therefore, Δ_b is the minimum bound for the control error. In this case we have proved the boundedness of the control error with a bound $2\Delta_b$. The interested reader can find the definition of a slightly more sophisticated adaptive system in [Mar84, MSF84], avoided in our exposition for the sake of simplicity, in which the bound for the control error approaches the limit Δ_b.

5.6.5 Real case with difference in structure

As defined in Section 4.2.1, the hypotheses on the process are as in the preceding case. Then the process dynamics is also described by equation (5.41).

According to hypothesis b_1 of this case, the AP model order can be less than that of the process equation and, consequently, the computation of adaptive predictive control will be performed according to equation (5.28), which involves the reduced dimension vectors $\hat{\theta}_r(k)$ and $\phi_r(k)$.

The analysis performed in Section 4.8 for this case has proved the existence of an adaptive system with an adaptation mechanism defined by equations (4.65)–(4.72) and (4.75)–(4.80) such that, according to Lemmas 4.6 and 4.7, there exists a time instant $k_f < +\infty$ for which the following convergence properties are satisfied:

$$\hat{\theta}_r(k) = \hat{\theta}_r(k-1) \qquad \forall k \geq k_f > 0 \tag{5.44}$$

$$|e(k|k)| < 2\Delta_b^n \max\{\|\phi(k-d), c\} \qquad \forall k \geq k_f > 0 \tag{5.45}$$

where Δ_b^n is defined in (4.80) as an upper bound for the sequence of the absolute values of the normalized perturbation signal $\{|\Delta^n(k)|\}$ defined in (4.72).

From these properties, the global stability result is stated in the following theorem.

Theorem 5.6: If the process described by equation (5.41) is of a linear and stable inverse nature, the application of the adaptive predictive control law of (5.28) and the adaptation mechanism defined by (4.65)–(4.72) and (4.75)–(4.80) guarantees the following properties under Assumption 5.1:

1. $\|\phi(k)\| < \Omega < +\infty \qquad \forall k \geq 0$
2. $\exists k_f < +\infty$ such that $\hat{\theta}_r(k) = \hat{\theta}_r(k-1) \qquad \forall k \geq k_f > 0$
3. $|\epsilon(k)| = |e(k|k)| < 2\Delta_b^n \max\{\|\phi(k-d), c\} \qquad \forall k \geq k_f + d$

provided that the following condition is verified:

$$\alpha > 2\Delta_b^n \tag{5.46}$$

α being the greatest of all possible α_1 considered in Lemma 5.2.

□

Proof: Lemma 4.7 has proved the existence of $k_f < +\infty$ such that

$$2\Delta_b^n > \frac{|e(k|k)|}{\max\{\|\phi(k-d), c\}} \qquad \forall k \geq k_f > 0 \tag{5.47}$$

and from condition (5.46):

$$\alpha > 2\Delta_b^n > \frac{|e(k|k)|}{\max\{\|\phi(k-d), c\}} \qquad \forall k \geq k_f > 0 \tag{5.48}$$

Given property 2 of this theorem, which is a result of Lemma 4.6, equation (5.48) implies that condition (5.29) of Theorem 5.3 is satisfied, which proves property

1 of this theorem. Property 3 is deduced from property 2, equation (5.31) and Lemma 4.7, thus concluding the proof of Theorem 5.6.

■

The result presented in Theorem 5.6 guarantees the global stability of APCS in the real case with differences in structure, within limits imposed by parameters α and Δ_b^n. α depends on the dynamic nature of the process, as stated in Lemma 5.2. Δ_b^n, an upper bound of $|\Delta^n(k)|$, depends basically on the norm of the unmodelled parameter vector $\|\theta_u\|$ as derived from equation (4.72) and on the absolute value of the perturbation signal $|\Delta(k)|$. Reducing the AP model order leads to increasing $\|\theta_u\|$ and thus to increasing Δ_b^n.

It is clear from (5.46) that, if α is large, the margin for Δ_b^n is also large, which allows a significant reduction in the AP model order. If α is small, a slight order reduction can lead to instability.

It can be seen from property 3 of Theorem 5.6 that reducing the AP model order increases the bound on the control error $\epsilon(k)$ and thus leads to deterioration of the tracking performance of the control system. Clearly, the allowed order reduction for the AP model is conditioned by the requirement of (5.46) and the desired bound for the tracking error $\epsilon(k)$. In the limit, when $\|\theta_u\| = 0$, we recover the unconditional global stability of the real case with no differences in structure with the tracking error bounded as in property 3 of Theorem 5.6. Also, if the perturbation signal reaches zero, we obtain the asymptotic stability of the ideal case stated in property 2 of Theorem 5.4

Finally, it is important to note that the stability condition of (5.46) could be relaxed by using a more sophisticated adaptive system, which, for the sake of simplicity, has not been considered here. The reader can find an adaptive system for which the term $2\Delta_b^n$ in (5.46) is substituted by $\Delta_b^n + \delta$ in [CMSF88], where δ is a positive arbitrarily small constant, thus approaching the most favourable stability condition for this case.

5.6.6 Time varying parameters

In the analysis carried out in Section 4.9 of the preceding chapter for the real case with time varying parameters, Lemma 4.8 proved that the normalized adaptive system considered there ensures that

$$\|\theta_r(k) - \hat{\theta}_r(k)\| - \|\theta_r(k) - \hat{\theta}_r(k-1)\| < 0$$

where $\theta_r(k)$ is the reduced process parameter vector and $\hat{\theta}_r(k-1)$ and $\hat{\theta}_r(k)$ are, respectively, the estimated AP model parameter vectors before and after the adaptation is performed. Thus, the estimated parameters tend to follow the evolution of the process parameters. The adaptation is performed as long as the *a posteriori* estimation error is beyond a certain bounded function of the perturbation signal.

Although taking different forms, Theorems 5.2 and 5.3 both state the conditions for APCS global stability for the real case with time varying parameters. Theorem 5.2 guarantees APCS global stability when the norm of the identification error is less than a certain value, which depends on the process dynamics. Theorem 5.3 states the conditions for stability based on the boundedness of the *a posteriori* estimation error and of the increments of the AP model parameters.

Both theorems admit a permanent variation in the AP model parameters and stability is guaranteed as long as inequalities (5.27) or (5.29) are satisfied. In both cases this will depend basically on the type of variation that the process parameters undergo. As previously discussed in Section 4.9, the parametric variations observed in the operation of industrial processes are usually due to the action of perturbations on or changes in the operating points, which could produce rapid changes in the parameters, or to continuous mild variations of the operating conditions, which could produce slow and sustained changes in the parameters. In any case, it seems reasonable to expect that the operation of the adaptive system is able to retain the above stability conditions in these circumstances.

Also, as already considered in Section 4.9, the efficiency of the adaptation will strengthen the predictive control strategy and will generally assist in the stabilization of the process and its dynamic behaviour. In this case the convergence results obtained for the real case with differences in structure are again valid. As we know, this last one includes as particular cases the real case with no differences in structure and the ideal case, along with their respective stability results.

5.7 APCS STABILITY OF PROCESSES OF A LINEAR AND STABLE NATURE

5.7.1 A general condition for stability

In this section we deal with the class of processes of a linear and stable nature, which is characterized by Definition 5.1, irrespective of their inverse stability or instability.

The case where the inverse is unstable is the one that renders the class of processes considered in this section not a subset of the class considered in the preceding section. When the inverse is unstable, the property stated in Lemma 5.2 is not true, and thus the stability results derived in the preceding section are not valid.

The class of processes considered here has the common property stated in Lemma 5.1. We will use it in this section to derive the stability results.

The stability results of the preceding section are still valid for those processes considered here that have stable inverse. Predictive or adaptive predictive control can be applied to them through the basic or the extended strategy. However, for stable processes with an unstable inverse, as has already been analyzed in Chapters 2 and 3, the application of the basic strategy is not possible as it can generate unbounded control actions.

There is a wide variety of forms of application of the predictive control extended strategy, as has been considered in the preceding section and analyzed in Chapter 3. The rigorous stability analysis for each case, in both the predictive and the adaptive predictive contexts, is a broad and complex open area for theoretical research. Therefore, in this section, following a similar didactic, intuitive and unified method as the one given in Section 5.4, while still being rigorous, we base our analysis on the implicit execution of the principle of predictive control as defined in (4.11) or (4.12) and on the general condition defined in the following assumption.

Assumption 5.2: The driving desired output $y_d(k+d)$ is physically realizable, $\forall k \geq 0$, and the time delay d is known.

□

From the theory developed in Chapters 3 and 4, it seems reasonable that, through the application of the extended strategy and with a proper choice of the prediction horizon and the AP model structure, Assumption 5.2 should be satisfied for processes of a linear and stable nature, which is actually confirmed in practical applications.

Sections 5.7.2 and 5.7.3 derive general stability results for this class of processes which will be valid for any method of application of the extended predictive control strategy that satisfies Assumption 5.2.

Appendix D presents a proof of APCS stability for stable processes, considering a particular but representative case of the extended predictive control strategy, without appealing to the assumption of the physical realizability of the DDT, but obtaining it as a result of the analysis.

5.7.2 Predictive control

The stability result in the context of predictive control is analogous to that obtained in the preceding section and is stated in the following theorem.

Theorem 5.7: Let a process be of a linear and stable nature described by an equation such as (5.1) and controlled by the predictive control law of (5.17). Under Assumption 5.2, the sequence $\{\|\phi(k)\|\}$ is bounded if there exists a time instant $k_f > 0$ for which the following condition is met:

$$\gamma > \|\theta(k) - \hat{\theta}\| \qquad \forall k \geq k_f \tag{5.49}$$

where

$$\gamma = \theta_{1m}\gamma_o$$

γ_o being the greatest of all possible γ_1 verifying condition 1 of Lemma 5.1 and $0 < \theta_{1m} \leq |\theta_1(k)|, \quad \forall k \geq 0$.

□

Proof: As in Theorem 5.1, from the process equation (5.1) and the control law of (5.17), we obtain equation (5.19):

$$\epsilon(k) = y(k) - y_d(k) = [\theta(k) - \hat{\theta}]^T \phi(k-d) + \Delta(k)$$

As for Corollary 5.1.3, by subtracting equation (5.16) from the process equation (5.1b), we obtain equation (5.25):

$$\epsilon(k) = y(k) - y_d(k) = \theta_1(k)[u(k-d) - u_d(k-d)] \qquad \forall k \geq d$$

From the above two equations we obtain

$$\theta_1(k)u(k-d) = [\theta(k) - \hat{\theta}]^T \phi(k-d) + \Delta(k) + \theta_1(k)u_d(k-d) \tag{5.50}$$

By using the triangular and Cauchy–Schwarz inequalities:

$$\theta_{1m}|u(k-d)| \leq \|\theta(k) - \hat{\theta}\|\|\phi(k-d)\| + |\Delta(k)| + |\theta_1(k)u_d(k-d)| \tag{5.51}$$

Let us suppose that the sequence $\{\|\phi(k)\|\}$ is unbounded. In such a case, using property 1 of Lemma 5.1 and equation (5.51), there will exist a subsequence $\{k_s\}$ of $\{k\}$ for which we may write

$$\gamma\|\phi(k_s-d)\| < \|\theta(k_s) - \hat{\theta}\|\|\phi(k_s-d)\| + |\Delta(k_s)| + |\theta_1(k_s)u_d(k_s-d)| + \gamma_2 \tag{5.52}$$

whcrc

$$\gamma = \gamma_o\theta_{1m}$$

By dividing both members of inequality (5.52) by $\|\phi(k_s - d)\|$:

$$\gamma < \|\theta(k_s) - \hat{\theta}\| + \frac{|\Delta(k_s)| + |\theta_1(k_s)u_d(k_s-d)| + \gamma_2}{\|\phi(k_s-d)\|} \tag{5.53}$$

$\|\phi(k_s - d)\| \to +\infty$ as $k_s \to +\infty$ according to property 2 of Lemma 5.1, and the sequence $\{|u_d(k-d)|\}$ is bounded due to the physical realizability of the DDT (Definition 5.3). Consequently, the last term on the right-hand side of (5.53) tends to 0 as $k_s \to +\infty$. This clearly contradicts condition (5.49) and thus Theorem 5.7 is proved.

■

From this theorem, corollaries similar to 5.1.1 and 5.1.2 of Theorem 5.1 can be derived which are stated without proof in the following.

Corollary 5.7.1: If condition (5.49) of Theorem 5.7 is satisfied, the sequence $\{|\epsilon(k)|\}$ is also bounded in the form

$$|\epsilon(k)| < \gamma\|\phi(k-d)\| + \Delta_b \qquad \forall k \geq d$$

□

Corollary 5.7.2: If the process parameter vector is constant and equal to the parameter vector of the predictive model, then

1. $\|\phi(k)\| < \Omega^2 \qquad \forall k \geq 0$

2. $\epsilon(k) = \Delta(k) \qquad \forall k \geq 0$

□

Likewise, the following corollary is derived.

Corollary 5.7.3: Under the stability conditions stated in Theorem 5.7, the DDT is bounded.

□

Proof: From Theorem 5.7, the sequence $\{\|\phi(k)\|\}$ is bounded, which implies that the sequences $\{|y(k)|\}$ and $\{|u(k)|\}$ are also bounded. The sequence $\{|u_d(k)|\}$ is bounded because the DDT is physically realizable. The parameter $\theta_1(k)$ is also bounded. Consequently, the boundedness of $\{|y_d(k)|\}$ is derived directly from (5.25), thus proving Corollary 5.7.3.

■

Remarks on the result of Theorem 5.7 and its corollaries are similar to those made in Section 5.5.2 for the result of Theorem 5.1 and its corollaries, excepting in this case the result of Corollary 5.7.3, which proves that the physical realizability of the DDT implies its boundedness.

5.7.3 Adaptive predictive control

The motivation and objectives for adaptive predictive control are, in this case, similar to those considered in Section 5.6.1 for the application of adaptive predictive control to processes of a linear and stable inverse nature.

Likewise, the stability result for predictive control stated in Theorem 5.7 can be directly transferred to the adaptive predictive context through the following theorem.

Theorem 5.8: If the process described by equation (5.1) is of a linear and stable nature, the application of the adaptive predictive control law of (5.26) guarantees, under Assumption 5.2, that the sequence $\{\|\phi(k)\|\}$ will be bounded if the following condition holds:

$$\gamma > \|\theta(k) - \hat{\theta}(k)\| \qquad \forall k \geq k_f > 0 \tag{5.54}$$

□

Proof: The proof of this theorem is equivalent to that of Theorem 5.7, and similar corollaries can be proved from it.

■

The main result of stability in the context of adaptive predictive control is also analogous to that obtained in the preceding section and is stated in the following theorem.

Theorem 5.9: Let a process be of a linear and stable nature described by an equation such as (5.1) and controlled by the adaptive predictive control law of (5.28). Under Assumption 5.2, the sequence $\{\|\phi(k)\|\}$ will be bounded if there exists a time instant $k_f > 0$ for which the following condition holds:

$$\gamma > \|\hat{\theta}_r(k) - \hat{\theta}_r(k-d)\| + \frac{|e(k|k)|}{\max\{\|\phi(k-d)\|, c\}} \qquad \forall k \geq k_f \tag{5.55}$$

where c is any positive constant.

□

Proof: Following similar arguments as given in the proof of Theorem 5.3, we may derive equation (5.31):

$$\epsilon(k) = [\hat{\theta}_r(k) - \hat{\theta}_r(k-d)]^T \phi_r(k-d) + e(k|k)$$

Recall equation (5.25):

$$\epsilon(k) = \theta_1(k)[u(k-d) - u_d(k-d)] \qquad \forall k \geq d$$

From the above two equations we obtain

$$\theta_1(k)u(k-d) = [\hat{\theta}_r(k) - \hat{\theta}_r(k-d)]^T \phi_r(k-d) + e(k|k) + \theta_1(k)u_d(k-d) \tag{5.56}$$

Using triangular and Cauchy–Schwarz inequalities in (5.56) and taking the fact that $\|\phi(k-d)\| \geq \|\phi_r(k-d)\|$ into account, we obtain

$$\begin{aligned}\theta_{1m}|u(k-d)| \leq \|\hat{\theta}_r(k) - \hat{\theta}_r(k-d)\| \, \|\phi(k-d)\| + |e(k|k)| \\ + |\theta_1(k)u_d(k-d)|\end{aligned} \tag{5.57}$$

Let us assume that the sequence $\{\|\phi(k)\|\}$ is unbounded. Then, using property 1 of Lemma 5.1 and equation (5.57) we may write

$$\begin{aligned}\gamma\|\phi(k_s-d)\| < \|\hat{\theta}_r(k_s) - \hat{\theta}_r(k_s-d)\| \, \|\phi(k_s-d)\| \\ + |e(k_s|k_s)| + |\theta_1(k_s)u_d(k_s-d)| + \gamma_2\end{aligned} \tag{5.58}$$

where

$$\gamma = \gamma_o \theta_{1m}$$

Dividing both sides of (5.58) by $\|\phi(k_s - d)\|$, we obtain

$$\gamma < \|\hat{\theta}_r(k_s) - \hat{\theta}_r(k_s - d)\| + \frac{|e(k_s|k_s)|}{\|\phi(k_s - d)\|} + \frac{|\theta_1(k_s)u_d(k_s - d)| + \gamma_2}{\|\phi(k_s - d)\|} \tag{5.59}$$

Given that, according to property 2 of Lemma 5.1, $\|\phi(k_s - d)\| \to +\infty$ as $k_s \to +\infty$, the last term on the right-hand side of (5.59) tends to zero because the sequence $\{u_d(k_s - d)|\}$ is bounded due to the physical realizability of the DDT. Consequently, inequality (5.59) contradicts condition (5.55) of this theorem, which is thus proved.

■

A stability analysis for the ideal case and the different real cases can be derived from the above theorem in terms that are completely analogous to the one carried out in the preceding section from Theorem 5.3 for processes of a linear and stable inverse nature, but in this case for processes of a linear and stable nature and based on Assumption 5.2.

5.8 APCS STABILITY OF UNSTABLE PROCESSES WITH AN UNSTABLE INVERSE

5.8.1 A general condition for stability

As has already been indicated in the introduction to this chapter, the classes of processes of a linear and stable inverse nature and of a linear and stable nature include nearly all industrial processes. However, with the intention of completing the theoretical material presented in this chapter, in this section we will analyze the APCS stability applied to the class of unstable processes with an unstable inverse in a similar manner to that carried out in the last two sections. General stability results will be derived both in the context of predictive control and in that of adaptive predictive control. These will be based on the dynamic nature of this kind of process and on the implicit execution of the principle of predictive control through (4.11) or (4.12), as well as on the following assumption.

Assumption 5.3: The driving desired output $y_d(k + d)$ is physically realizable and bounded, $\forall k \geq 0$, and the time delay d is known.

□

As has already been mentioned in the introduction, both the definition and the application of the extended strategy that are able to satisfy this assumption and its theoretical verification are open problems and are outside the scope of this book.

5.8.2 Predictive control

The stability result in this case is stated in the following theorem.

Theorem 5.10: Let an unstable process with unstable inverse be described by an equation such as (5.1) and controlled by the predictive control law of (5.17). Under Assumption 5.3, the sequence $\{\|\phi(k)\|\}$ will be bounded if there exists a time instant $k_f > 0$ for which the following condition holds:

$$\min\{\alpha, \gamma\} > \|\theta(k) - \hat{\theta}\| \qquad \forall k \geq k_f \tag{5.60}$$

where α and γ are positive constants analogous to those defined in Theorems 5.1 and 5.7 respectively.

□

Proof: Assume that $\{\|\phi(k)\|\}$ is unbounded. Then, there will exist one of the two sequences $\{k_{s1}\}$ and $\{k_{s2}\}$ for which properties 1 and 2, respectively, of Lemma 5.3 hold. Property 1 is related to Lemma 5.1 and states that $\|\phi(k_{s1} - d)\| \to +\infty$ as $k_{s1} \to +\infty$. Property 2 is related to Lemma 5.2 and states that $\|\phi(k_{s2} - d)\| \to +\infty$ as $k_{s2} \to +\infty$.

Taking the fact that the DDT is physically realizable into account, we may use similar arguments as those given in the proof of Theorem 5.7 in order to prove the contradiction between the existence of such a sequence as $\{k_{s1}\}$ and condition (5.60). Taking also the fact that the DDT is bounded into account, we may use similar arguments to those given in Theorem 5.1 in order to prove the contradiction between the existence of $\{k_{s2}\}$ and condition (5.60). Both contradictions conclude the proof of Theorem 5.10.

■

From this theorem, corollaries such as 5.1.1 and 5.1.2 of Theorem 5.1 can be derived. Likewise, we can draw similar comments to those made in Section 5.5.2 for the result of Theorem 5.1 and its corollaries, except that in this case, according to Assumption 5.3, the DDT is assumed bounded and physically realizable.

5.8.3 Adaptive predictive control

From Theorem 5.10 we can derive the following theorem for the adaptive predictive context.

Theorem 5.11: If the process described by equation (5.1) is unstable and has an unstable inverse, the application of the adaptive predictive control law of (5.26) guarantees, under Assumption 5.3, that the sequence $\{\|\phi(k)\|\}$ will be bounded if the following condition is verified:

$$\min\{\alpha,\gamma\} > \|\theta(k)-\hat{\theta}(k)\| \qquad \forall k \geq k_f \tag{5.61}$$

□

Proof: The proof of this theorem is equivalent to that of Theorem 5.10, and similar corollaries can be derived from it.

■

The main result of stability in the context of adaptive predictive control is stated in the following theorem.

Theorem 5.12: Let an unstable process with an unstable inverse be described by an equation such as (5.1) and controlled by the adaptive predictive control law of (5.28). Under Assumption 5.3, the sequence $\{\|\phi(k)\|\}$ is bounded if there exists a time instant $k_f > 0$ for which the following condition is met:

$$\min\{\alpha,\gamma\} > \|\hat{\theta}_r(k)-\hat{\theta}_r(k-d)\| + \frac{|e(k|k)|}{\max\{\|\phi(k-d)\|, c\}} \qquad \forall k \geq k_f \tag{5.62}$$

where c is any positive constant.

□

Proof: As considered in the proof of Theorem 5.10, if the sequence $\{\|\phi(k)\|\}$ is unbounded, one of the sequences $\{k_{s1}\}$ or $\{k_{s2}\}$ may exist. For the case where $\{k_{s1}\}$ exists, using the assumption that the DDT is physically realizable, similar arguments to those given in the proof of Theorem 5.9 prove that condition (5.62) cannot hold. For the case of where $\{k_{s2}\}$ exists, taking into account the fact that the DDT is bounded, we see that similar arguments to those given in the proof of Theorem 5.3 also prove that condition (5.62) cannot hold. Both cases establish the contradiction between the unboundedness of $\{\|\phi(k)\|\}$ and condition (5.62), thus proving Theorem 5.12.

■

From this theorem, a stability analysis can be derived for the ideal case and the different real cases in terms that are completely analogous to the one performed in Section 5.6 from Theorem 5.3 for processes of a linear and stable inverse nature, but in this case for unstable processes with an unstable inverse and based on Assumption 5.3.

5.9 CONCLUSIONS

The results derived in this chapter form a theoretical body of stability both in the context of predictive control and adaptive predictive control.

The formulation started from the dynamic nature of the processes, which has been divided into three classes, namely: (1) processes of a linear and stable inverse nature; (2) processes of a linear and stable nature; and (3) unstable processes with an unstable inverse.

For each of these classes, a stability condition has been established based on the characteristics of the boundedness and/or physical realizability of the driving desired trajectory (DDT). In the first class, where the process is stable, the corresponding condition may be satisfied using the basic or the extended strategy. In the second class, where the process inverse may be unstable, the satisfaction of the condition requires the use of the extended strategy. This condition has unified the analysis performed for each class of processes, and has given general validity to the stability results derived, irrespective of the specific control strategy used.

The analysis has progressed from the simplest formulation (ideal case) to a formulation defined by hypotheses that describe an industrial environment. First, this chapter has proved that the stability of predictive control depends on a measure of the modelling error that has been mathematically formulated in relation to the process dynamics. On the other hand, the result was intuitive and could reasonably be expected. The basic motivation for adaptive predictive control has been to overcome the stability problem due to the modelling error. In the ideal case, the expected result was the one that has been proven, that is to say asymptotic stability guaranteeing that the process output will follow the desired output.

As expected, asymptotic stability was not possible in the real cases due to the existence of the unknown and unpredictable perturbation vector. Nevertheless, when there is no difference in structure, we aim the control error to approach the perturbation vector as closely as possible. The results obtained here proved the stability for this case in terms of the boundedness of the control error, approaching the corresponding bound of the perturbation signal.

Where there is a difference in structure, it seems logical that stability may not always be guaranteed and that there must exist a stability limit in terms of the model order reduction. The theory presented in this chapter has derived the mathematical formulation for this limit and its relationship with the process dynamics. When the model order reduction is compatible with this limit, adaptive predictive control guarantees the stability result and the control error may be minimized by appropriate selection of the adaptation mechanism.

In the case of time varying parameters, the theory developed here has established the stability condition that, given the usual nature of parametric changes within the industrial context, should reasonably be satisfied by the adaptation mechanisms formulated in Chapter 4. These adaptation mechanisms guarantee tracking of the process parameters by the AP model parameters until stability is reached, which has been repeatedly confirmed by practice.

Those areas of theoretical analysis which are still an open subject for research have been indicated, but it is important to emphasize that the results arising from the theory presented here have clear implications in a wide practical context.

In summary, the results presented in this chapter provide theoretical support for the APCS methodology as a whole, that is, both for predictive control and for adaptive predictive control. Profound understanding of this theoretical body clearly contributes to better application of the methodology. Productive interaction between theory and practice may lead to accurate control and optimization of the process operation. These are the subjects of the following parts of this book.

Part 2

THE APPLICATIONS

Chapter 6

MULTIVARIABLE CONTROL OF A BINARY DISTILLATION COLUMN

6.1 INTRODUCTION

With this chapter we open the part of the book that is devoted to the application of adaptive predictive control systems (APCS). This and the following chapter present the application of APCS using the basic strategy of predictive control, while the extended predictive control strategy will be illustrated in Chapters 8 and 9.

A particular purpose of this chapter is to emphasize the key issues involved in the practical application of APCS to real processes. These practical issues are illustrated by actual application to single-input, single-output and multivariable control of a pilot scale binary distillation column.

This project of APCS application was carried out over the first months of 1976 at the Department of Chemical Engineering of the University of Alberta and the results were first published in [Mar76a, MS84]. At that time, several authors had criticized the control theory for chemical processes in particular and that for industrial processes in general [Fos73, KST76, LW76], stressing the existence of a gap between theory and practice. In this critique the multivariable control of a distillation column was given as a typical example of the difficulties found in the practical application of modern control theory. The examination of the different applications of available methodologies to the multivariable control of distillation columns [RS76, Edg76] emphasized the inability of these methods to respond satisfactorily to existing problems. Clearly, the reasons were not unconnected to the complex dynamic nature of these problems. In fact, a distillation column presents the following characteristics: it is a very non-linear process; it is multivariable with strong interactions, and it exhibits pure time delays between its input and output variables. Consequently, the application of APCS to the control of a distillation column was a real and challenging project with which to assess its capabilities as an advanced control methodology.

6.2 PRACTICAL ISSUES IN APCS IMPLEMENTATION

First in this section, we will briefly review the mathematical formulation of the APCS basic strategy in a multivariable setting. Later, we will emphasize the practical issues involved in the implementation of the method.

6.2.1 Mathematical formulation

The dynamic relation between the inputs and outputs of a large class of multivariable processes can be described around their steady state values by multi-input, multi-output (MIMO) difference equations of the form

$$\begin{aligned} Y(k) = {} & \sum_{i=1}^{h} A_i(k)Y(k-i-r) + \sum_{i=1}^{f} B_i(k)U(k-i-r) \\ & + \sum_{i=1}^{g} C_i(k)W(k-i-r) + \Delta(k) \end{aligned} \tag{6.1}$$

where $Y(k-i-r)$, $U(k-i-r)$ and $W(k-i-r)$ are, respectively, the increments at time $k-i-r$ of the measured output, input and measurable disturbance vectors with respect to the steady state values. This equation may be derived from equation (1.10) of Chapter 1 by recursive substitutions, already considered in Section 4.2.2. The following paragraph completes the description of equation (6.1).

The dimensions of Y and U are assumed to be equal to n and the scalar components of U are assumed to be the degrees of freedom available for controlling the process, that is, the variables that can be manipulated independently. If U had a larger dimension, the extra components could be included in vector W as non-manipulable measurable inputs. $A_i(k)$, $B_i(k)$ and $C_i(k)$ are time variant matrices of appropriate dimensions which determine the dynamics of the process. $\Delta(k)$ is the perturbation vector and r is an integer defined as $r = \min d_{ij}$ ($i = 1, \ldots, n$; $j = 1, \ldots, n$), where the d_{ij} are integers that represent the estimated pure time delays (expressed in number of sampling periods) between output y_i and input u_j. It can be seen that the time delay related to the measurable disturbance vector W is also equal to r. This is to assume that the delays associated with the process inputs and the measurable disturbances are such that any effect on the process output due to measurable disturbances can be compensated for by the process input vector. If these perturbations cannot be compensated for, then they are included in $\Delta(k)$.

The $n \times n$ matrix $D = [d_{ij}]$ is defined as the *delay matrix*. It will be shown later in this section that this matrix plays a significant practical role in the implementation of APCS for MIMO systems with time delays, and it is related to the interactor matrix concept discussed in [WF76]. In the application of the basic strategy of predictive control, it is generally assumed that the delay matrix is known.

An adaptive predictive (AP) model is used to calculate an *a priori* estimation of the process output $Y(k)$ and to predict, at time k, the process output for time $k+r+1$ as follows:

$$\begin{aligned}\hat{Y}(k|k-1) = &\sum_{i=1}^{h} \hat{A}_i(k-1)Y(k-i-r) + \sum_{i=1}^{f} \hat{B}_i(k-1)U(k-i-r)\\ &+ \sum_{i=1}^{g} \hat{C}_i(k-1)W(k-i-r)\end{aligned} \tag{6.2}$$

$$\begin{aligned}\hat{Y}(k+r+1|k) = &\sum_{i=1}^{h} \hat{A}_i(k)Y(k-i+1) + \sum_{i=1}^{f} \hat{B}_i(k)U(k-i+1)\\ &+ \sum_{i=1}^{g} \hat{C}_i(k)W(k-i+1)\end{aligned} \tag{6.3}$$

The adaptation mechanism uses the error of the *a priori* estimation, $Y(k) - \hat{Y}(k|k-1)$, to adjust the AP model parameter matrices $\hat{A}_i$, $\hat{B}_i$ and $\hat{C}_i$ at each time k according to the design philosophy developed in Chapter 4. On the other hand, following the basic strategy of predictive control described in Chapter 2, the control vector that renders the predicted output $\hat{Y}(k+r+1|k)$ equal to the driving desired output $Y_d(k+r+1)$ is given by

$$\begin{aligned}U(k) = &\hat{B}_1(k)^{-1}Y_d(k+r+1) - \hat{B}_1(k)^{-1}[\sum_{i=1}^{h} \hat{A}_i(k)\hat{Y}(k-i+1)\\ &+ \sum_{i=2}^{f} \hat{B}_i(k)\hat{U}(k-i+1) + \sum_{i=1}^{g} \hat{C}_i(k)W(k-i+1)]\end{aligned} \tag{6.4}$$

where the adaptation mechanism must ensure that $\hat{B}_1(k)$ is non-singular.

The actual implementation of such an adaptive predictive control strategy requires the execution of the following steps at each control instant:

- adjustment (or recursive estimation) of the AP model parameters;
- generation of the driving desired output by the driver block;
- computation of the control signal.

6.2.2 Practical issues

For the actual implementation of the above steps on real processes, a number of practical issues need to be considered concerning the AP model, time delays, the adaptation algorithm and the driver block. These practical issues and their possible solutions are discussed next.

Incremental choice of the I/O variables in the AP model

An initial practical consideration about the use of the AP model of (6.2) arises from the fact that, given the non-linear, time varying nature of the process, exact *a priori* information on the steady state values may not be available. Therefore, the input/output vectors Y, U and W in the AP model cannot be computed as deviation variables around their steady state operating values. There are basically two methods for overcoming this problem easily.

One possibility consists of adding a term to the AP model to compensate for the non-availability or the erroneous knowledge of the steady state values. For example, if in a scalar first-order process model the unknown steady state values for the output and input are denoted by y_{ss} and u_{ss} respectively, then the process can be represented as follows:

$$y(k) - y_{ss} = a[y(k-1) - y_{ss}] + b[u(k-1) - u_{ss}]$$

The corresponding AP model for this example will be

$$\hat{y}(k) = \hat{a}(k-1)y(k-1) + \hat{b}(k-1)u(k-1) + \hat{s}(k-1)$$

$\hat{s}(k-1)$ will be adjusted on-line by the adaptation mechanism. This accounts for the lack of knowledge about the true value of $s = y_{ss} - ay_{ss} - bu_{ss}$.

Another solution is to choose $Y(k)$, $U(k)$ and $W(k)$ as incremental values of the measured output, input and measurable disturbance vectors between the sampling instants k and $k-\gamma$, where γ is a suitably chosen positive integer. In this way the input/output representation defined by equation (6.1) is independent of the steady state values, and the corresponding incremental vectors entering the AP model (6.2) are easily computed. This is the approach considered in the application to the distillation column that is presented in this chapter. Some advantages of this approach for predictive control are discussed in Appendix B (Section B.1.3).

The delay matrix and the non-singularity of $B_1(k)$

The delay matrix D will determine the singularity or non-singularity of the process matrix $B_1(k)$ given in (6.1). Since the zero elements in $B_1(k)$ correspond to those components d_{ij} in D that are greater than the minimum delay r, we may establish the following rule in order to determine the singularity or non-singularity of $B_1(k)$ from the value of the delay matrix:

- If, by substituting zero values for the $d_{ij} > r$ and any non-zero real numbers for the $d_{ij} = r$, the set of matrices derived from D is singular, then $B_1(k)$ will be singular. If this is not the case, then $B_1(k)$ will most probably be non-singular.

When matrix $B_1(k)$ is non-singular, there is a provision in the adaptation mechanism of APCS [MFS81] which ensures that $\hat{B}_1(k)$ will also be non-singular and thus the control law (6.4) will be implementable. When $B_1(k)$ is singular, a practical strategy for avoiding it consists of transforming the original system by means of an appropriate change in the delay between the input/output variables in such a way that a new delay matrix results which does not imply the singularity of $B_1(k)$. This procedure is illustrated in the following example.

Consider a two-input/two-output system represented by

$$\begin{pmatrix} y_1(k) \\ y_2(k) \end{pmatrix} = \begin{pmatrix} a_1 & 0 \\ 0 & 0 \end{pmatrix} \begin{pmatrix} y_1(k-1) \\ y_2(k-1) \end{pmatrix} + \begin{pmatrix} 0 & 0 \\ 0 & a_2 \end{pmatrix} \begin{pmatrix} y_1(k-2) \\ y_2(k-2) \end{pmatrix}$$

$$+ \begin{pmatrix} b_1 & b_2 \\ 0 & 0 \end{pmatrix} \begin{pmatrix} u_1(k-1) \\ u_2(k-1) \end{pmatrix} + \begin{pmatrix} 0 & 0 \\ 0 & b_4 \end{pmatrix} \begin{pmatrix} u_1(k-2) \\ u_2(k-2) \end{pmatrix}$$

$$+ \begin{pmatrix} 0 & 0 \\ b_3 & 0 \end{pmatrix} \begin{pmatrix} u_1(k-3) \\ u_2(k-3) \end{pmatrix}$$

The corresponding delay matrix for this MIMO system is

$$D = \begin{pmatrix} 0 & 0 \\ 2 & 1 \end{pmatrix}$$

Since in this case we have $r = 0$, D clearly implies the singularity of B_1 according to the above rule.

We may consider a simple transformation by shifting the time delay in the first component of the process output vector, thus introducing a new output vector Y' defined as $y'_1(k) = y_1(k-1)$ and $y'_2(k) = y_2(k)$. The transformed process equation is now

$$\begin{pmatrix} y_1'(k) \\ y_2'(k) \end{pmatrix} = \begin{pmatrix} a_1 & 0 \\ 0 & 0 \end{pmatrix} \begin{pmatrix} y_1'(k-1) \\ y_2'(k-1) \end{pmatrix} + \begin{pmatrix} 0 & 0 \\ 0 & a_2 \end{pmatrix} \begin{pmatrix} y_1'(k-2) \\ y_2'(k-2) \end{pmatrix}$$

$$+ \begin{pmatrix} b_1 & b_2 \\ 0 & b_4 \end{pmatrix} \begin{pmatrix} u_1(k-2) \\ u_2(k-2) \end{pmatrix} + \begin{pmatrix} 0 & 0 \\ b_3 & 0 \end{pmatrix} \begin{pmatrix} u_1(k-3) \\ u_2(k-3) \end{pmatrix}$$

The corresponding delay matrix for this system is:

$$D' = \begin{pmatrix} 1 & 1 \\ 2 & 1 \end{pmatrix}$$

which does not imply the singularity of B_1'. The input vector $U(k)$ can now be computed at time k using the corresponding AP model from the following equation:

$$\begin{pmatrix} y_{d_1}'(k+2) \\ y_{d_2}'(k+2) \end{pmatrix} = \begin{pmatrix} \hat{a}_1(k) & 0 \\ 0 & 0 \end{pmatrix} \begin{pmatrix} y_1'(k+1) \\ y_2'(k+1) \end{pmatrix} + \begin{pmatrix} 0 & 0 \\ 0 & \hat{a}_2(k) \end{pmatrix} \begin{pmatrix} y_1'(k) \\ y_2'(k) \end{pmatrix}$$

$$+ \begin{pmatrix} \hat{b}_1(k) & \hat{b}_2(k) \\ 0 & \hat{b}_4(k) \end{pmatrix} \begin{pmatrix} u_1(k) \\ u_2(k) \end{pmatrix} + \begin{pmatrix} 0 & 0 \\ \hat{b}_3(k) & 0 \end{pmatrix} \begin{pmatrix} u_1(k-1) \\ u_2(k-1) \end{pmatrix}$$

Notice that the first term $\hat{A}_1'(k)Y'(k+1)$ on the right-hand side of the above equation can be computed since $y_1'(k+1)$ is already known at time k. Note also that, in order to apply this strategy, the columns of matrix $\hat{A}_1'(k)$ corresponding to the non-transformed output variables must be zero. If necessary, this could be accomplished easily by successive backward substitution of the original system into the above equation.

The strategy illustrated above by means of a simple example is applicable in general and may require delay changes in the output and/or the input variables to obtain the non-singularity of B_1'(k).

Another way of analyzing the problem associated with the singularity of $B_1(k)$ is as follows. The MIMO system can be decomposed into a set of n multi-input, single-output (MISO) systems. The control vector can in some cases be computed by considering the AP model of each of these MISO systems in a particular sequential order so as to satisfy the set of n conditions imposed on the control vector by the set of n MISO AP models. The following example illustrates this procedure.

Consider the two MISO systems of the process discussed in the preceding example:

$$y_1(k) = a_1 y_1(k-1) + b_1 u_1(k-1) + b_2 u_2(k-1)$$
$$y_2(k) = a_2 y_2(k-2) + b_3 u_1(k-3) + b_4 u_2(k-2)$$

The control inputs $u_2(k)$ and $u_1(k)$ are then computed sequentially from the following AP models:

$$y_{d_2}(k+2) = \hat{a}_2(k) y_2(k) + \hat{b}_3(k) u_1(k-1) + \hat{b}_4(k) u_2(k)$$
$$y_{d_1}(k+1) = \hat{a}_1(k) y_1(k) + \hat{b}_1(k) u_1(k) + \hat{b}_2(k) u_2(k)$$

Note that in practice this strategy is a particular case of the more general delay change or transformation strategy considered earlier, where the delay transformation is only carried out on the output variables.

Choice of the AP model order

A practical decision that must be taken in the implementation of APCS concerns the choice of the integers h, f and g, which determine the structure of the AP model in equation (6.2). As we have seen in Chapters 4 and 5, APCS stability theory ensures a satisfactory performance even in the presence of a reduced order AP model. We may recall that the key point in achieving this performance is the formulation of a normalized adaptive system. The use of a normalized system is always recommended in practice and, accordingly, it is possible to neglect many of the dynamic terms in the process equation, thus considering a small AP model order while maintaining the stability requirement.

In spite of the above recommendation, the application described in this chapter does not use a normalized adaptive system. The reason is historical: the project was carried out before the introduction of this type of adaptive system to APCS. Regardless, APCS was applied with a reduced order AP model, leading to satisfactory results, as we will see later in this chapter. This may be justified because in physical processes described by an equation such as (6.1), the parameters in the matrices A_i, B_i and C_i may decrease rapidly as i increases. Therefore, within the basic strategy used here, neglecting these terms may have little effect on the one-step or $(r+1)$-step ahead prediction of APCS considered in equation (6.3).

This argument may explain, as practice has confirmed [Mar78, MFS81], that in the implementation of APCS it is possible in many cases to choose AP model structures of lower order than that of the process and, moreover, that the conditions imposed by the theory to guarantee stability may be relaxed in the applications.

Choice of a driver block and a recursive estimation law

As pointed out in preceding chapters, the driver block must generate a driving desired trajectory (DDT) such that: (i) it is physically realizable; (ii) it takes into account the previous and current 'state' of the process; (iii) it ensures that the generated trajectory satisfies a specified performance index, resulting, for example, in the desired output trajectory approaching the steady state value with the desired dynamics.

Requirements (i) and (ii) will ensure that the desired output is compatible with the control limits in the process and that the driver block is not simply an autonomous entity. A particular driver block design already considered in Chapter 2 is used in the distillation column application considered in this chapter and will serve as an illustrative example.

In adaptive control, any controller design method can be combined with a suitable recursive estimation scheme. It is recommended that in APCS the adaptation mechanism be designed by considering the stability approach presented in Chapter 4 and by the specific type of application as well as the user's knowledge of and familiarity with a particular algorithm. The emphasis in the APCS approach is to estimate parameters so that prediction error is minimized. In this chapter, a particular gradient parameter identification technique will be used for the adaptation mechanism.

6.3 EXPERIMENTAL EQUIPMENT

A schematic diagram of the experimental set-up used in this and other studies [SSW77] is shown in Figure 6.1 and is briefly described below.

The 22.86 cm diameter column contains eight bubble cap trays (four caps per tray) and has a 30.48 cm tray spacing. The reboiler is of the thermosyphon type and the liquid overhead product is withdrawn from a total condenser, with the column operating at atmospheric pressure. Methanol-water feed of about 46 percent by weight of methanol is introduced into the column at the fourth plate. Further details on the column characteristics can be found in [WB73].

There exist two output variables: the composition of the top product and that of the bottom product respectively. Also there exist two control variables: steam and reflux flow rates respectively. Another input variable with a significant influence on the outputs is the feed flow rate, which can be assumed to be a measurable disturbance. Typical operating conditions for the column are given in Table 6.1.

The control of the column is referred to as single-input/single-output (SISO) when only one output is controlled by manipulating one input. The top composition is usually controlled by the reflux flow and the bottom composition is controlled by the steam flow. When both compositions are controlled simultaneously by manipulating both reflux and steam flows, the controlled is referred to as multi-input/multi-output (MIMO).

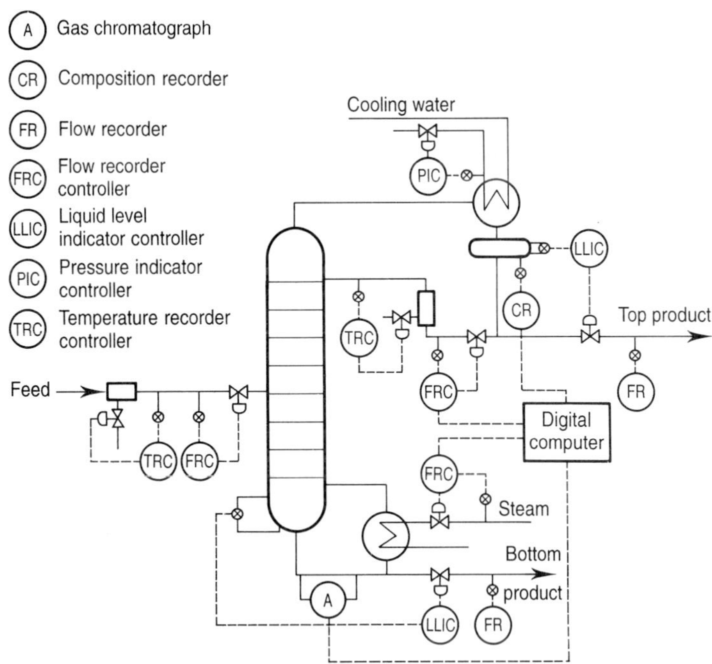

Fig. 6.1. Distillation column and control loops.

TABLE 6.1: Typical operating conditions			
Feed flow	18.0 g/s	Bottom flow	8.9 g/s
Reflux flow	18.1 g/s	Top flow	9.1 g/s
Steam flow	15.5 g/s	Feed composition	45.7 wt %
Top composition	96.0 wt %	Feed inlet temperature	71.8 ° C
Bottom composition	1.1 wt %	Reflux inlet temperature	63.9 ° C

Continuous measurement of the composition of the top product is accomplished by a capacitance cell. This method is adequate for solutions of high methanol content, but gives erratic results for low methanol concentrations. Consequently, an alternative method was used for the measurement of the bottom composition. This consisted of a Beckman series 4 Industrial Gas Chromatograph (GC) with an automatic liquid sampling valve and a product circulation system. The consecutive analysis of periodic samples of the bottom product provides the composition measurement. The use of this method poses three difficulties for the SISO bottom composition control and for the multivariable control:

1. Each GC analysis requires approximately 256 seconds. Consequently, this imposes a minimum value of 256 seconds for the sampling period.

2. The GC analysis also imposes a measurement time delay of one sampling period.

3. The accuracy of measurement is estimated to be ± 0.5%.

Figure 6.2 shows steady state measurements of the bottom composition and illustrates the high level of measurement noise.

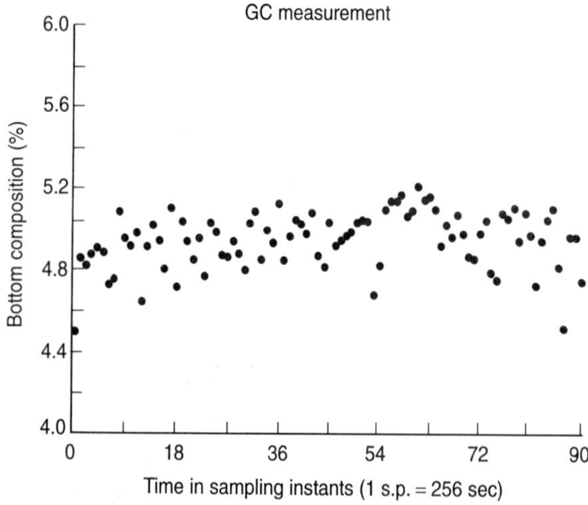

Fig. 6.2. Steady state measurements of the bottom composition.

The column was interfaced with an IBM 1800 digital computer, which allowed extensive data acquisition and provided the means for implementing control algorithms. All the significant inputs to the column were monitored and/or controlled by means of control loops, as is shown in Figure 6.1. Control of the top and bottom compositions was accomplished using the IBM 1800 computer. The required control actions (i.e. the reflux and steam flow rates) were cascaded to the setpoints of the flow controllers.

6.4 ADAPTIVE PREDICTIVE CONTROL: SEQUENCE OF OPERATIONS AND DESIGN PARAMETERS

In this section we describe the sequence of operations used in the implementation of adaptive predictive control to the distillation column. We also specify the main parameters selected by the designer.

6.4.1 SISO top composition control

In this case the top composition is the output controlled by using the reflux flow rate as the control variable. The sampling period is 64 seconds. The time delay between the reflux flow and the top composition is observed to be one sampling period. The sequence of operations at each sampling instant is as follows:

1. Measurement of the process output $y_p(k)$.

2. Computation of the incremental process output $y(k)$ by

$$y(k) = y_p(k) - y_p(k-2) \tag{6.5}$$

3. A second-order AP model structure with time delay 1 is chosen for this example. No measurable perturbations are considered. Accordingly, the computation of the *a priori* estimate $\hat{y}(k|k-1)$ is

$$\begin{aligned}\hat{y}(k|k-1) = \sum_{i=1}^{2} \hat{a}_i(k-1)y(k-1-i) \\ + \sum_{i=1}^{3} \hat{b}_i(k-1)u(k-1-i)\end{aligned} \tag{6.6}$$

   and the error of this estimation is given by

$$e(k|k-1) = y(k) - \hat{y}(k|k-1) \tag{6.7}$$

   The values $u(k-1-i)$ are obtained by

$$u(k-1-i) = u_p(k-1-i) - u_p(k-3-i) \tag{6.8}$$

where $u_p(k-l-i)$ is the reflux flow rate applied to the process at $k-1-i$.

Note that in (6.6) the number of parameters related to the process output is equal to the order of the AP model, and the number of $\hat{b}_i$ parameters is equal to the AP model order plus the time delay expressed in integer numbers of sample intervals.

4. Computation of the parameters $\hat{a}_i(k)$ and $\hat{b}_i(k)$ by gradient parameter algorithm of the form

$$\begin{aligned} \hat{a}_i(k) &= \beta_{ai}\alpha(k)e(k|k-1)y(k-1-i) + \hat{a}_i(k-1) \\ \hat{b}_i(k) &= \beta_{bi}\alpha(k)e(k|k-1)u(k-1-i) + \hat{b}_i(k-1) \\ \alpha(k) &= \frac{1}{1 + \sum_{i=1}^{2} \beta_{ai} y(k-1-i)^2 + \sum_{i=1}^{3} \beta_{bi} u(k-1-i)^2} \end{aligned} \tag{6.9}$$

The values of the coefficients β_{ai} and β_{bi} were chosen equal to 10 and 0.1 respectively. This choice took into account the approximate ratio between the absolute values of the parameters of a well-adjusted AP model.

In all the experiments the initial parameters of the AP model were chosen in a straightforward manner. For example, in some cases they were first initialized to (arbitrary) zero values and were found to converge to some appropriate values at the end of the experiment. These values could be used as initial values for subsequent runs.

When little or no knowledge about the process is available, the system may be run in an identification mode during the first few sampling instants, that is, only the above four steps need to be performed. For the experiments that will be described later, the system was generally operating in identification mode during the first two sampling instants only.

5. Computation of the desired incremental output $y_d(k+2)$ by

$$\begin{aligned} y_{pd}(k+2) &= \sum_{i=1}^{2} f_i y_p(k+1-i) + \sum_{i=1}^{3} h_i v(k+1-i) \\ y_d(k+2) &= y_{pd}(k+2) - y_p(k) \end{aligned} \tag{6.10}$$

where $y_{pd}(k+2)$ and $v(k+1-i)$ are the desired top composition and the desired steady state value (setpoint) at $k+2$ and $k+1-i$ respectively. f_i and h_i are parameters of a model with the desired dynamics. In this way $y_{pd}(k+2)$ is the computed value at time $k+2$ of a trajectory that, starting from the previously measured top composition value $y_p(k+1-i)$, will approach the desired steady state value $v(k)$ according to the desired dynamics specified by the parameters f_i and h_i. Therefore, $y_{pd}(k+2)$ is the desired value of the top composition at time $k+2$ and $y_d(k+2)$ is the corresponding desired incremental process output. Note that this illustrates one method of selecting or designing the driver block. However, as discussed in Chapter 2, this driver block design is not to be recommended for processes with an unstable inverse, which is not the case considered here.

The above parameters f_i and h_i were chosen as a result of the discretization of a second-order model with a natural frequency of 0.0084 rad/sec, damping ratio and static gain equal to 1 and a time-delay of one sampling period.

6. Computation of the control signal $u_p(k)$ by

$$u(k) = \frac{y_d(k+2) - \sum_{i=1}^{2} \hat{a}_i(k) y(k+1-i) - \sum_{i=2}^{3} \hat{b}_i(k) u(k+1-i)}{\hat{b}_1(k)} \tag{6.11}$$

$$u_p(k) = u(k) + u_p(k-2)$$

7. Application of control constraints: The control signal can be limit checked in an absolute and incremental manner, imposing limits on $u_p(k)$ and $u(k)$ respectively.

6.4.2 SISO bottom composition control

The single-input/single-output control of the bottom composition is accomplished using steam flow rate as control variable and a sampling period of 256 seconds. Though the effect of steam flow on the bottom composition is immediate, there is a measurement time delay of one sampling period due to the CG analysis. Thus, there is a time delay of one sampling period in both top and bottom composition loops, and hence the sequence of operations in this case is identical to the one described above.

The design parameters are the same as for the top composition control, with the exception of the natural frequency of the driver block, chosen as 0.0056 rads/sec.

6.4.3 MIMO control

In this case we have a multivariable system with two inputs u_1 (reflux flow rate) and u_2 (steam flow rate) and two outputs y_1 (top composition) and y_2 (bottom composition).

Due to the time required for the GC analysis, the sampling period was chosen to be 256 seconds. Because of this large sampling period, there was no time delay between the top composition and the steam and reflux flow rates. There was a measurement time delay of one sampling period between the bottom composition and the steam flow rate, while the time delay between the bottom composition and the reflux flow rate was observed to be two sampling intervals. Therefore, the delay matrix was

$$D = \begin{pmatrix} d_{11} & d_{12} \\ d_{21} & d_{22} \end{pmatrix} = \begin{pmatrix} 0 & 0 \\ 2 & 1 \end{pmatrix}$$

As discussed in Section 6.2, this kind of delay matrix would imply the singularity of matrix $B_1(k)$. To avoid this problem, a delay change or transformation in $y_1(k)$ identical to the one used in the illustrative example of Section 6.2 was made. As already explained in Section 6.2, this transformation renders the transformed matrix $B_1'(k)$ non-singular.

The specific sequence of operations used in this case is similar to the one described before for the SISO cases, except for recognition of the fact that the system to be controlled is a MIMO system with a change in the delay in $y_1(k)$. The complete sequence of operations for this case has been described in [Mar76a].

The multivariable AP model has a second-order structure, and the initial values of its parameters were also chosen in a straightforward manner. All the values of the coefficients β related to the top composition were chosen equal to 1, and those related to the bottom were set to 0.1.

The top and bottom desired outputs were generated by two separate driver blocks, identical to the scalar form previously presented for the SISO cases, the parameters of which corresponded to those of a discretized second-order model, with and without one sampling time delay respectively, a natural frequency of 0.0056 rad/sec, and a damping ratio and static gain both equal to 1.

The compensation of the disturbance effect was found to be satisfactory without including the terms related to the measurable disturbance (feed flow) in the AP model for the top composition y_1. In one experiment these terms were included for the bottom composition y_2. Thus the AP model was conveniently modified to include a new variable $w(k)$ for the feed flow and two parameters $\hat{c}_{12}(k)$ and $\hat{c}_{22}(k)$ to take its influence on the bottom composition into account.

6.5 EXPERIMENTAL RESULTS AND DISCUSSION

Some representative experiments conducted for SISO APCS control of the top composition are summarized in Table 6.2, while Figures 6.3 – 6.4 present the results obtained. In the same way, SISO control of the bottom composition is summarized in Table 6.3 and the results are presented in Figure 6.5.

TABLE 6.2: SISO top composition control experiments

Figure	Setpoint changes	Other details
6.3	From 93.0 to 96.0% at k = 30	• Initial values of AP model parameters equal to 0 • Absolute control limits during the 20 first instants
6.4a	From 96.0 to 97.0% at k = 23 From 97.0 to 96.0% at k = 65	• Compared to a well-tuned PI controller $K_P = 4.72$ and $K_I = 0.772$
6.4b	From 96.0 to 94.5% at k = 23 From 94.5 to 96.0% at k = 64	• Compared to the same PI controller

The details of the MIMO experiments are given in Table 6.4, Figures 6.6 – 6.8 displaying the results obtained.

An overall view of the time histories of the variables and parameters illustrated in these diagrams gives a general idea of the excellent behaviour of the APCS control of the pilot distillation column. In order to analyze this behaviour further, the remainder of this section discusses the experimental results in more detail, focusing on the characteristics of the plant, the simplicity of the strategy and the practical issues of the implementation, as well as the overall performance.

TABLE 6.3: SISO bottom composition control experiments

Figure	Setpoint and load changes	Other details
6.5	Setpoint changes: From 5 to 3% at k = 22 From 3 to 5% at k = 44 From 5 to 7% at k = 138 Load (feedflow) changes: From 18 to 21 g/s at k = 67 From 21 to 18 g/s at k = 93 From 18 to 15 g/s at k = 120	• Duration: 10 h 40 min • Incremental control limited to ± 2.1 g/s • PI control with $K_P = -0.15$ and $K_I = -0.0024$ • APCS control started at transient state with bottom composition at 6.8%

TABLE 6.4: MIMO control experiments

Figure	Setpoint and load changes	Other details
6.6	Top composition: From 96.0 to 97.0% at k = 29 Bottom composition: From 3 to 5% at k = 55	• Duration: 6 h 24 min • Control started with the plant at equilibrium and composition values 96.5% (top) and 1% (bottom)
6.7	Top composition: From 97.0 to 96.0% at k = 125 Bottom composition: From 5 to 2% at k = 125 Load (feedflow) changes: From 18 to 21 g/s at k = 7 From 21 to 18 g/s at k = 34 From 18 to 15 g/s at k = 66 From 15 to 18 g/s at k = 88	• Duration: 10h 40 min • Incremental control limited to ± 2.9 g/s
6.8	Stochastic load (feedflow) (mean 18 g/s) (standard deviation 0.9 g/s) Mean changes: 21 g/s at k = 85 18 g/s at k = 101 15 g/s at k = 115 18 g/s at k = 129	• Duration: 10 h 40 min • The AP model compensation action starts at k = 41 • Initial values of parameters $\hat{c}_{12}(0) = \hat{c}_{22}(0) = 0$ • Incremental control limited to ± 3.5 g/s

6.5.1 Characteristics of the plant

Distillation columns have often been cited as examples of the control difficulties experienced in chemical plants [Fos73]. They also are some of the most common units used in a petrochemical operation and are high energy users. Thus, a demonstration of improved performance in their operation is of significant practical value. Control of the distillation column using conventional PID or multiloop PID strategies has not generally been satisfactory due to the following:

1. The distillation column is a process with important interactions, as can clearly be seen in Figure 6.4, where the response of the uncontrolled bottom composition is shown to be influenced by a setpoint change performed at the top composition.

2. The distillation column is a non-linear plant. The following facts demonstrate the high non-linearity of the plant:

   (a) Figures 6.4 and 6.5 show the asymmetric behaviour of a PI controller when the system is subjected to positive and negative setpoint and perturbation changes.

   (b) When comparing the responses under APCS control in Figure 6.4a and b, it can be seen that the gain of the plant differs depending on the sign of the setpoint change.

   (c) The non-linearity and high order of the plant is also clearly apparent from modelling studies performed on this column. A non-linear ordinary differential equation model of the same column, based on mass and energy balance, results in thirty differential equations containing 140 variables [Sim76].

6.5.2 Simplicity of the adaptive predictive control strategy

The following design variables and strategies have to be selected in the actual implementation of APCS: choice of the AP model order and type (with incremental variables and/or transformed models to handle time delays, etc.); choice of the recursive estimation law; design of the driver block; selection of suitable control limits; and choice of the initial parameter values for the AP model. The different choices made in this application are now discussed, with special focus being placed on their simplicity and their influence on the overall performance of APCS.

Choice of the AP model order

Although the dynamic input/output relations of the column are described by non-linear equations of a high-order structure, AP models of first and second order were used in order to solve the (short-term) adaptive prediction problem in a practical way. As demonstrated in the experimental runs, satisfactory control was obtained at all times with these simple choices of the AP model.

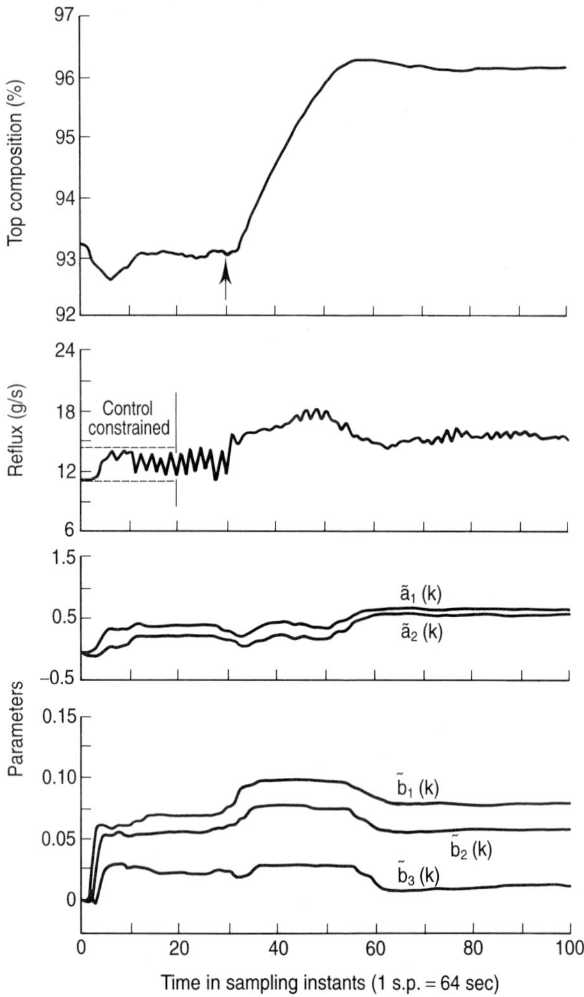

Fig. 6.3. Startup without any knowledge of the plant parameters.

Choice of the incremental input/output variables

The choice of incremental input/output variables in the AP model rendered the APCS implementation extremely simple and eliminated the need for any previous knowledge of the steady state values of the plant, thus allowing automatic control in the full operation range of the column.

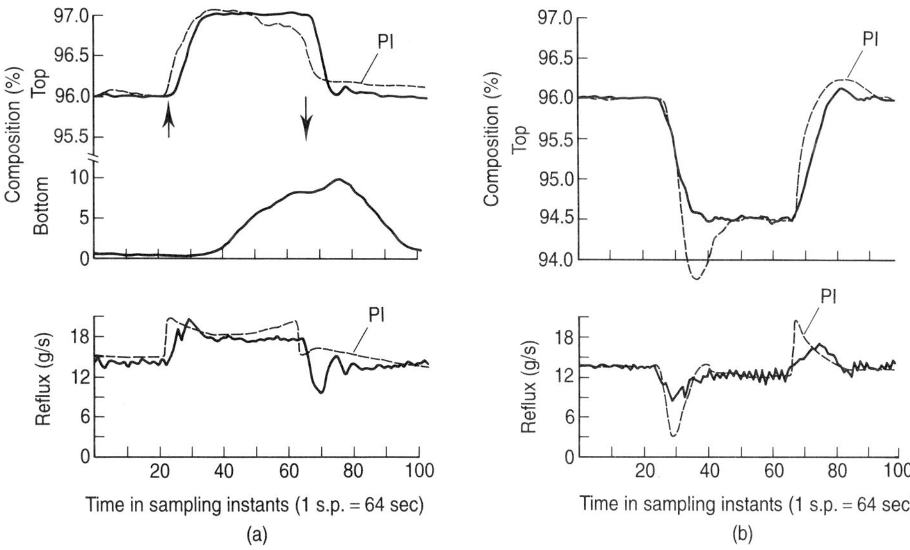

Fig. 6.4. SISO control of the top composition.

Use of the delay transformation for MIMO control

The delay change made on $y_1(k)$ avoided, in a simple way, the singularity of matrix $B_1(k)$ for the MIMO APCS strategy without deterioration of the control pcrformancc. Our approach in handling timc dclays for MIMO adaptive predictive control, as outlined in Section 6.2, is made with special emphasis on simplicity and practicality.

Choice of the adaptation mechanismm

The purpose of the estimation algorithm in the APCS implementation is to minimize the prediction error, that is, the error between the predicted and the measured process outputs. Note that this feature does not require a convergent estimation of the process parameters and hence no excitation signal in the process input is necessary. In this application, a gradient parameter algorithm was used as the adaptation mechanism. Chapter 4 analyzed the convergence of similar gradient parameter algorithms in the presence of bounded noise and disturbances, including criteria for conditional adaptation. These kind of criteria are of practical use in APCS implementation [Mar76a] and help to strengthen the robustness and stability of the overall control strategy, as analyzed in Chapter 5.

In the present application, although continuous adaptation was carried out in the presence of noise and disturbances, the algorithm gave satisfactory results. Experimental runs and the corresponding parameter trajectories shown in Figures 6.3 and 6.5 clearly demonstrate the stable behaviour of this algorithm. In particular, Figure 6.5 shows a more than ten-hour run with several setpoint changes and disturbances being introduced into the process. In all cases the parameters change smoothly and these changes coincide with a change of setpoint or the input of a disturbance into the process. The fast convergence property of the algorithm can clearly be seen in Figure 6.3.

Simplicity of the driver block design

The absence of a desired output trajectory in the control strategy, which would drive the process output from the preceding process outputs to the desired steady state value, could result in abrupt control signals, which are undesirable. For instance, in the experiment shown in Figure 6.7, the bottom composition measurement at sampling instant $k = 10$ is approximately 7.7% as compared with the desired (setpoint) steady state value of 5%, due to the effect of the feed flow disturbance. If the control strategy were to attempt to return the bottom composition to the desired steady state (5%) immediately, then the resulting control action generated could be extremely abrupt. This kind of control action would make the solution of the prediction problem more difficult in such a non-linear plant, since large control actions could result in large variations in the parameters of a linearized representation of the process. The driver block design used in this application is simple and practicable, and generates a desired output trajectory that, starting from the previous measured outputs, approaches the desired steady state by taking the desired process dynamics into account, leading to bumpless transfer of the process output variable from its current state to its desired state in a satisfactory manner.

Use of limits on the control signal

Control limits (incremental and absolute) were used for various purposes in this application. The incremental (over two sampling periods) control signal cannot detect and correct for the effects of the intersample ripples in the control action. Such ripples can be observed in Figures 6.3 and 6.4b. However, they can be eliminated by the introduction of incremental (over one sampling period) control limits. These limits were used in all the experiments described in Tables 6.3 and 6.4, with satisfactory results.

Choice of the initial parameter values of the AP model

In the absence of any knowledge of the plant parameters, control limits can be imposed on the control inputs to prevent APCS from producing an initial

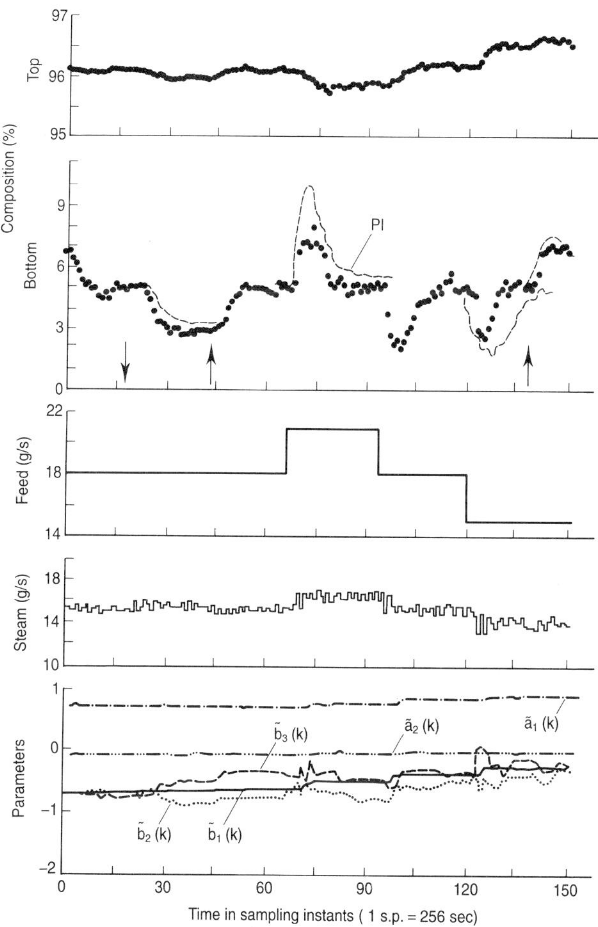

Fig. 6.5. SISO control of the bottom composition.

sequence of large control signals. For example, in Figure 6.3 the initial values of the AP model parameters were set to zero and APCS was applied by imposing suitable absolute control limits which constrained the input to remain within a narrow band. Following a short identification period, these control limits can be changed or relaxed somewhat. The fast convergence property of the algorithm, together with the control constraints imposed during the startup period (see Figure 6.3), lead us to the conclusion that any suitably chosen initial values should give good control under the APCS strategy.

6.5.3 Performance of the adaptive predictive control strategy

The experimental evaluation of the APCS strategy is now discussed. It is worthwhile noting that this implementation was performed on a pilot-scale unit using industrial-type instrumentation and was executed in an environment (in terms of noise and disturbances) that is, in many respects, typical of an industrial situation.

SISO and MIMO control

Some significant issues summarizing the performance of APCS are the following:

(a) In both the SISO and MIMO experiments, the response of APCS to setpoint changes and step feed flow disturbances is fast and smooth, without any offset or excessive control action. Note that all the experiments involving bottom composition control (Figures 6.5–6.8) were carried out with noisy GC measurements (see Figure 6.2). The feed flow disturbance has a greater effect on the bottom than on the top composition, as can be seen in Figure 6.7 for instance. In addition, due to the measurement time delay of one sampling period, the significant effect of step feed flow disturbances on the bottom composition is not detected by the control system until two sampling periods after the disturbance affects the process. Consequently, the corresponding corrective action does not affect the plant output until the third measurement. Figure 6.5 (SISO case) and Figure 6.7 (MIMO case) show how this corrective action returns the bottom composition value to the setpoint.

(b) The control system is able to adapt itself immediately to changes in the plant operating conditions, that is, due to multiple and simultaneous setpoint changes and load disturbances.

(c) In the multivariable case, APCS can manage the inherent system interactions in a satisfactory manner, as demonstrated in the following examples.

In the SISO control of top and bottom compositions, these composition values decrease as reflux decreases and steam increases. In the multivariable case, when simultaneous setpoint changes are made, as shown in Figure 6.7, the two control variables decrease to achieve the desired setpoint tracking.

With the bottom composition under open loop or no control (constant steam flow), an increase from 96% to 97% in the top composition results in a significant increase in the bottom composition, as can clearly be seen in Figure 6.4a. In the multivariable case, this interaction is compensated for by a suitable increase in steam and reflux, as occurs in the experiment illustrated in Figure 6.6.

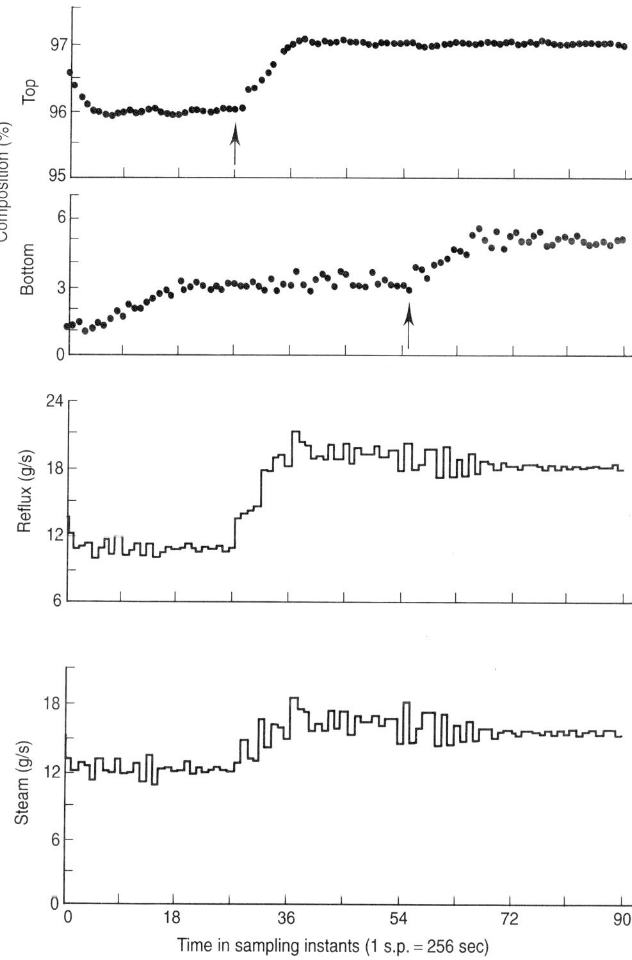

Fig. 6.6. MIMO control: setpoint changes.

Regulatory performance of stochastic disturbances

To evaluate the regulatory performance of the MIMO APCS controller, the distillation column was subjected to a stochastic-type disturbance in the feed flowrate of a mean of 18 g/sec and a standard deviation of 0.9 g/sec. The disturbance sequence and the resulting regulation of the top and the bottom composition can be seen in Figure 6.8. Appreciation of the severity of such a disturbance on the column can be achieved by examining the instantaneous effect of a positive and a negative step change in feed flowrate, as shown in Figures 6.5 and 6.7.

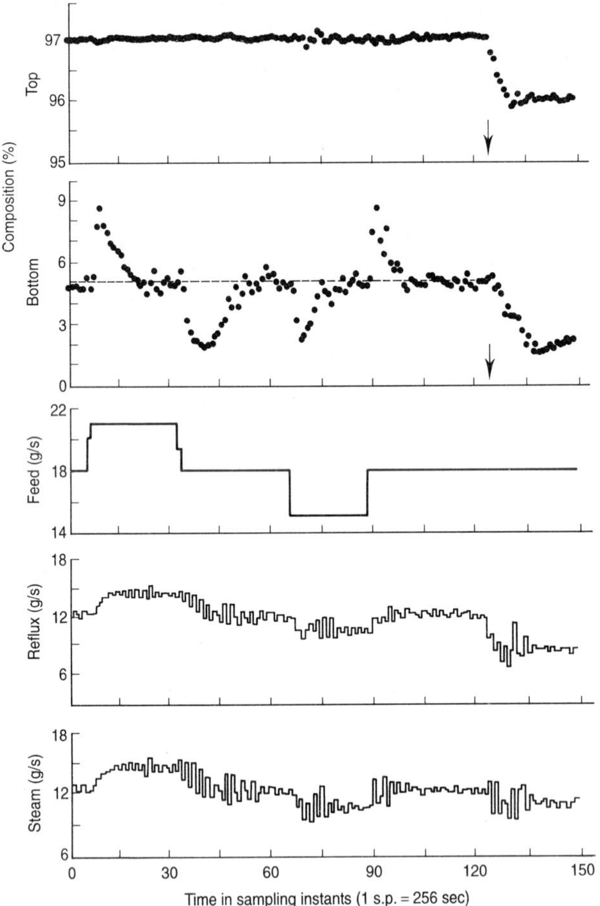

Fig. 6.7. MIMO control: simultaneous load and setpoint changes.

The other aspect of this regulatory performance that should be considered is the fact that there is a delay of three sample intervals between the time at which disturbance occurs and that at which corrective control action takes effect. This implies that initial transients observed in the bottom composition are unavoidable unless the disturbance is introduced within the AP model and is thus taken into account in the computation of the control action.

The effect of considering the disturbance in the AP model is demonstrated in Figure 6.8. The measured disturbance, appearing in equations (6.2)–(6.4), was taken into account from sampling instant $k = 41$ with initial parameter values for $\hat{c}_{12}(0) = 0$ and $\hat{c}_{22}(0) = 0$. Note that other parameters $\hat{c}_{ij}$ were fixed at zero and therefore that the AP model considered the disturbance for the bottom composition loop only. After approximately thirty sampling intervals, this method of generating the control action is able to compensate for the stochastic disturbance, resulting in a satisfactory regulation. Examination of Figure 6.8 also reveals that when a non-stationary change in the feed flowrate is introduced (for instance, mean changes from 18 g/sec to 21 g/sec at $k = 85$) the resulting control action cannot completely compensate for this disturbance because of the incremental control limits imposed on the steam flowrate. However, the regulation of both outputs y_1 and y_2 under such extremely severe conditions is excellent.

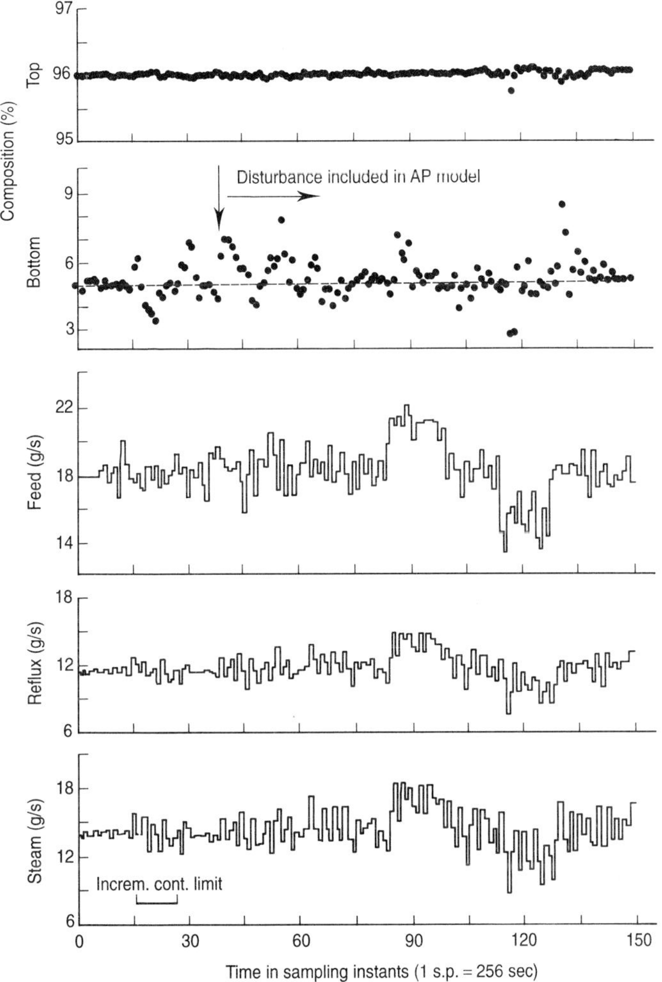

Fig. 6.8. Stochastic disturbances and AP model compensation.

The advantage of MIMO regulation can be appreciated by comparing Figures 6.5 and 6.8. In Figure 6.5 notice how changes in the steam flowrate (interaction) affect the top composition (uncontrolled). In comparison, satisfactory regulatory control of the top composition is observed under MIMO APCS control in the presence of a severe stochastic-type disturbance.

Comparison with conventional PI control

A clear qualitative difference between APCS and PI controllers is that the former is self-tuning whereas the latter is a fixed-gain controller. The tuning of PI involves a time consuming trial and error procedure. A comparison of complexity between the two methods is not an issue because either strategy can easily be executed by a digital computer.

In the SISO case, the APCS easily outperforms a well-tuned PI controller (Figures 6.4 and 6.5). The advantage of applying an adaptive predictive controller to a system that is characterized by asymmetric dynamics and time varying parameters, in comparison with a fixed-gain PI controller, is shown clearly in Figure 6.4a and b. The performance of APCS under consecutive setpoint changes and disturbances is compared in Figure 6.5 with four separate PI experiments, each run being started from steady state. The purpose is to illustrate the improvement over well-tuned PI controllers and the maintenance of good control by APCS under changing conditions.

The MIMO APCS performance proved to be excellent and can be compared with the results of a classical multiloop PI scheme applied to the same distillation column [CSW83]. It is important to point out that the tuning of a multiloop PI scheme poses an extra difficulty: the final parameter tuning depends on the sequence in which the loops are closed and it also requires several iterations for the fine tuning of each loop by an extremely time consuming and tedious process.

6.6 CONCLUSIONS

This chapter has presented an application of the basic APCS strategy to the SISO and MIMO control of a binary distillation column. The experiments were carried out in an environment that is, in many respects, typical of an industrial situation. The plant is basically non-linear, multivariable, has time delays and a high-order structure. The control system was tested under severe conditions such as the presence of step and stochastic feed flow disturbances, significant measurement noise and setpoint changes, acting alone or simultaneously. The experimental evaluation of this method reveals (i) the simplicity of the algorithm, and (ii) the excellent performance obtained in all cases.

The simplicity of the resulting control strategy is due to the basic APCS concepts. Their implementation allows simple rationalization of such practical issues as: the lack of knowledge of the steady state values of the plant, the presence of time delays in the multivariable case, the choice of the model order and the selection of suitable control limits.

The experimental results presented here easily outperformed those obtained using classical control techniques and highlighted the potential of adaptive predictive control as a solution to industrial control problems.

Chapter 7

DESIGN OF AN AUTOPILOT FOR THE F-8 AIRCRAFT

7.1 INTRODUCTION

With the purpose of further illustration of the application of APCS to real time processes using the basic strategy of predictive control, this chapter describes the design and experimental evaluation of an automatic pilot for the pitch angle dynamic control of NASA's F-8 supersonic aircraft. This application is also of historical interest in being the first real time application of adaptive predictive control systems, and was carried out in 1975 using the facilities of the Charles Stark Draper Laboratory in Cambridge, Massachusetts, USA.

The main difficulty in the application of control systems to the longitudinal dynamics (pitch angle) of aircraft arises from the large variation in the plant parameters occurring during the course of normal flight operations. This variable nature of aircraft dynamics represented a challenge for the application of any control system, and particularly for the application of adaptive control systems. The difficulty of this problem was emphasized in a NASA research programme for the adaptive control of the F-8, the results of which were published in 1977 [Bry77, ADDG77, AK77, DM77, DDDW77, Ell77, SH77]. In fact, the purpose of the programme was not to develop new control methods, but, rather, to apply existing methods to the problem. The results of the programme showed up the inability of the methodologies employed to solve this problem. In this context, the application described in this chapter was a real challenge for APCS, which was implemented on the same F-8 aircraft utilized in the aforementioned NASA programme.

In this chapter we will first discuss the interest and difficulty of the pitch angle control and the design objectives of the APCS automatic pilot. Later, we will describe the way in which the basic predictive control strategy was used. This strategy was implemented through a computer program that controlled the flight of a high-fidelity aircraft hybrid simulator. In order to evaluate the performance of the APCS, automatic pilot control tests were performed over the whole spectrum of the different aircraft flight regimes.

Special interest is devoted to the case in which the aircraft exhibits unstable natural behaviour. This case will illustrate the ability of the basic strategy of predictive control, within the adaptive context, to handle unstable processes, as guaranteed by the stability theory developed in preceding chapters.

7.2 MANUAL AND AUTOMATIC PITCH ANGLE CONTROL IN AN AIRCRAFT

Due to stability requirements, aircraft design traditionally requires that the centre of mass be located between the aerodynamic centre and the elevator as represented in the free body diagram of Figure 7.1. In this way, the torque originated by the lift force, applied in the aerodynamic centre, and the weight, applied at the centre of mass, is compensated by the torque produced by the elevator position.

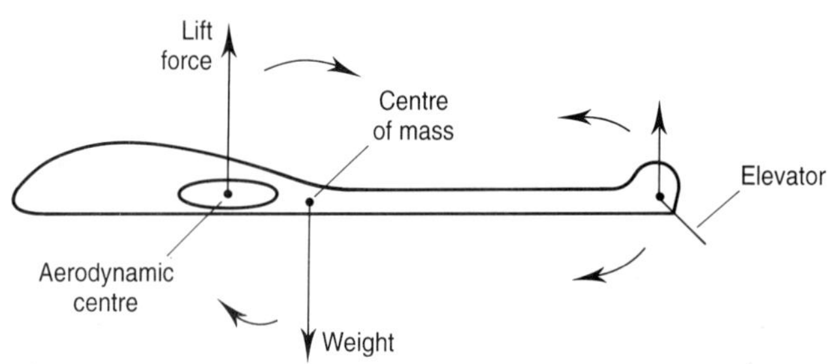

Fig. 7.1. Aircraft free body diagram.

Thus, the stability obtained is due to an increment in the air resistance to the aircraft movement. For each stable flight condition, a certain elevator position corresponds to a determined pitch angle. If the elevator position deviates from equilibrium, the aforementioned torques will not be compensated for and the aircraft will begin to change its pitch angle.

The interest in accurate pitch angle control is basic as it allows excellent control of the aircraft's speed and altitude. The pitch angle dynamic control in aircraft is basically controlled manually: when the pilot moves the rudder, he or she is, by means of a servomechanism, moving the elevator position, which acts as the control input to the aircraft, as illustrated in Figure 7.2a. In supersonic aircraft, this manual control requires considerable skill on the part of the pilot, since the aircraft's dynamic response is highly non-linear and mainly depends on the different flight conditions. The left-hand sides of Figures 7.5–7.10 illustrate typical responses of the pitch angle to pilot step commands on the rudder. In these diagrams we can see how the elevator follows the pilot's commands directly and the corresponding pitch rate and pitch angle responses under different flight conditions. Due to the complex dynamic nature of the aircraft, when the pilot wishes to change the pitch angle, he (she) will attempt to do so slowly in order to prevent oscillations, which may be extremely dangerous.

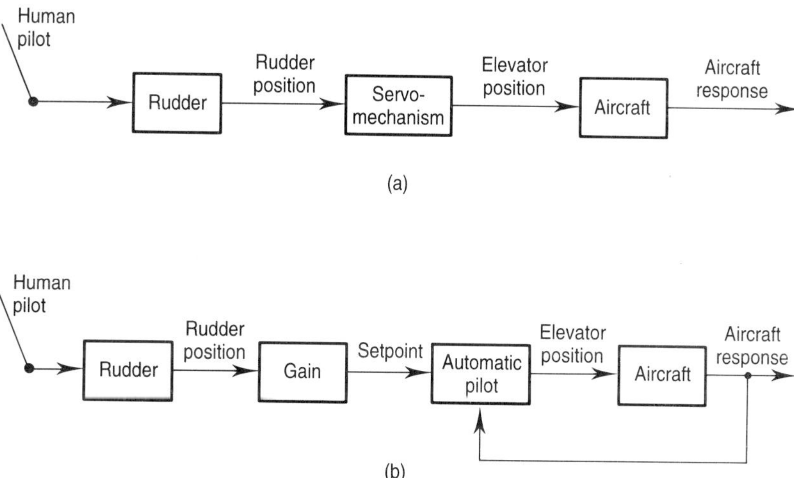

Fig. 7.2. Manual (a) and automatic (b) aircraft control.

The automatic pilot was designed to operate between the human pilot and the aircraft, as illustrated in Figure 7.2b. Moving the rudder, the human pilot enters, through a simple gain, the desired setpoint to the automatic pilot, which uses the measurements of the aircraft's response to generate the control signal, that is, the elevator position. The objective of an ideal automatic pilot would be to allow the human pilot to execute changes in the pitch angle easily and with the desired dynamics, irrespective of flight conditions and avoiding any kind of oscillations.

In the next section, we describe the way in which adaptive predictive control was used for the autopilot design.

7.3 APCS DESIGN OF THE AUTOPILOT

7.3.1 Objectives

As indicated in the preceding section, the role assigned to the automatic pilot is to translate the pilot's commands on the rudder into the desired control performance. In order to do this, the *pitch rate* was selected as the *process output* to be controlled and the *elevator position* as the *control input*, and the specific practical objectives of the autopilot's design were set as illustrated in Figure 7.3 and defined in the following points:

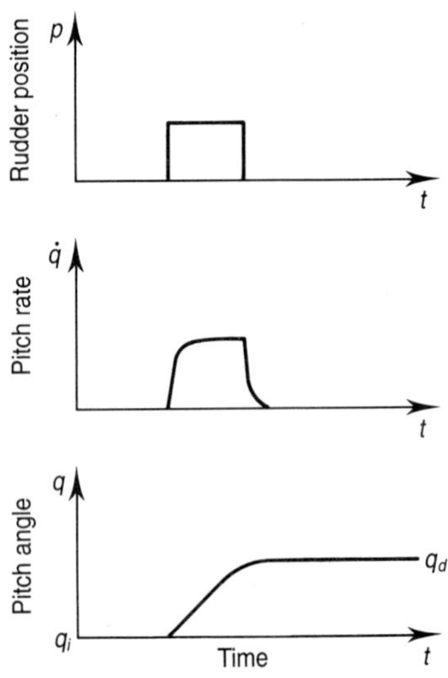

Fig. 7.3. Desired behaviour for the APCS autopilot.

1. When the rudder is at equilibrium position, the aircraft must keep a constant pitch angle.

2. If the pilot wishes to change the position of the pitch angle, he (she) will move the rudder from its equilibrium position. This will result in the pitch rate, following a desired trajectory, reaching a value that is proportional

to the rudder displacement. In this way, the pitch angle will begin to change according to the value of the pitch rate.

3. When the plane reaches the final desired pitch angle, the pilot will allow the rudder to return to its equilibrium position. This will result in the pitch rate, following a desired trajectory, returning to zero. Consequently, the new position of the pitch angle will be maintained, and the change will have occurred without oscillations and at a rate that is controlled by the pilot.

7.3.2 APCS autopilot structure

In order to achieve the objectives listed above, the design for the APCS autopilot corresponds to the block diagram represented in Figure 7.4.

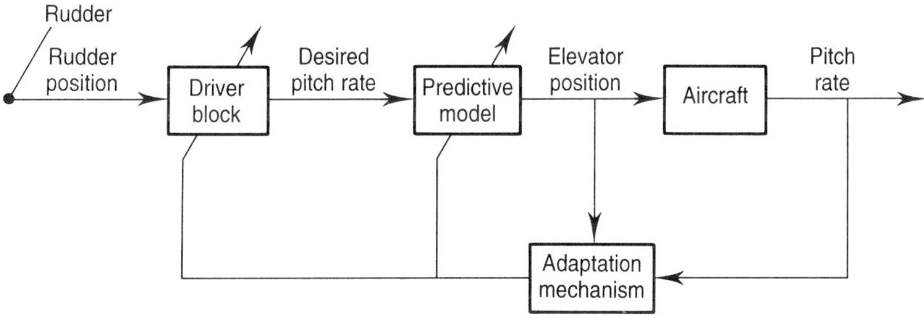

Fig. 7.4. APCS autopilot block diagram.

The main aspects concerning the different blocks of this diagram are described in the following.

Process dynamics and choice of the adaptive predictive model

Although the dynamic behaviour of the F-8 aircraft is non-linear, there is a dynamic relation between the elevator position and the pitch rate which, at each sampling instant k, can be described approximately by a linear time varying difference equation of the form

$$y(k+1) = \sum_{i=1}^{n} a_i(k) y(k+1-i) + \sum_{i=1}^{m} b_i(k) u(k+1-i) + \Delta(k) \qquad (7.1)$$

where the variables $y(k)$ and $u(k)$ are the increments at instant k with respect to instant $k-1$ of the pitch rate and the elevator position, respectively, and the time varying parameters $a_i(k)$ and $b_i(k)$ describe the complex process dynamics. The perturbation term $\Delta(k)$, as considered in preceding chapters, corresponds to the influence of measurement noise and disturbances not included in the dynamics being modelled.

Consequently, considering this dynamic description of the process, the adaptive predictive model adopted in the APCS application was selected as follows:

$$\hat{y}(k+1|k) = \sum_{i=1}^{\hat{n}} \hat{a}_i(k) y(k+1-i) + \sum_{i=1}^{\hat{m}} \hat{b}_i(k) u(k+1-i) \qquad (7.2)$$

This adaptive predictive model is completely defined by choosing its dimensions $\hat{n}$ and $\hat{m}$ and the initial values of the parameters $\hat{a}_i(0)$ and $\hat{b}_i(0)$. Their values for each instant k will be properly calculated by the adaptation mechanism.

Driver block and control action computation

In accordance with the block diagram of Figure 7.4, the driver block receives the rudder position signal as input which is generated directly by the human pilot when displacing the rudder. This position signal, according to the design objective, is indicative, through a proportionality constant, of the pitch rate that the human pilot requires.

The driver block also receives information on the actual value of the pitch rate at each sampling instant. From the value of the pitch rate desired by the pilot and its actual value, at each sampling instant k the driver block will generate a desired pitch rate increment $y_d(k+1|k)$ for the next sampling instant $k+1$. This increment is calculated by the driver block and basically follows the design described in Chapter 2, so that the pitch rate approaches the value desired by the pilot following a driving desired trajectory. However, it must be noted that this driving desired trajectory, as it is always generated from the pitch rate measured values, will be influenced by the disturbances and noise that may affect this variable, and so we cannot expect it to be a perfectly predefined curve. On the contrary, as we will observe in the experimental results given later, this desired trajectory will be irregular in many cases.

Once the driver block has calculated the desired pitch rate increment, the predictive model of (7.2) is used to calculate the increment for the elevator position $u(k)$ which renders the predicted pitch rate increment equal to the desired one, according to the principle of predictive control. Consequently, substituting $\hat{y}(k+1|k)$ by $y_d(k+1|k)$ in (7.2) and solving for $u(k)$, we obtain

$$u(k) = \frac{y_d(k+1|k) - \sum_{i=1}^{\hat{n}} \hat{a}_i(k) y(k+1-i) - \sum_{i=2}^{\hat{m}} \hat{b}_i(k) u(k+1-i)}{\hat{b}_1(k)} \tag{7.3}$$

After being limited by the corresponding physical constraints for the control action, the incremental signal $u(k)$ given by (7.3) determines the new position that will be applied to the elevator.

Adaptation mechanism

As can be seen in the block diagram of Figure 7.4, the adaptation mechanism will act on the following two levels:

1. On the first level and at each instant k, from the error produced in the prediction of the pitch rate increment, the adaptation mechanism adjusts the value of the parameters $\hat{a}_i(k)$ and $\hat{b}_i(k)$ of the predictive model by means of an algorithm such as those formulated in Chapter 4.

2. On the second level, the adaptation mechanism informs the driver block of the measured value of the pitch rate, so that the driver block can generate a physically realizable driving desired trajectory for the variable under control.

Sequence of operations

The APCS structure described above results in the following operation sequence for the calculation of the control action at each sampling instant k:

1. measurement of the pitch rate and calculation of the corresponding increment $y(k)$;

2. calculation of the prediction error associated with the increment $y(k)$;

3. adjustment of the value of the parameters $\hat{a}_i(k)$ and $\hat{b}_i(k)$ of the predictive model;

4. calculation of the desired increment $y_d(k+1|k)$;

5. calculation of the incremental control action $u(k)$;

6. application of the control limits to the incremental action $u(k)$ to determine the position signal that will be applied to the aircraft elevator.

7.4 EXPERIMENTS AND RESULTS

The APCS autopilot was tested in the high-fidelity hybrid simulator of the F-8 aircraft available in the Charles Stark Draper Laboratory (CSDL). To evaluate the performance of the APCS automatic pilot, control tests were performed over the whole spectrum of the different aircraft flight regimes. Also, the performance of the automatic pilot was tested when the aircraft dynamics was rendered both neutrally stable and inherently unstable in the degree of freedom under control. After summarizing the main characteristics of the F-8 and its hybrid simulation, we present some illustrative examples of the experimental results obtained.

7.4.1 The F-8 aircraft and its hybrid simulation

The F-8 is a single engine, supersonic, shipboard day fighter. In addition to the conventional controls (elevator, ailerons, rudder and throttle), this aircraft has a two-position wing to allow high lift at slow speed without a large fuselage angle, variable leading edge for increased subsonic maneuverability, a speedbrake to allow rapid deceleration from high speed and an afterburner for increased thrust.

The CSDL hybrid simulation of the F-8 aircraft was a real time, non-linear, six degrees of freedom, rigid body simulation with constant mass. The equations of motion were programmed on two Beckman 2200 analog computers and one Xerox Data Systems 9300 digital computer. The entire flight regime of the F-8 was covered, from stall to Mach 1.9, from sea level to 15000 meters, including ground effects for take-off and landing. Solution of the flight equations was obtained with no small-angle/small-perturbation assumptions being made. The allocation of computational tasks between the analog and the digital computers was so done to assure the good dynamic response required to model this high performance aircraft accurately.

7.4.2 Autopilot's adaptive behaviour

Four similar experiments with four different flight conditions have been performed in order to test the adaptive behaviour of the autopilot. The following circumstances were present in all these experiments:

(a) The sampling period was 0.05 seconds.

(b) The desired dynamics chosen for the driver block was that of a second-order system, with a natural frequency of 5 rad/sec, a damping ratio of 0.7 and a static gain of -2.28. The dynamics of this model was considered as the most desirable one to produce a well-damped pitch rate and fast acceleration responses.

(c) The AP model used had a third-order structure.

(d) The AP model parameters always had the same initial values, which were chosen realistically.

(e) From each trim flight condition, the rudder was moved to generate a 1.33-degree step for five seconds.

The four different flight conditions are summarized in Table 7.1

TABLE 7.1: Flight conditions in tests

Case	1	2	3	4
h : Altitude (m)	9144	9144	1524	1524
M: Mach number	1.2	0.5	0.7	0.33
p : Pressure (N/m^2)	30260	5218	28919	6463
v : Speed (m/s)	363	151	234	110

The results for test cases 1–4 are illustrated in Figures 7.5–7.8 respectively. The natural response of the aircraft (i.e. autopilot off) is shown on the left-hand side of each diagram, while the controlled response (autopilot on) is shown on the right-hand side.

7.4.3 Autopilot behaviour for neutrally stable and inherently unstable aircraft

Figure 7.9 shows an experiment similar to those previously described, except that the centre of mass was shifted over the aerodynamic centre, thus making the aircraft neutrally stable in pitch. The AP model in this case had a second-order structure, and the sampling period was 0.1 seconds. The flight conditions were: $h = 1761$ m, $M = 0.68$, $p = 3420$ N/m^2 and $v = 230$ m/s, which are close to those of case 3 of Table 7.1.

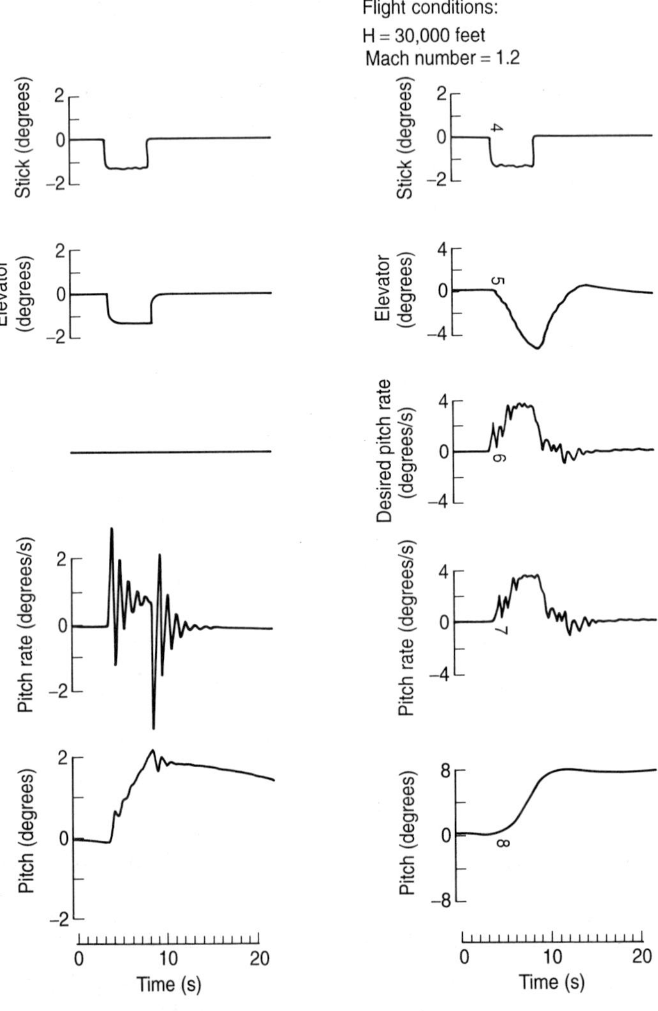

Fig. 7.5. Autopilot's adaptive behaviour. Case 1.

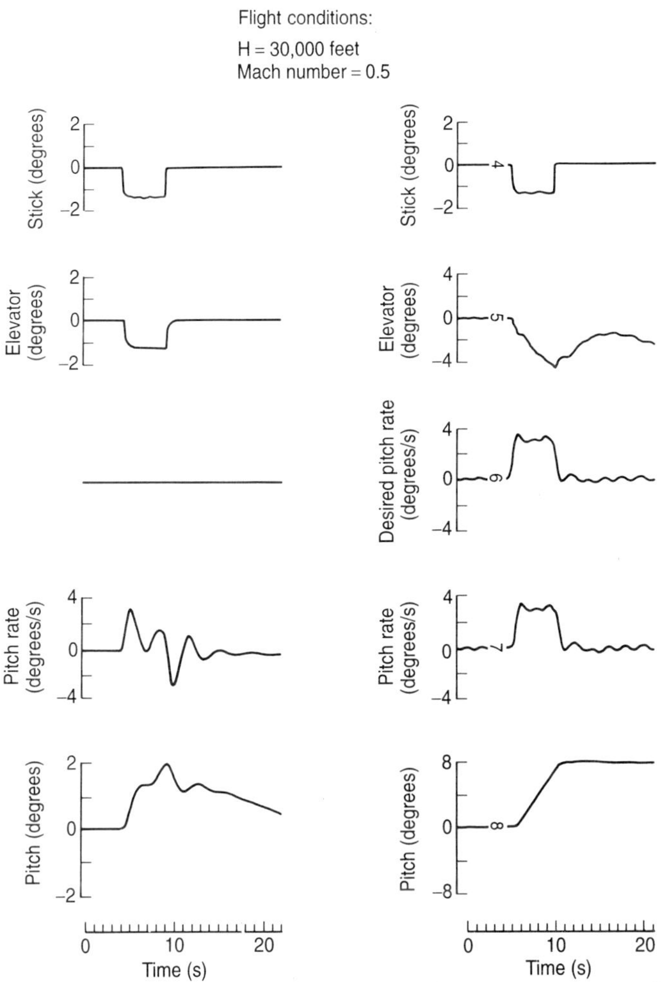

Fig. 7.6. Autopilot's adaptive behaviour. Case 2.

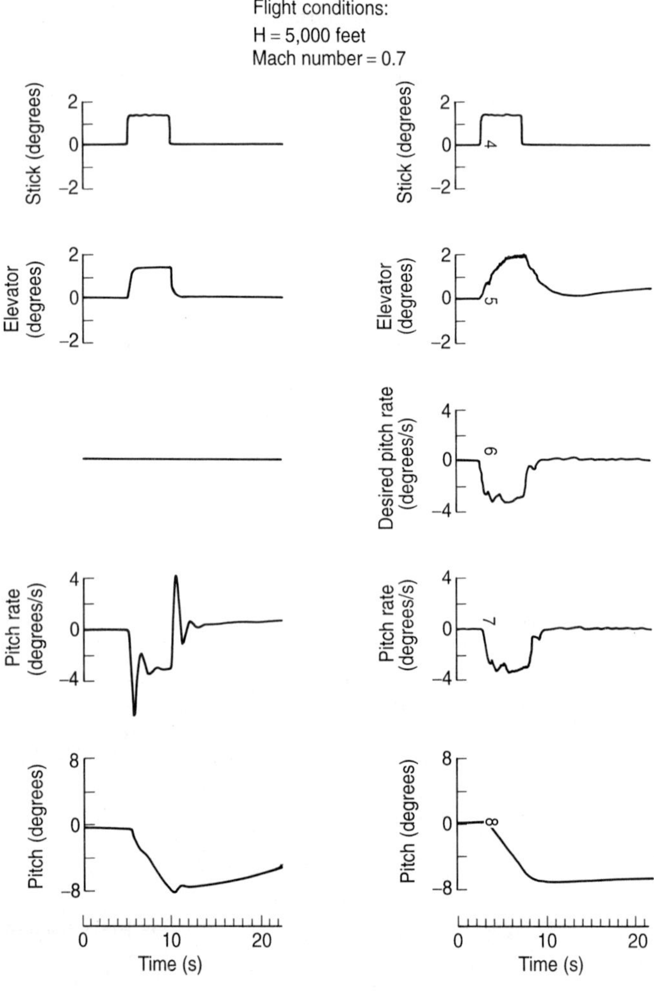

Fig. 7.7. Autopilot's adaptive behaviour. Case 3.

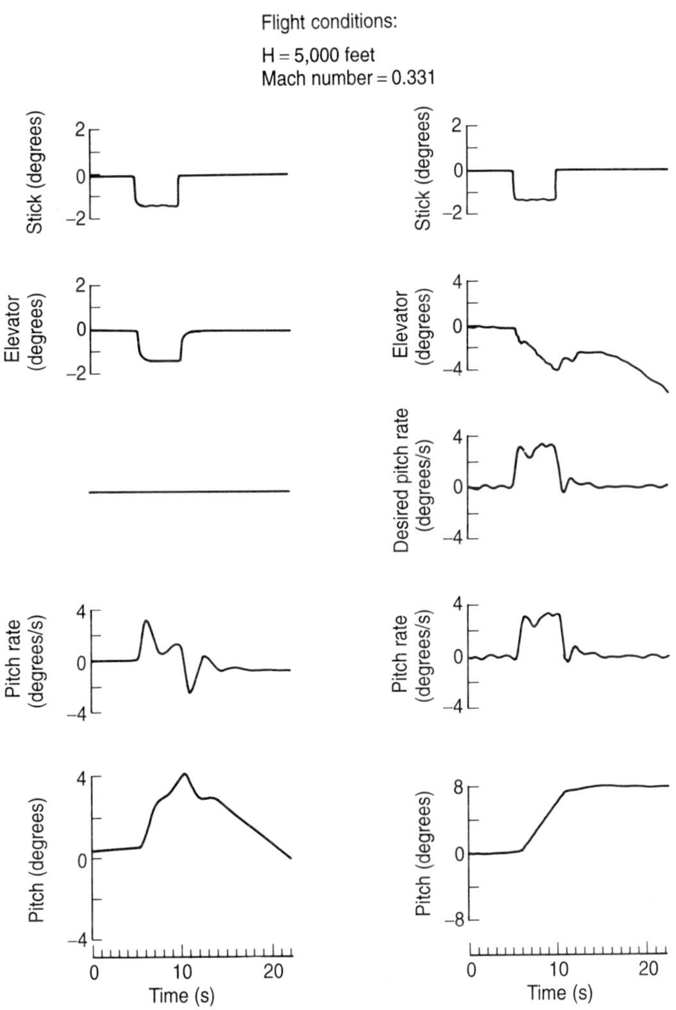

Fig. 7.8. Autopilot's adaptive behaviour. Case 4.

Figure 7.10 shows the response of the aircraft with the centre of mass displaced from its nominal position to a position such that the aerodynamic centre is placed between the centre of mass and the elevator. The new distance between the aerodynamic centre and the centre of mass was one-half of that of the design and in the opposite direction. This means that the aircraft was inherently unstable in pitch.

The first pitch angle change shown in Figure 7.10 was performed under the following flight conditions: $h = 3048$ m, $M = 0.4$, $p = 7804$ N/m^2 and $v = 131$ m/s. The second pitch change corresponded to: $h = 762$ m, $M = 0.74$, $p = 38830$ N/m^2 and $v = 251$ m/s.

The AP model had a third-order structure and the sampling period was 0.1 seconds. The desired dynamics used in the driver block for the experiments illustrated in Figures 7.9 and 7.10 was the same as that used in the preceding experiments.

The oscillations of the desired output produced by the driver block design were in this case accentuated by the instability of the aircraft itself. However, these oscillations have a positive effect, in that they help to drive the identification process of the AP model parameters.

7.5 DISCUSSION OF THE RESULTS

We will discuss the most significant practical issues observed from the above experiments. This analysis will consider the following aspects:

- control of the stable aircraft;
- parameter identification;
- control of the neutrally stable and unstable aircraft.

7.5.1 Control of the stable aircraft

As concluded from the results of all the experiments performed, the autopilot remained stable and, in spite of the widely different flight conditions, all the pitch responses were excellent and almost identical.

A closer look reveals the following aspects:

1. The F-8 aircraft pitch is basically a non-linear process. The aircraft's natural responses to the same rudder (elevator) command under four different flight conditions are shown on the left-hand sides of Figures 7.5–7.8.

These natural responses show the high non-linearity of the process (note the different scales used in the diagrams). Furthermore, under every flight condition, the dynamics of the natural response depends significantly upon the sign and the magnitude of the rudder command.

2. The pitch rate always follows the driving desired output generated by the driver block closely. This can be observed in all those diagrams showing the autopilot operation.

3. The driving desired trajectory (DDT) is in accordance with the desired dynamics. The DDT follows the desired dynamics well enough to maintain a well-damped pitch rate and fast acceleration responses.

4. The aircraft pitch control becomes extremely easy for the human pilot. In fact, he (she) can control the pitch at will. To pass from one pitch position to another, he (she) only needs to displace the rudder from its equilibrium position in the appropriate direction, until the aircraft attains the new desired pitch angle. When the pilot allows the rudder to return to its equilibrium, the aircraft will remain at the new pitch position without oscillations. During the transition, oscillations are avoided and the pitch rate is proportional to the displacement of the rudder.

5. The trim (steady state) values are adapted automatically by the autopilot and knowledge of them is not required. In the experiments presented here, once the pitch angle is changed from its steady state value, the flight conditions start to vary. Therefore, as can be seen from the graphs, when the new pitch angle position is reached, the elevator position (control signal generated by the autopilot) does not return to the initial steady state position. In order to hold the desired pitch angle position in spite of the continuous change in flight conditions, the elevator continues to move until a new steady state is attained.

6. The autopilot retains its desired behaviour in all types of flight operation. From our experience with the simulation, and as is apparent from the results presented here, we may note that the autopilot adapts immediately to any flight condition and can be used in any kind of flight operation.

7.5.2 Parameter identification

The following aspects are emphasized:

1. No previous knowledge of the dynamics of the aircraft or its variations

over time is necessary. Indeed, the only information required by the autopilot is the measurement of the pitch rate.

2. The AP model may use structures of different order. Second- and third-order structures were used with satisfactory results in both cases, as shown in Figure 7.9 and other diagrams.

3. The values of the AP model parameters changed in accordance with the non-linearity of the plant. Though not shown here, rapid changes and fluctuations were observed in the evolution of the AP model parameters when the aircraft was out of steady state position.

7.5.3 Control of the neutrally stable and unstable aircraft

The results obtained when the aircraft was made neutrally stable and unstable lead us to the following conclusions:

1. The control system adapts automatically to the new characteristics of the process, following its desired dynamic behaviour. As has already been mentioned in this chapter, stability considerations require that the centre of mass lies between the aerodynamic centre and the elevator. This results in increased air resistance to the aircraft's motion and creates a torque that is compensated for by the torque generated by the elevator position. When the centre of mass coincides with the aerodynamic centre, the aircraft is considered to be neutrally stable and the torque to be compensated for is close to zero or is negligible. This is in accordance with the results of the experiment illustrated on the right-hand side of Figure 7.9 in which it is shown that, in order to obtain an identical desired pitch response, the elevator position needs negligible changes in comparison with those of previously illustrated experiments.

2. The experiment illustrated in Figure 7.10 shows that the proposed control scheme is able to control unstable processes, as guaranteed by the theory. In this example, the aircraft has been controlled to pass rapidly from a high altitude and low speed to a low altitude and high speed position. The aircraft's behaviour maintains its desired dynamics in accordance with the human pilot's inputs from the rudder. But the control applied to the process (elevator position) now has quite a different shape and logic as compared with the results shown when the plant was stable.

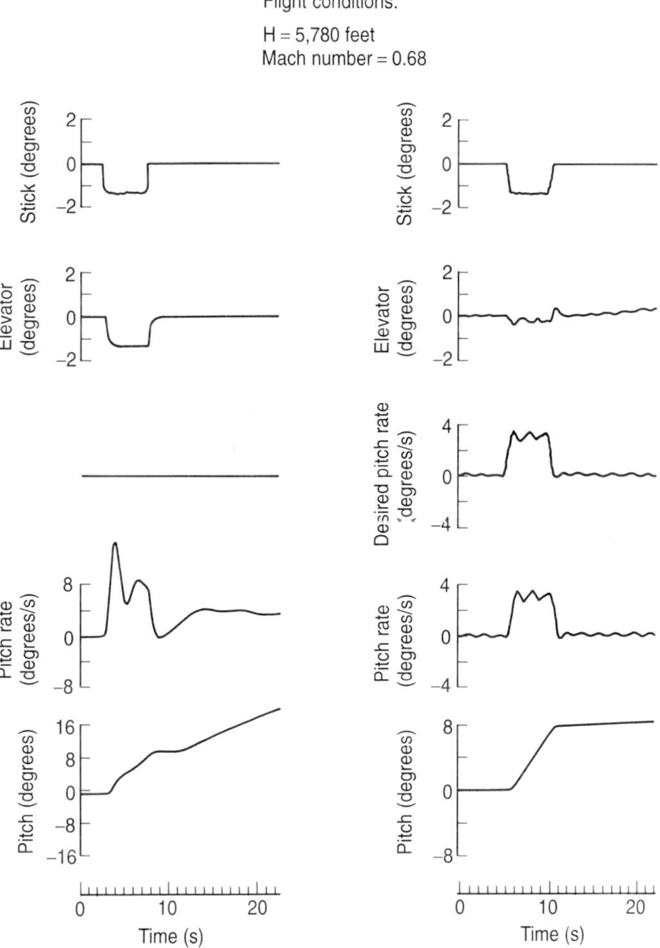

Fig. 7.9. Autopilot behaviour for neutrally stable aircraft.

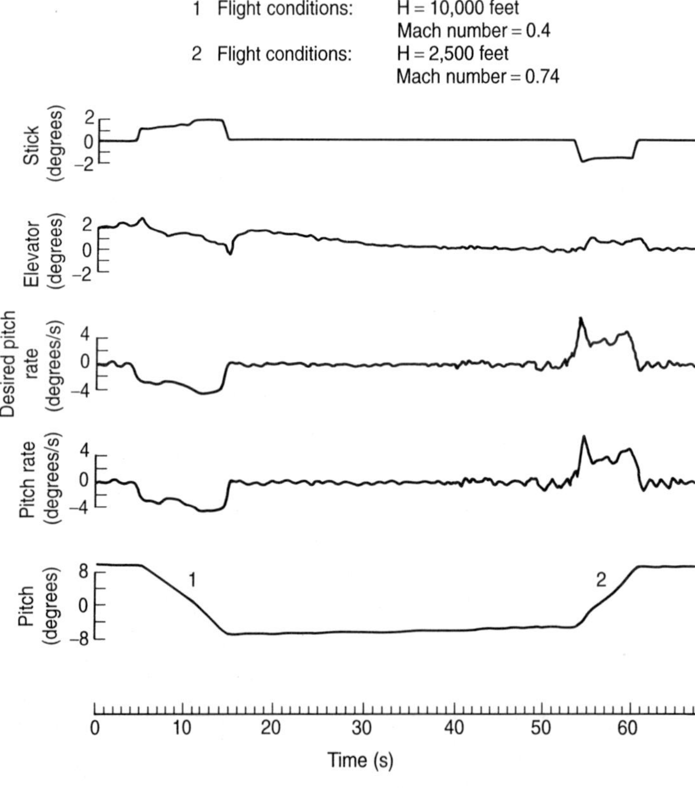

Fig. 7.10. Autopilot behaviour for unstable aircraft.

7.6 CONCLUSIONS

This chapter has presented the application of APCS, through the basic strategy of predictive control, to the design of an automatic pilot for the pitch angle of the NASA F-8 supersonic aircraft. The results obtained illustrate the ability of APCS to adapt to different flight conditions and its satisfactory performance in all situations, despite the non-linearity of the process. The desired behaviour was preserved even when the aircraft dynamics was made unstable by displacing the centre of mass over and beyond the aerodynamic centre. The APCS autopilot allows the human pilot a high performance pitch angle control. This control is simple and easy in all flight operations since it standardizes the aircraft's dynamic response to the pilot's actions.

The results presented here solved timely and satisfactorily the problem of the automatic dynamic control of the longitudinal characteristics (pitch rate) in a supersonic aircraft. By extension, they show that adaptive predictive control is well suited to the control of non-linear and time varying processes, and that it can also influence and vary the process design. In fact, the use of this type of automatic pilot may influence the design of aircraft itself, since it allows the separation of the design from stringent stability requirements. Shifting the centre of mass of the aircraft to make it coincide with the aerodynamic centre causes the air resistance to decrease significantly. Therefore, the possibility of controlling the aircraft with neutrally stable or unstable dynamics may imply significant advantages both from the perspective of fuel saving and from that of maneuverability.

Chapter 8

CONTROL OF A BLEACH PLANT IN A PULP FACTORY

8.1 INTRODUCTION

This chapter presents the application of APCS to the control of the bleach plant in Port Mellon, British Columbia, Canada, owned by the company CANFOR Ltd. Differently from those presented in the two preceding chapters, this application uses the extended strategy of predictive control.

When this project was carried out in 1984, most of the published papers on the applications of adaptive control dealt with computer simulation results and only a few included results on experimental pilot plants. This project was the first application of adaptive predictive control to an industrial production process and its results were presented in [MD85, DMZ89].

The practical interest of its realization was twofold. First, the process presented serious control problems related to both its industrial environment and to its dynamic and operational characteristics. Among them we can emphasize its high non-linearity, the existence of time delays, intermittent load changes and, in consequence, changes in the point of operation, as well as frequent perturbations acting significantly over the process. Second, the control performance in the bleach process is of considerable economic relevance to the overall pulp production process.

In this chapter, after describing the control problem in a bleach plant, we will summarize the main issues involved in the application of the extended strategy of predictive control and its practical implementation. Then we will present and discuss the experimental results.

8.2 THE CONTROL PROBLEM IN A BLEACH PLANT

The objective of using bleach is to eliminate the residual lignin and other colouring compounds in the pulp in order to make it suitable for production of white paper. Bleaching is performed by following a sequence of processes, the first and most important of which is chloring, whose simplified diagram is shown in Figure 8.1.

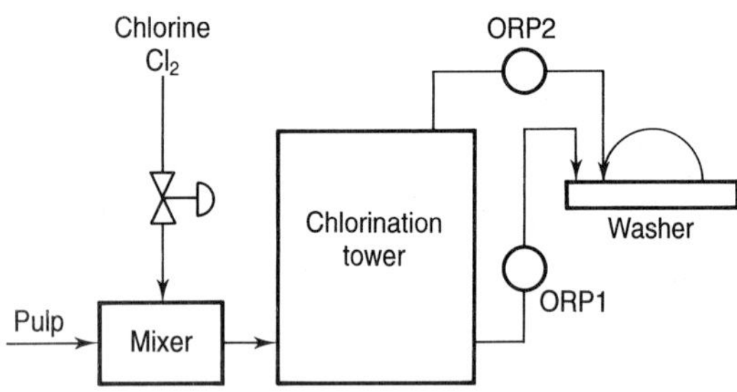

Fig. 8.1. Chloring process.

Pulp is fed into the mixing tower, where chlorine is added. Chloring is produced as the result of two chemical reactions: a fast substitution followed by a slow oxidation. The first reaction is completed after a few minutes, while the second one takes approximately one hour. The reaction rate is affected by the amount of lignin in the incoming pulp, as well as by the chlorine concentration, the wood species and the stock temperature. Thus the dynamics is complex and may vary over time. Moreover, surges of black liquor carryover may be frequent and are very detrimental as they consume a large amount of chlorine. Thus this is a difficult control problem in which the incentive for good control is high for economic reasons.

Efficient performance of the chloring process requires a rather tight control of the chlorine consumption, as underchlorination as well as overchlorination are detrimental to pulp quality and result in the production of off-grade pulp. Thus, the variable to be controlled is the residual chlorine in the tower, which is required to be kept close to zero using the chlorine flow as the manipulated control variable.

The residual chlorine is measured using two ORP sensors. One of them (ORP2) is placed after the tower, supplying the information about the correct operation of the process. In fact, it is the measurement from the ORP2 sensor that is desired to remain as close to zero as possible. However, because of the long time required to perform the oxidation reaction in the tower, it is recommended that the residual chlorine be measured after the substitution reaction has been achieved, which is done by sensor ORP1, and that this measurement be used as the feedback signal for control.

The PID control scheme that had been implemented in this process is shown in block diagram form on Figure 8.2. Using the signal from the ORP1 sensor, the PID manipulates the setpoint of an inner loop which controls the chlorine flow.

Note the presence of the linearizing block that compensates for the non-linear characteristics of the ORP1 sensor. In addition, there is feedforward from the fiber production flow, which is estimated from the flow and concentration sensors. This PID scheme was implemented in FORTRAN on the mill PDP 11/44 process computer. However, it was not generally used and the process was controlled directly by the operator. The presence of a time delay of about two minutes from the chlorine addition to the ORP1 sensor and the frequent changes in the conditions of operation prevented satisfactory performance of the PID control scheme by manual tuning. However, the application of an automatic PID tuning method, aided by a gain scheduling to compensate for production rate changes, improved the robustness of the PID scheme, as is described in [DMZ89].

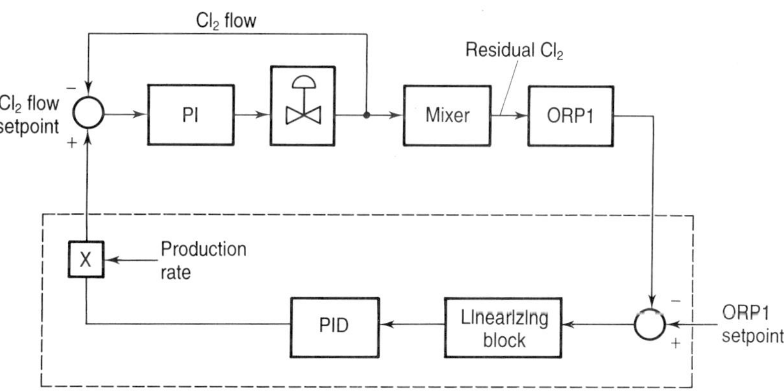

Fig. 8.2. Residual chloring PID control scheme.

8.3 APPLICATION OF THE EXTENDED STRATEGY OF PREDICTIVE CONTROL

As explained in Chapter 3, the extended strategy of predictive control considers an interval $[k, k+\lambda]$ at time instant k and uses the AP model to predict a process output sequence as a function of a control input sequence. Then, the driver block selects, as the projected desired trajectory (PDT), the trajectory predicted by the AP model in response to a control sequence such that both the trajectory and the control sequence satisfy certain performance criterion. The control applied at instant k is the first term of this control sequence.

In the implementation described in this chapter we use the same kind of incremental AP model as that considered in Chapter 6, but it is now extended over the prediction interval $[k, k+\lambda]$ in the form

$$\hat{y}(k+j|k) = \sum_{i=1}^{\hat{n}} \hat{a}_i(k)\hat{y}(k+j-i|k) + \sum_{i=1}^{\hat{m}} \hat{b}_i(k)\hat{u}(k+j-i|k) \quad (j = 1, \ldots, \lambda) \tag{8.1}$$

where $\hat{a}_i(k)$ and $\hat{b}_i(k)$ are the AP model parameters updated at time k, and $\hat{y}(\cdot|k)$ and $\hat{u}(\cdot|k)$ are, respectively, the incremental predicted output and control sequences, which verify

$$\begin{aligned} \hat{y}(k+1-i|k) &= y(k+1-i) = y_p(k+1-i) - y_p(k-i) \qquad i = 1, \ldots, \hat{n} \\ \hat{u}(k+1-i|k) &= u(k+1-i) = u_p(k+1-i) - u_p(k-i) \qquad i = 1, \ldots, \hat{m} \end{aligned} \tag{8.2}$$

where y_p and u_p are the process output and control respectively.

The specific performance criterion used in this implementation is the one considered in Section 3.4 of Chapter 3 under the incremental formulation presented and analyzed in Section C.1.3 of Appendix C. This performance criterion is defined by the two conditions given below.

Condition 8.1: $\hat{y}_p(k+\lambda|k) = y_{pr}(k+\lambda|k)$, where:

(i) $\hat{y}_p(k+\lambda|k)$ is the output predicted at instant k for instant $k+\lambda$;

(ii) $y_{pr}(k+\lambda|k)$ is a desired value for the output at instant $k+\lambda$ that is explicitly computed at instant k as belonging to a reference trajectory defined by

$$\begin{aligned} y_{pr}(k+j|k) &= \sum_{i=1}^{p} \alpha_i y_{pr}(k+j-i|k) + \sum_{i=1}^{q} \beta_i y_{sp}(k+j-i) \qquad (j = 1, \ldots, \lambda) \\ y_{pr}(k+1-l|k) &= y_p(k+1-l) \qquad (l = 1, \ldots, p) \end{aligned} \tag{8.3}$$

p, q, α_i and β_i being chosen to guarantee that, starting from the present value of the measured process output, this trajectory approaches the setpoint y_{sp} in the desired manner. □

Condition 8.2: The control sequence is constant in the prediction interval:

$$\hat{u}(k+j|k) = 0 \qquad \forall j = 1, \ldots, \lambda - 1$$

□

Using Condition 8.2 in equation (8.1) recursively from the initial conditions of (8.2), we may write

$$\hat{y}(k+j|k) = \sum_{i=1}^{\hat{n}} \hat{e}_i^{(j)}(k)y(k+1-i) + \sum_{i=2}^{\hat{m}} \hat{g}_i^{(j)}(k)u(k+1-i) + \hat{g}_1^{(j)}(k)\hat{u}(k|k) \qquad (j=1,\ldots,\lambda) \tag{8.4}$$

where $\hat{e}_i^{(j)}(k)$ and $\hat{g}_i^{(j)}(k)$ are obtained from the AP model parameters $\hat{a}_i(k)$ and $\hat{b}_i(k)$ by means of the algorithms defined in Chapter 3, equation (3.7).

By summing up the λ equations of (8.4) and using (8.2), the predicted process output for $k+\lambda$ can be computed in the form

$$\hat{y}_P(k+\lambda|k) - y_P(k) = \sum_{i=1}^{\hat{n}} \hat{\eta}_i^{(\lambda)}(k)y(k+1-i) + \sum_{i=2}^{\hat{m}} \hat{\gamma}_i^{(\lambda)}(k)u(k+1-i) + \hat{h}^{(\lambda)}(k)\hat{u}(k|k) \tag{8.5}$$

where

$$\hat{\eta}_i^{(\lambda)}(k) = \sum_{j=1}^{\lambda} \hat{e}_i^{(j)}(k); \qquad \hat{\gamma}_i^{(\lambda)}(k) = \sum_{j=1}^{\lambda} \hat{g}_i^{(j)}(k)$$
$$\hat{h}^{(\lambda)}(k) = \sum_{j=1}^{\lambda} \hat{g}_1^{(j)}(k) \tag{8.6}$$

Using Condition 8.1 in equation (8.5), we have

$$u(k) = \hat{u}(k|k) = \frac{y_{Pr}(k+\lambda|k) - y_P(k)}{\hat{h}^{(\lambda)}(k)} - \frac{\sum_{i=1}^{\hat{n}} \hat{\eta}_i^{(\lambda)}(k)y(k+1-i) + \sum_{i=2}^{\hat{m}} \hat{\gamma}_i^{(\lambda)}(k)u(k+1-i)}{\hat{h}^{(\lambda)}(k)} \tag{8.7}$$

and the control $u_P(k)$ to be applied to the process at instant k is

$$u_P(k) = u(k) + u_P(k-1) \tag{8.8}$$

This control law appears as a simple generalization, for $\lambda > 1$, of the APCS control law used in Chapters 6 and 7 for $\lambda = 1$, $y_{Pr}(k+\lambda|k)$ being the value of the projected desired trajectory (PDT) for instant $k+\lambda$.

8.4 IMPLEMENTATION OF APCS

8.4.1 Structure of the implementation

The block diagram of Figure 8.3 summarizes the structure of the implementation of APCS.

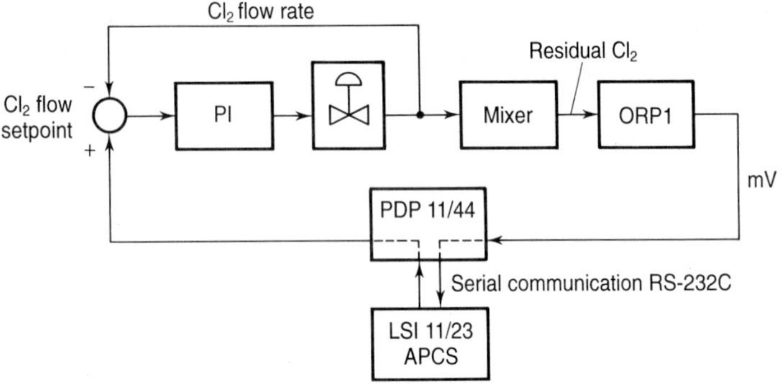

Fig. 8.3. APCS control structure.

The input to APCS (process output to be controlled) was the measurement of the ORP1 sensor (in mV), and the APCS control signal was the setpoint for the chlorine flow rate loop (also in mV). Therefore, no linearization of the ORP1 signal or use of the production rate measurement was considered in the implementation of APCS, as was the case for the PID control scheme previously considered.

APCS was implemented in FORTRAN on an LSI 11/23 microcomputer under the real time operating system RT–11. The LSI 11/23 communicated with the PDP 11/44 mill process computer through a serial communication line RS-232.

The main program, written at the background level, initialized all the variables required for the execution of APCS and called a control routine. This routine was a general purpose program prepared for application to processes with different characteristics by setting the values of a number of so-called *structure variables*. Among these variables, the operator selects the design parameters associated with the control strategy, such as the AP model order and initial parameters, the prediction horizon λ, the coefficients defining the reference trajectory, the parameters involved in the adaptation algorithms and the limits for the control signal. After being called for the first time, the completion routine reschedules itself according to the chosen sampling period. The time between consecutive executions of the completion routine is employed by the main program to allow the operator to change the values of the structure variables.

8.4.2 Design issues

In the implementation, the prediction horizon λ was equal to 5 and two different sampling periods were used: thirty seconds and one minute. Two different AP model structures were used. In the experiments where the sampling period was equal to thirty seconds, the AP model equation was:

$$\hat{y}(k+j|k) = \sum_{i=1}^{2} \hat{a}_i(k)\hat{y}(k+j-i|k) + \sum_{i=1}^{2} \hat{b}_i^*(k)\hat{u}(k+j-i-2|k) \qquad (j=1,\ldots,5) \tag{8.9}$$

This is a second-order model with a time delay of two sampling periods. This was approximately the time delay observed in the process under control. This model is equivalent to the one given in (8.1) (having no pure time delays) when we consider $\hat{m} = 4$ with

$$\hat{b}_1(k) = \hat{b}_2(k) = 0, \quad \hat{b}_3(k) = \hat{b}_1^*(k), \quad \hat{b}_4(k) = \hat{b}_2^*(k) \qquad \forall k \geq 0$$

The initial values given to the AP model parameters were

$$\hat{a}_1(0) = 1; \quad \hat{a}_2(0) = -0.2; \quad \hat{b}_1^*(0) = \hat{b}_2^*(0) = 0.1$$

In the experiments where the sampling period was equal to one minute, the AP model equation was:

$$\hat{y}(k+j|k) = \sum_{i=1}^{2} \hat{a}_i(k)\hat{y}(k+j-i|k) + \sum_{i=1}^{3} \hat{b}_i(k)\hat{u}(k+j-i|k) \qquad (j=1,\ldots,5) \tag{8.10}$$

The initial values of the AP model parameters in this case were

$$\hat{a}_1(0) = 1; \quad \hat{a}_2(0) = -0.1; \quad \hat{b}_1(0) = 0.01; \quad \hat{b}_2(0) = 0.1; \quad \hat{b}_3(0) = 0.05$$

The number of $\hat{b}_i$ parameters in this AP model was set equal to three in order to make provision for a possible variation in the time delay of the process. If the time delay were smaller than one minute, then the parameter $\hat{b}_1$ could have some significant value, but if the time delay were to become greater than one minute, then the parameter $\hat{b}_1$ should be identified equal to zero by the adaptation mechanism. The extended strategy of predictive control used in this implementation is able to cope with unknown or time varying time delays provided that the prediction horizon λ is long enough and that a number of $\hat{b}_i$ parameters in the AP model are set up to allow the adaptation mechanism to accommodate to the time delay by estimating the corresponding first leading b_i parameters as zero [Rod82, RM83].

In this implementation, the adaptation algorithms were similar to those designed in Chapter 4 and previously used in the applications of Chapters 6 and 7.

The reference trajectory was designed by discretizing a second-order model with static gain and damping ratio equal to 1 and time constant equal to 1.5 times the sampling period, thus ensuring a smooth and sufficiently fast evolution towards the setpoint.

8.5 EXPERIMENTAL RESULTS AND DISCUSSION

APCS was implemented in the bleach plant for a ten-day trial period. In this section we present four different runs, whose responses are plotted in Figures 8.4–8.7. The most significant circumstances of these runs are summarized in Table 8.1. Among them, we may emphasize two failures that occurred in the transmission of data between the two computers during the experiments. The response of APCS in the presence of this kind of emergency condition is also presented.

In the following we point out some features of APCS observed from the tests concerning, respectively, the tracking and regulatory performance, the ability to handle a process with an unstable inverse and time variant time delay, the robustness against failures in the data transmission and the simplicity of startup and use.

8.5.1 Tracking and regulatory performance

Since APCS did not use a linearizing block to compensate for the non-linear characteristics of the ORP sensor or the production rate (fiber flow rate), the process, as seen by APCS, was basically non-linear. Also, it had time delays and was subject to severe and unknown disturbances.

A global look at the experimental results confirms that the APCS performance was satisfactory in tracking setpoint changes and as a regulator of fixed setpoint. Figures 8.4 and 8.5 show how the response of APCS to setpoint changes and/or unknown load changes was fast and smooth, without offset and excessive control action. However, this response was inevitably affected by the size and character of the unknown disturbances. In fact, two different kinds of unknown disturbance were acting on the process. One was the production rate, shown in Figures 8.4

and 8.5, and the other was the quality of the fiber, which is not shown and is difficult to measure. In some cases the changes in production rate were drastic; for instance in those shown during the second setpoint change in Figure 8.4 and the setpoint change in Figure 8.5. These changes had a considerable effect on the operating conditions of the plant, but, nevertheless, APCS maintained its satisfactory performance. Some observations that illustrate the logic of the control action are as follows:

TABLE 8.1: APCS experiments

Figure	Setpoint and load changes	Other details
8.4	From 1051 to 1048 From 1048 to 1053 Fiber flow rate changes shown in Fig. 8.4	• Sampling period: 30 sec • Duration: 67.5 min
8.5	From 1055 to 1053 Fiber flow rate changes shown in Fig. 8.5	• Sampling period: 1 min • Duration: 3 h 45 min • Control action starts at instant 5
8.6	Setpoint constant at 1062	• Sampling period: 30 sec • A failure in transmission indicated by an arrow in Fig. 8.6 gave an ORP1 value of 1008 while the actual value was approximately 1062
8.7	Setpoint constant at 1050	• Sampling period: 1 min • Control action starts at instant 6 • A failure in transmission indicated by an arrow in Fig. 8.7 made the control signal equal to 0

(a) Changes in the production rate always lead to logical changes in the control signal (chlorine flow rate). This can clearly be seen in the first fifteen minutes

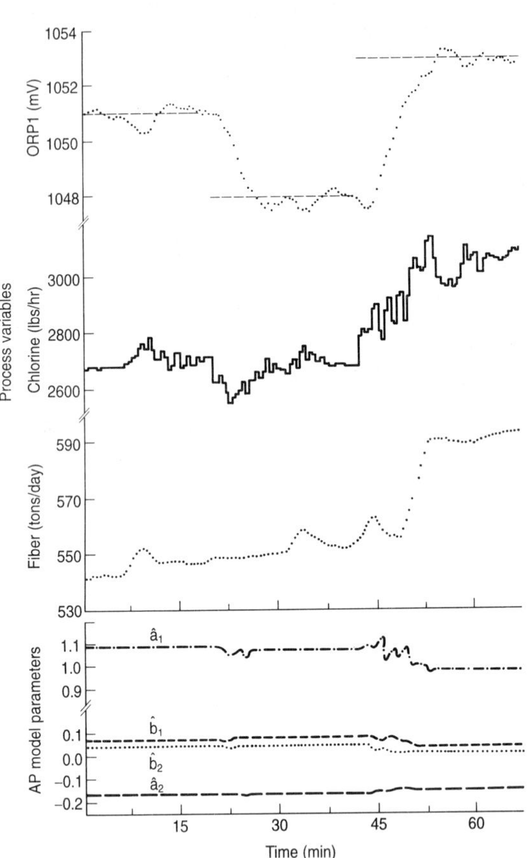

Fig. 8.4. APCS control experiment.

of APCS operation shown in Figure 8.4. An increase in production rate is compensated for by an increase in chlorine flow rate. In the same sense, a decrease in production rate is compensated for by a decrease in the control signal, as observed in the second half hour of operation shown in Figure 8.5. This correspondence between production rate and control action allows the estimation of the changes in production rate even when it is not shown, as later we will comment.

(b) The regulatory performance in the interval between 1.5 and 3 hours shown in Figure 8.5 shows a steady but slow increase in production rate, while the chlorine flow rate is steadily and slowly decreasing instead of increasing. The value of ORP1 is maintained slightly over the setpoint, although it constantly tends to it. This can probably be explained by the other unknown distur-

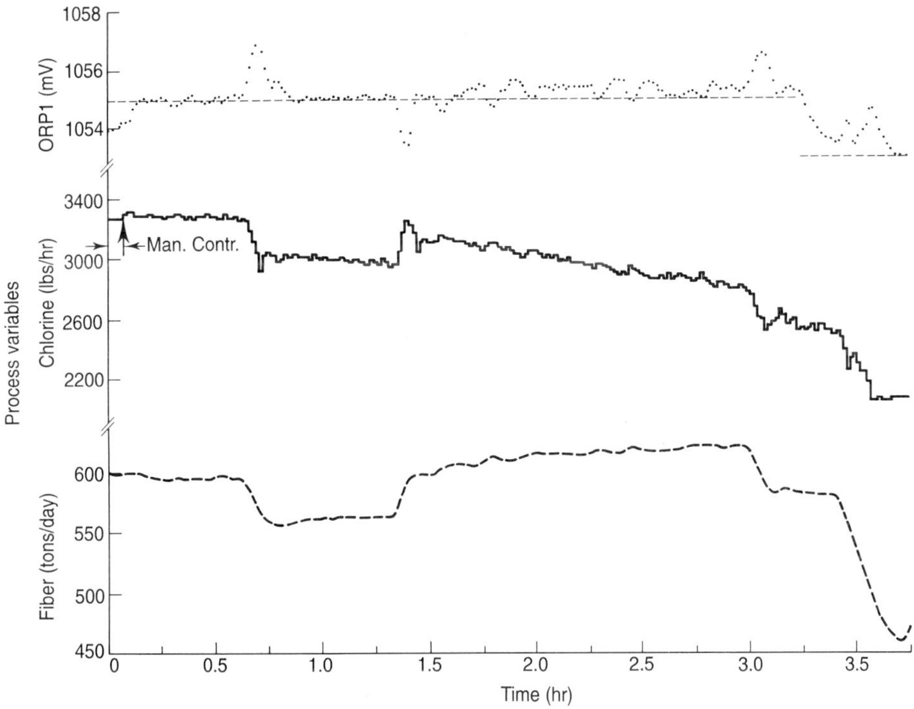

Fig. 8.5. APCS control experiment.

bance, that is, by the change in the fiber quality.

(c) Figure 8.4 shows how a decrease of three units in the setpoint value of ORP1 does not seem to decrease the required chlorine flow rate at steady state. This can probably be explained by the simultaneous increase in production rate. The level of chlorine flow rate is definitely much higher in order to maintain ORP1 at the third setpoint value shown in Figure 8.4. This increase is also motivated by the corresponding increase in production rate. The non-linearity of the plant is probably the third factor that determines the experimental results obtained. It must be noted that the increase in production rate would tend to lead to a decrease in the ORP1 signal. Therefore, APCS drives the ORP1 signal to a higher level satisfactorily, compensating in addition for the said tendency caused by an unknown disturbance.

8.5.2 Inverse instability and time delay compensation

The driver block used in this application is capable of generating a stable control signal even when the identified process has an unstable inverse and a variable time delay. These properties of the driver block design under consideration have been

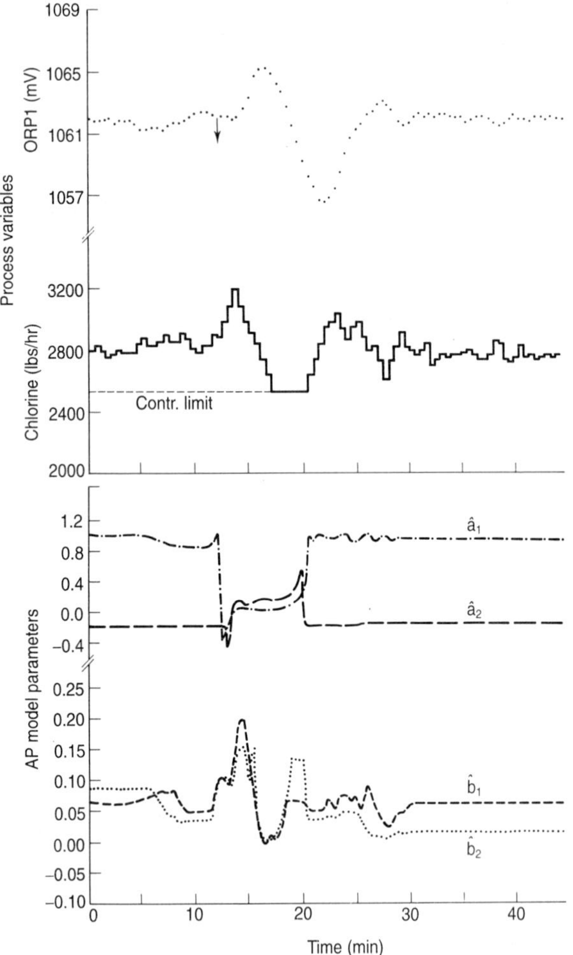

Fig. 8.6. APCS control experiment.

analyzed previously and proved in [Rod82, RM83]. They are also noticeable in the present implementation, as can be deduced from the following observations:

(a) As shown over the first seven minutes of Figure 8.6, the AP model parameter $\hat{b}_2^*$ was larger in terms of absolute value than the parameter $\hat{b}_1^*$, which implied that the AP model had an unstable inverse. This kind of situation had also been observed on several occasions during the trial period. It is clear that a driver block with a prediction horizon of $\lambda = 1$ would have generated an unstable control law. However, under the present driver block design with $\lambda = 5$, the control law remained satisfactorily stable, as shown in the diagrams.

(b) The time delay observed in the process was around one minute, but it could oscillate between forty-five seconds and one minute-fifteen seconds. These variations corresponded inversely to the variations in production rate. In Figure 8.7, after minute 46, we can observe an increase in the value of ORP1 and a significant and sustained decrease in the value of the control signal that returns the value of ORP1 to the setpoint. Analyzing this fact, it is easy to deduce that a significant decrease in the production rate has occurred, which must imply an increase in the process time delay. If the time delay exceeds one minute, the AP model parameter $\hat{b}_1$ should become equal to zero according to the chosen sampling period and the AP model structure. Precisely this is observed in Figure 8.7, since the parameter $\hat{b}_1$ becomes zero at minute 49 and remains at that value for the duration of the experiment shown in this diagram. Therefore, the variation in the AP model parameters tracks the process time delay as the control signal continues to be satisfactory.

8.5.3 Robustness against failure in the data transmision

The APCS robustness was drastically tested when failures in the transmission of the input/output data occurred accidentally. Figure 8.6 shows the response of APCS when, at minute 12 (sampling instant 24), a failure in the data transmission gave an ORP1 value of 1008, while the actual value was approximately 1062. Figure 8.7 shows the response of APCS when, at minute 25 (sampling instant 25), the control signal transmitted to the plant was equal to zero, while the computed control signal was equal to 3424. In both Figures 8.5 and 8.6 the accidents are indicated by an arrow. In order to analyze the effect of these failures comparatively, it must be taken into account that, although the scale of times of sampling instants is the same, the sampling period in Figure 8.7 was twice that of Figure 8.6. The analysis of these accidents shows the following:

(a) The failure in the transmission of the process output value (Figure 8.6) clearly does not affect the actual value of the process output, but it severely disturbs the values of all the parameters of the AP model, since they are directly updated from the *a priori* estimation error $e(k|k-1) = y(k) - \hat{y}(k|k-1)$. As a consequence, an erroneous AP model disrupts the APCS operation, producing an erratic control signal. However, the subsequent operation of the APCS estimation system was able to return the AP model parameters to adequate values in approximately nine minutes (eighteen sampling periods). The process output returned to the setpoint, and the satisfactory performance of APCS was restored in approximately twelve minutes (twenty-four sampling periods). The maximum deviation of ORP1 from the setpoint was slightly greater than five units.

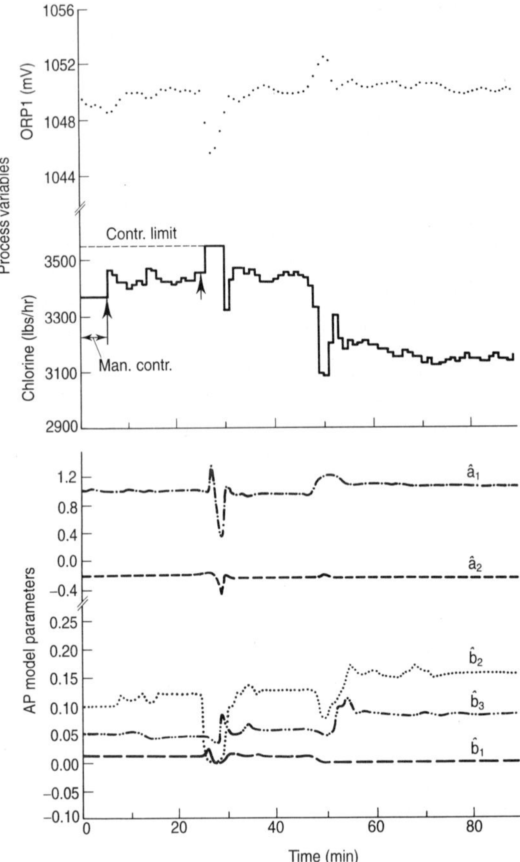

Fig. 8.7. APCS control experiment.

(b) The failure in the transmission of the control signal (Figure 8.7) affects the process output considerably and clearly deceives the APCS estimation system. However, the effect on the AP model parameters values is not so extreme as that observed in Figure 8.6, since in this case the effect is noticed through an erroneous process input and not directly through the *a priori* estimation error. It can be seen that the effect of the failure on parameters $\hat{a}_1$ and $\hat{a}_2$ is not so grave and they recover their previous values quite well in a few sampling periods. As a result, the correct value of the control signal is rapidly recovered, and it returns the value of ORP1 to the setpoint, as the estimation system simultaneously and adequately corrects the AP model parameter values. The maximum deviation of ORP1 from the setpoint was also approximately five units, but in this case the process output came back to the setpoint in approximately six minutes (six sampling periods).

(c) It can be observed in Figures 8.6 and 8.7 that the control limits, set at the beginning of these experiments by the operator, contributed to the robustness of the APCS operation in the previously analyzed severe conditions. Clearly, this experimental application of APCS did not take elemental security procedures into account which, by detecting and reacting to the transmission failure, could have minimized the effect of such a failure. The emphasis was placed on the study of the performance of the adaptive predictive controller itself.

8.5.4 Simplicity of startup and use

The change from manual control (from the PDP 11/44) to APCS automatic control (from the LSI 11/23) was always smooth and efficient. This kind of change, illustrated in Figures 8.5 and 8.7, was made with a realistic choice of initial values for the AP model parameters. However, this choice was made in a simple manner and, as demonstrated in the experiment shown in Figure 8.6, where the parameter identification process was disrupted by the process output transmision failure, if this choice had been very erratic, it would only have affected the initial performance of APCS.

Under the previously described implementation, APCS was simple to use and performed well under different choices of structure variable. For instance, two AP model structures, corresponding to two different control periods, were chosen by the simple consideration of the previously observed dynamics of the process. All the other structure variables were also chosen in a simple manner. However, it is clear that this simplicity of usage requires training and a certain familiarity with the system.

8.6. CONCLUSIONS

This chapter has presented the first application of APCS to an industrial production unit. This application was carried out in a bleach plant within a pulp factory and used the extended strategy of predictive control.

Despite the fact that the APCS control did not use any process signal linearization block, and was subjected to significant unknown disturbances due to changes in the fiber flow rate and in fiber quality, the results obtained were satisfactory. Likewise, APCS proved to be robust when failures in the transmission of input/output data occurred. These results emphasized the simplicity of use of the system and its capacity for application to production processes in environments with highly adverse conditions. Thus, APCS demonstrated its capacity as a reliable, robust and efficient adaptive industrial controller.

Chapter 9

ACTIVE CONTROL OF FLEXIBLE STRUCTURES

9.1 INTRODUCTION

With this chapter we close the part of the book devoted to the application of adaptive predictive control systems (APCS). The area selected, active control of structures, is an emerging but promising one and it also allows us to illustrate the application of APCS through a state space formulation.

In the context of civil structural engineering, the idea of using automatic control systems to reduce the vibration induced in structures such as buildings, bridges, towers and others by loads such as earthquakes, wind, moving traffic, etc., has been considered increasingly over the last two decades. In the first step, traditional methods were invoked from control theory to formulate control algorithms, and were tested by numerical simulations. In the second step, experimental projects have been conducted which implement active controllers in pilot structures. A third step concerns installations on full-scale structures. While there exist some real world structures equipped with active controls, this is still an open issue.

This chapter describes the implementation of the extended strategy of predictive control in experimental building structures subjected to seismic actions. This application was part of a project conducted at the National Center for Earthquake Engineering Research, State University of New York at Buffalo, USA, in 1986.

First, we will describe some basic issues that characterize the problem of active structural control. Second, we will describe the experimental setup. Third, we will summarize the formulation of the extended strategy of predictive control for this problem, which uses state space modelling. Finally, the experimental results will be presented and discussed.

9.2 SOME ASPECTS OF ACTIVE CONTROL

Structures are mechanical systems with flexible components exhibiting dynamic responses (displacements, velocities, etc.) when subjected to exciting loads (forces or other actions). Structural systems are found in many engineering areas such as, among others, mechanics, robotics, automotive, naval, aerospace and civil. Although having their own specific features, all these structural systems have a common need to improve on their performance when subjected to undesirable loads.

Traditional ways of enhancing structural behaviour against disturbing actions have mainly been oriented towards the design of specific parts of the structure to modify internal characteristics such as stiffness, damping, etc. As a complement or alternative to these design procedures, passive systems are devices that are added to the structure in order to absorb energy, thus reducing the vibration induced in the structure by the excitation. Typical passive systems belong to the broad family of vibration isolators and dampers which are being used in structural systems in different practical applications. Alternatively, active systems attempt to reduce the structural motion by the action of actuators driven by automatic controllers with feedback from the structural response.

In civil engineering, structures are usually massive, and static features are very important in determining their behaviour and in designing them to resist environmental loads such as earthquakes, wind and other disturbances. However, a constant trend has been to design lighter and more challenging structures while increasing the safety and comfort of their occupants and contents in the face of these loads. This trend has been pushed forward by the availability of new lighter materials and construction procedures, together with more powerful computational methods, thus making the dynamic features of the structures more significant and apparent.

The design of passive systems for civil structures has progressed significantly, especially in seismic isolation devices and energy dissipators. A recent state of art of the development and worldwide actual applications of these systems can be found in [SC94]. The main practical advantage of passive systems is that they do not rely on external energy and are almost maintenance free, while their main drawback is that they supply only a reactive action that is adequate for narrow band frequency ranges only.

Active control of civil engineering structures was first envisaged as an almost speculative idea about twenty years ago [Yao72], and proposed equipping the structures with sensors and actuators in a closed loop with a feedback control strategy. Since then, considerable attention has been devoted to this concept, progressing from theory through small prototype tests to full-scale implementation.

In comparison with passive systems, active systems require an external energy supply and more complex instrumentation, design and maintenance, but they can offer significant advantages such as: (1) the active action is based on a control strategy that can improve the behaviour of the structure as a dynamic system; (2) active systems are more flexible since they are more insensitive to specific operating conditions such as the particular site characteristics of the structure and the external excitations; (3) different objectives can be considered within the control strategy, for instance reducing displacements to enhance safety or reducing accelerations to improve comfort in the case of structures with human users.

As with any other application of automatic control, the development of active structural systems has been progressing as technology has been supplying sensors and actuators and control theory has been borrowing appropriate strategies. On the technological side, there is a wide availability of transducers to measure accelerations, velocities, displacements and other variables representing the structural response that can be adequate for feedback control. More difficult is the issue of the design of actuators that are able to supply forces with sufficient intensity and speed to modify the response of massive structures. Some of the features of a number of actuators proposed and tested over the last years are summarized in [SC94]. For instance, aerodynamic appendages have been considered to reduce wind induced motion of tall buildings [SS81]. Pulse generators applying intensive short forces by air ejectors have been also suggested [MMDC88]. Both deflectors and pulse generators have been tested on small-scale models but no further applications have been reported.

More attention has been given to the so-called active mass dampers (AMD) and active bracing systems (ABS) respectively. An AMD is essentially made up by installing a mass whose movement is driven by a servo controller on top of a building (or tower), with the aim of reducing the movement of the main structure. Intensive research and laboratory tests have resulted in some installations in real buildings, mainly in Japan [Iem94, AFMY92]. An ABS essentially consists of a set of prestressed tendons or braces connected to the structure whose tensions are controlled by electrohydraulic servovalves. One of the interesting possibilities of ABS is that tendons and braces are common elements in many structures and thus the control system being installed can start from the structural members already in existence. The main drawbacks are still associated with the complexity of the design and with reliability. Early experiments were conducted on small-scale models [Roo80, CRS88, RCSR89, CLSR89]. Experiments with an ABS installed on full-scale building structures have been also reported [SRWL91].

So-called semiactive systems have been proposed in an attempt to reduce the energy requirements of the active actuators, which can be high for large structures. The basic common characteristic of these systems is that they operate as passive reactive systems but with variable parameters allowing on line tuning. For instance, systems with active variable stiffness [KM92] and active variable dampers have been discussed and tested. With the same saving energy purpose, a combination of passive and active systems working cooperatively has been suggested. In this context, one of the schemes with more interesting perspectives is made up of base isolators complemented with active actuators applying forces to the base of a building against earthquakes [KLS87, IK90, YDL92, FSF93, BRRM95].

In the case of control strategies, optimal control has been the most frequently invoked methodology. Optimal control was appealing since the question of controlling structural motion can be stated in terms of minimizing a standard linear quadratic function involving energy and control action, where the mathematical formulation is well standardized [KS72, SW77, Soo90]. On the other hand, unlike industrial applications such as those considered in the preceding chapters, a time invariant model of the structural system is available *a priori* in many cases. However, the solution requires significant computing effort and a dubious selection of weighting parameters of the performance criterion, especially for structures with a high number of degrees of freedom, which motivated an interest in other control methods that give simpler formulations with more easily tuned parameters. In this context, predictive control has been shown to be an effective strategy for structural control [RBM87, RCSR89, LR89, LARR94]. In fact, predictive control results in a simple computational scheme with parameters having clear physical meaning. Another additional advantage is the easy handling of time delays related to the actuators in the control system. The following sections will illustrate these features.

9.3 EXPERIMENTAL SETUP FOR ASEISMIC CONTROL

9.3.1 Description

Figures 9.1 and 9.2 present the two structures used in this application schematically: three- and six-storey building structures respectively. Figure 9.2 also shows a block diagram with the basic elements of the complete experimental control loop set up at the Department of Civil Engineering, University of New York at Buffalo, USA.

The structure illustrated in Figure 9.1 is a single-bay three-storey steel frame equipped with an active tendon controller between the ground and the first floor. Each floor is installed with blocks such that the total system weighs 3.5 tons. The structure given in Figure 9.2 is a 1:4 scale, three-bay, six-storey metal model equipped with two active tendon controllers placed between the ground and the first floors and between the second and the third floors respectively. The structure and the additional mass placed on the different floors weighs 19.18 tons.

For the experiments, the structures are rigidly bolted through a rigid foundation to a concrete block placed on a shaking table. Table movement is driven by a computer control system to produce real time base accelerations reproducing actual or synthetic earthquakes. The structures are equipped with lateral braces between floors so that the structure moves horizontally along the direction of the shaking table movement. Measured feedback signals to the control system are the displacements and velocities of each floor relative to the ground. These signals are sampled and converted to discrete time values by A/D converters. These

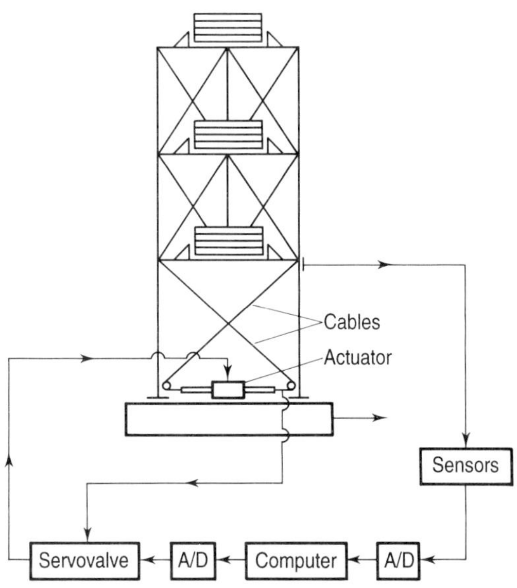

Fig. 9.1. Three-storey building structure and experimental control loop.

values are used by a computer to calculate the desired values of the control actions, which are translated into analog control signals by D/A converters. These control signals feed the servovalves of the tendon controllers.

Figure 9.3 shows a diagram of a tendon controller, which is composed of a horizontal rigid frame, a hydraulic actuator with a piston rod, a servovalve and four cables. The cables are braced to the upper floor by one of their ends while the other ends are attached to the rigid frame through four pulleys. The frame is connected to the piston rod of the hydraulic actuator whose motion is commanded by the servovalve proportional to the difference between the analog signal from the D/A converter and the signal obtained by measuring the actual displacement of the piston/rod. In this way, the tensions of the cables are actively modified, which results in horizontal control forces on the structure, as we will describe in the following.

9.3.2 Tendon controller forces and dynamic model

In this section we analyze how the active tendons generate the horizontal control

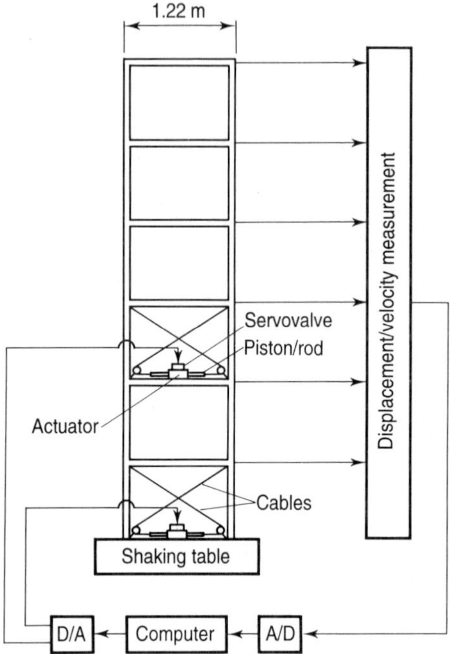

Fig. 9.2. Six-storey building structure and experimental control loop.

forces and we also derive the dynamic model of the controlled structure. We first study the six-storey structure in detail since this is the more general. Then we will consider the three-storey structure as a particular case.

Six-storey structure

Figure 9.4 illustrates the tendon controller 1, placed between the ground and the first floor, at instant t when the first floor relative displacement is $d_1(t)$. Tensions in the cables are given by

$$\begin{aligned} T_{1q} &= T_o - k_t \left[d_1(t) \cos\alpha_1 + u_{o1}(t)\right] \\ T_{1d} &= T_o + k_t \left[d_1(t) \cos\alpha_1 + u_{o1}(t)\right] \end{aligned} \tag{9.1}$$

where T_o is a pretension initially given to the cables to prevent tension release during control application; u_{o1} is the displacement of the rigid frame; and k_t is the

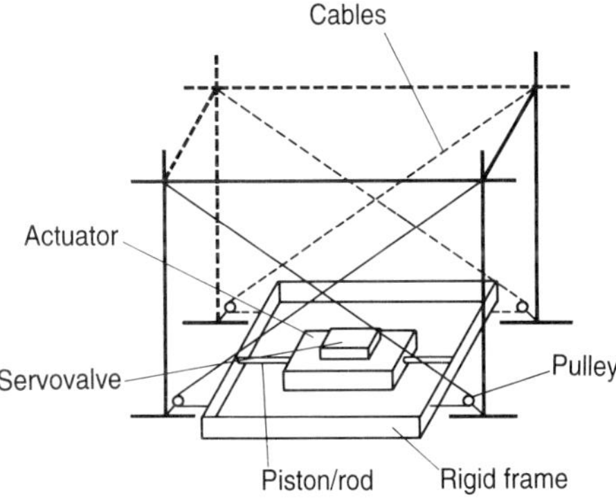

Fig. 9.3. Active tendon controller.

stiffness of the cables which determines the force associated with cable elongation. Then the horizontal control force f_1 at floor 1 results from the subtraction of the horizontal components of T_{1q} and T_{1d} both at the front and the back of the tendon controller. Therefore,

$$f_1(t) = -4\, k_t \, \cos\alpha_1 \, [d_1(t) \, \cos\alpha_1 + u_{o1}(t)] \tag{9.2}$$

In the same way, a force $-f_1(t)$ is generated on the ground, but it is absorbed by the ground itself.

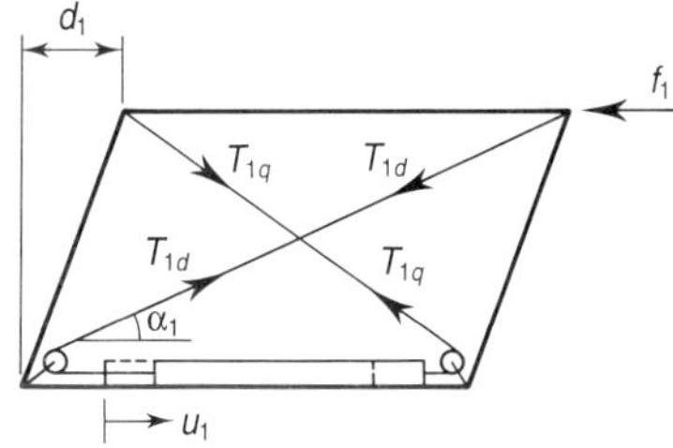

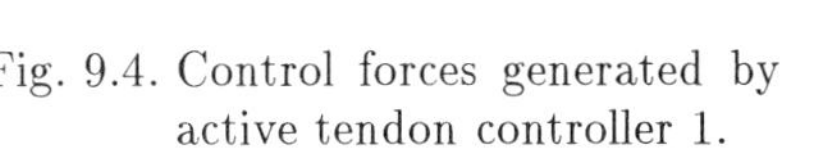
Fig. 9.4. Control forces generated by active tendon controller 1.

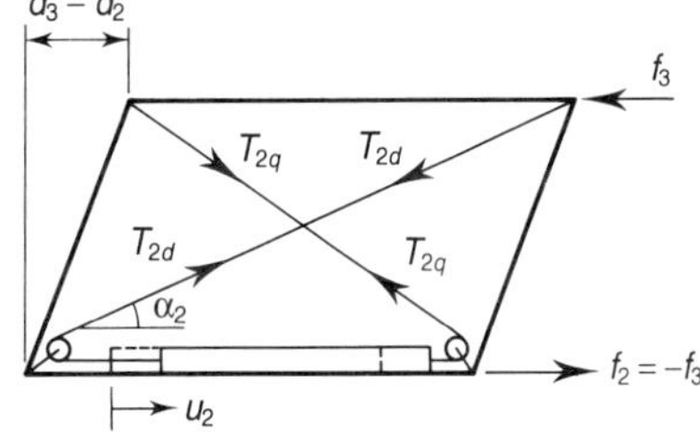

Fig. 9.5. Control forces generated by active tendon controller 2.

The same analysis for tendon controller 2, placed between the second and third floors (see Figure 9.5), shows the presence of control forces at floors 2 and

3, verifying:

$$\begin{aligned} f_2(t) &= 4\,k_t\,\cos\alpha_2\,[(d_3(t)-d_2(t))\,\cos\alpha_2 + u_{o2}(t)] \\ f_3(t) &= -f_2(t) \end{aligned} \tag{9.3}$$

where d_2 and d_3 are the displacements of the second and third floors, respectively, and u_{o2} is the displacement of the rigid frame of tendon controller 2.

The dynamic behaviour of a structure such as the one considered here may be described [CP75] by an equation of the form

$$M\ddot{d}(t) = f_e(t) + f_r(t) \tag{9.4}$$

where M is the mass matrix (in this case it is diagonal with the values of the masses of the floors in its diagonal terms), $\ddot{d}$ is the vector containing the acceleration of each floor, f_e is the vector of the external forces acting on the structure and f_r is the vector of the internal forces reacting to motion. Assuming linear behaviour, f_r may be described by

$$f_r(t) = -C\dot{d}(t) - Kd(t) \tag{9.5}$$

where C and K are, respectively, the matrices describing the damping and stiffness features of the structure, while $\dot{d}$ and d are the vectors whose components are the velocity and displacement of each floor respectively.

Finally, for f_e we may distinguish two terms. One describes the exciting forces acting on the structure due to the acceleration of the ground motion, which is defined by

$$-M\,j\,\ddot{d}_o(t)$$

where j is the vector with all its components equal to 1, and $\ddot{d}_o$ is the acceleration of the ground motion. The other term in f_e accounts for the control forces generated by the tendon controllers defined by the vector

$$f^T = [f_1, f_2, f_3, 0, 0, 0]$$

where f_1, f_2, f_3 are the forces applied at floors 1, 2, 3 given by (9.2) and (9.3). Therefore,

$$f_e(t) = -Mj\ddot{d}_o(t) + f(t) \tag{9.6}$$

From (9.2) and (9.3), f may be written in the form

$$f = K_p\,d + L\,u_o \tag{9.7}$$

where

$$K_p = \begin{bmatrix} -4\,k_t\,\cos^2\alpha_1 & 0 & 0 & 0 & 0 & 0 \\ 0 & -4\,k_t\,\cos^2\alpha_2 & 4\,k_t\,\cos^2\alpha_2 & 0 & 0 & 0 \\ 0 & 4\,k_t\,\cos^2\alpha_2 & -4\,k_t\,\cos^2\alpha_2 & 0 & 0 & 0 \\ 0 & 0 & 0 & 0 & 0 & 0 \\ 0 & 0 & 0 & 0 & 0 & 0 \\ 0 & 0 & 0 & 0 & 0 & 0 \end{bmatrix}$$

$$L = \begin{bmatrix} -4\,k_t\,\cos^2\alpha_1 & 0 \\ 0 & 4\,k_t\,\cos^2\alpha_2 \\ 0 & -4\,k_t\,\cos^2\alpha_2 \\ 0 & 0 \\ 0 & 0 \\ 0 & 0 \end{bmatrix} \tag{9.8}$$

$$u_o^T = [u_{o1}, u_{o2}]$$

Using (9.4)–(9.7) we may write

$$M\,\ddot{d}(t) + C\,\dot{d}(t) + K'\,d(t) = L\,u_o(t) - M\,j\,\ddot{d}_o(t) \tag{9.9}$$

where

$$K' = K + K_p$$

Equation (9.9) describes the motion of the structure under the control of active cables. We may notice the double effect of the cables: they supply a passive control action $K_p\,d$ and an active control $L\,u_o$. The first one is equivalent to a change in the stiffness matrix of the structure, which is now K'. The active control action is associated with the displacement u_o of the rigid frame of the tendon controllers as commanded in the closed loop by the control system.

Three-storey structure

In this study the two upper floors of the structure were made rigid by diagonal braces. In this way the structure behaved as a single-degree-of-freedom system whose horizontal movement is characterized by the displacement d of the first floor. By means of an analysis like the one performed for the six-storey structure, we can see that this displacement is described by the equation

$$\ddot{d}(t) + 2\xi\omega\dot{d}(t) + \omega^2 d(t) = -\frac{4k_t \cos\alpha}{m} u_o(t) - \ddot{d}_o(t) \tag{9.10}$$

where m is the mass, ξ is the damping and ω is the natural frequency of the system; u_o is now the (scalar) displacement of the rigid frame of the tendon controller.

9.3.3 State model and control problem

For the purpose of formulating the control problem, it is convenient to express the above dynamic models in a state space setting. Defining the state vector $x^T = [d, \dot{d}]$, equation (9.9) can be rewritten in the form

$$\dot{x}(t) = F\,x(t) + G\,u_o(t) + w(t) \tag{9.11}$$

where

$$F = \begin{pmatrix} 0 & I \\ -M^{-1}K' & -M^{-1}C \end{pmatrix}; \quad G = \begin{pmatrix} 0 \\ M^{-1}L \end{pmatrix}; \quad w = \begin{pmatrix} 0 \\ -j \end{pmatrix} \ddot{d}_o \tag{9.12}$$

For the design of the active controller, the full state is available since displacement and velocity sensors are located at each floor. The seismic acceleration is not measured, so it has to be considered as an unknown disturbance. It is convenient to complete the above model with a time delay to compensate for the inertia of the active tendon controller. This can be modelled assuming that the actuator displacement vector u_o is related to the control vector u supplied by the controller in the form $u_o(t) = u(t - \tau)$, τ being the time delay.

For the three-storey structure, we may write the corresponding two dimensional state equation

$$\dot{x}(t) = F\,x(t) + G\,u(t - \tau) + w(t) \tag{9.13}$$

with

$$F = \begin{pmatrix} 0 & 1 \\ -\omega^2 & -2\xi\omega \end{pmatrix}; \quad G = \begin{pmatrix} 0 \\ \dfrac{-4k_t \cos\alpha}{m} \end{pmatrix}; \quad w = \begin{pmatrix} 0 \\ -1 \end{pmatrix} \ddot{d}_o \tag{9.14}$$

For both structural systems, the control problem basically lies in the formulation of the control law to give u as a feedback of the state vector x.

Before describing the implementation of predictive control, we point out one feature that distinguishes this implementation from those described in preceding chapters. In this case, the controller has to react immediately and efficiently against the earthquake excitation when it begins to shake the structure and the high intensity and frequency, the short duration and the unknown character of the exciting forces make the use of an on-line parameter adaptation impracticable, which we have used successfully in the preceding applications in this book. Therefore, the approach taken in this case has been to formulate the predictive control law with fixed parameters, which are obtained after determining the matrices involved in the above state models. On the other hand, although the state models previously formulated are linear, they capture the structure's dynamics to a reasonable approximation and have been widely used in the area of structural control [Soo90].

There are analytical tools in the literature on structural dynamics with which to determine the structure's parameter matrices [CP75]. Another approach, with an experimental emphasis, to obtaining structural parameters is essentially based on recording known excitations and their corresponding responses and using any appropriate identification technique [KN86, Saf89]. We followed this approach in the application considered here. The structure was subjected to a series of white noise ground accelerations by the shaking table and the responses of the different floors were recorded. From the input/output data, standardized modal analysis techniques [RSLW89] led to the frequency and damping parameters of the modes of vibration. From these parameters, the mass, damping and stiffness matrices were finally obtained.

9.4 IMPLEMENTATION OF PREDICTIVE CONTROL

Since the models described above are given in state space form, we will follow the procedure described in Appendix B, Section B.3 for the formulation of a state feedback predictive controller. First, we will introduce the predictive model and then we will formulate the control law.

9.4.1 Predictive model

The models of (9.11) and (9.14) must be discretized due to the digital nature of the control system. The usual procedure for this, as outlined in Appendix A, assumes that the control input $u(k)$ is constant between sampling instants k and $k+1$. Using this fact and assuming no disturbance in the model due to the fact that the seismic excitation is unknown, the following discrete time model is obtained:

$$x(k+1) = Ax(k) + Bu(k-r) \tag{9.15}$$

where

$$A = \exp(TF); \qquad B = F^{-1}(A-I)G$$

I is the identity matrix; T is the sampling period; and r is an integer defined as $r = \tau/T$, which implies the assumption that the time delay τ can be measured as an integer multiple of the sampling period. This is a common assumption regarding time delay in digital control implementations, but equation (9.15) can easily be modified to take into account the non-integer nature of the time delay [FPW90].

As in Appendix B, Section B.3, it seems natural to consider the model of (9.15) to predict the structural state over the prediction interval $[k, k+\lambda+r]$ in the form

$$\begin{array}{c}\hat{x}(k+j|k) = A\hat{x}(k+j-1|k) + B\hat{u}(k+j-1-r|k) \\ (j = 1, 2, \ldots, \lambda + r)\end{array} \tag{9.16}$$

where $\hat{x}(k+j|k)$ denotes the $n \times 1$ state vector predicted at instant k for instant $k+j$ and $\hat{u}(\cdot|k)$ denotes the sequence of $p \times 1$ control vectors on the prediction interval, which verify

$$\hat{x}(k|k) = x(k); \quad \hat{u}(k-j|k) = u(k-j) \qquad (j = 1, \ldots, r)$$

9.4.2 Control law

For Q, R symmetric weighting matrices properly chosen, consider the problem of the minimization of the simplified linear quadratic performance index

$$J_k = \frac{1}{2}\hat{x}(k+r+\lambda|k)^T Q\hat{x}(k+r+\lambda|k) + \frac{1}{2}\hat{u}(k|k)^T R\hat{u}(k|k) \tag{9.17}$$

along with the condition for the control sequence

$$\hat{u}(k|k) = \hat{u}(k+1|k) = \ldots = \hat{u}(k+\lambda-1|k) \tag{9.18}$$

As detailed in Section B.3 of Appendix B, the solution of this problem leads to the following control law:

$$u(k) = \hat{u}(k|k) = -D_1 x(k) - D_2 U_k \tag{9.19}$$

where

$$\begin{aligned} U_k &= [u(k-1)^T, u(k-2)^T, \ldots, u(k-r)^T]^T \\ D_1 &= D_3 Z \\ D_2 &= D_3 T \\ D_3 &= (N^T Q N + R)^{-1} N^T Q \end{aligned} \tag{9.20}$$

with

$$\begin{aligned} Z &= A^{r+\lambda} \\ T &= [\,A^{\lambda}B \quad A^{\lambda-1}B \quad \ldots \quad A^{r+\lambda-2}B \quad A^{r+\lambda-1}B\,] \\ N &= A^{\lambda-1}B + A^{\lambda-2}B + \ldots + AB + B \end{aligned} \tag{9.21}$$

9.5 EXPERIMENTAL RESULTS AND DISCUSSION

The experiments consist of shaking the structures to simulate seismic actions in order to test the performance of the active tendon system driven by the predictive control law of (9.19). As previously mentioned, before running the tests, the characteristics of the structures were identified by using experimental modal analysis techniques. Using the identified structural parameters and the selected sampling period T, the matrices involved in the control law were calculated and kept constant during the experiments. The design parameters to be selected for the controller were the prediction horizon λ, the time delay r and the weighting matrices Q, R.

9.5.1 Three-storey structure

The parameters of the single degree of freedom model of (9.10) were identified experimentally [CRS88], the following values being obtained: $m = 2922.7$ kg, $\omega = 21.79$ rad/sec, $\xi = 1.24\%$, $k_t = 371950.8$ N/m and $\alpha = 36$ degrees. The sampling period was $T = 0.01$ sec and the time delay was $r = 2$. The weighting matrix Q was chosen as

$$Q = \begin{pmatrix} 1 & 0 \\ 0 & 0 \end{pmatrix}$$

which implies weighting the structural displacement, while R is now a scalar defined as $R = \beta k_t$, β being the weighting factor on the control action. Substituting $R = \beta k_t$ into the performance index of (9.17), we observe the presence of the term $\beta k_t \hat{u}(k|k)^2$. Since the energy required to produce a cable elongation u is proportional to $k_t u^2$, the factor β has the physical meaning of weighting the energy applied to the actuator.

The results of two series of experiments with the shaking table are now described, in which the base of the structure has been subjected to a band-limited white noise and to an earthquake acceleration respectively. Different tests were performed for different values of the prediction horizon parameter λ and the weighting factor β assigned to the control action.

White noise excitation experiments

The shaking table has been programmed to supply a 0–8 Hz banded white noise acceleration to the base of the structure for about three minutes. The amplitude of the transfer function between acceleration of the first floor and that of the base has been obtained in the frequency range under consideration by means of a spectrum analyzer and has been used to illustrate the results of the different control tests. Figure 9.6 plots the amplitude as a function of the frequency of the excitation for the structure under predictive control for different values of parameters λ and β. Table 9.1 gives the peak value of the transfer function and its corresponding frequency for the plots given in Figure 9.6. For the sake of comparison, it also gives the peak value for the uncontrolled structure.

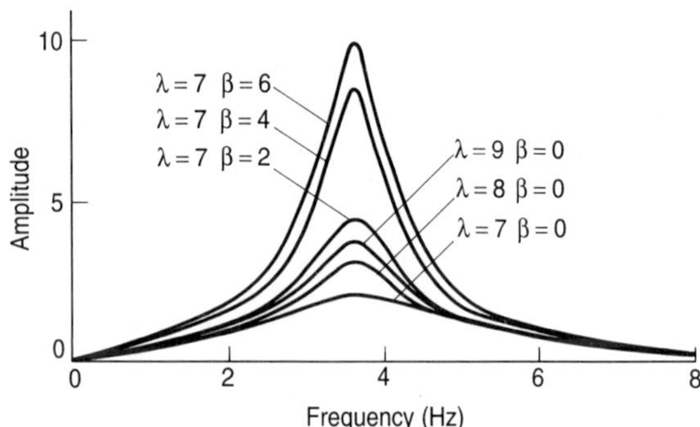

Fig. 9.6. Transfer function magnitude with predictive control.

TABLE 9.1: White noise excitation experiments

λ	β	Transfer function peak	Frequency (Hz)
No control	No control	154.00	3.52
7	6	9.99	3.60
7	4	8.35	3.60
7	2	4.63	3.68
7	0	2.65	3.76
8	0	3.18	3.68
9	0	4.21	3.68

The influence of λ and β on control performance can be analyzed from these tests. By comparing the transfer functions for $\lambda = 7, 8$ and 9 (with $\beta = 0$), more reduction in the response is observed for smaller values of λ. By comparing the cases for λ fixed equal to 7 with values of $\beta = 0, 2, 4$ and 6, more reduction is observed as smaller values of β are used. These observations are in accordance with the meanings of λ and β. In fact, the predictive control law has been derived from the minimization of the performance index of (9.17). This minimization requires the response predicted for instant $k + \lambda + r$ to be near the equilibrium state. Thus, the smaller the value of λ, the more demanding the performance criterion becomes, which results in a more drastic control action with more reduction in the response. On the other hand, weighting of the control action in the performance index of (9.17) is imposed by $R = \beta k_t$. Consequently, for a fixed prediction horizon, a smaller control action is generated when the value of β is increased.

In all cases only a slight increase in the effective frequency of the controlled system is observed with respect to the natural frequency of the structure as shown in Table 9.1. The damping effect of the control is noticeable in all the cases. In fact, the peak amplitude of the transfer function is reduced from 154 for the uncontrolled case to 9.99 for the control case with $\lambda = 7$ and $\beta = 6$.

Earthquake excitation experiments

The structure has been excited by the shaking table simulating the ground acceleration of the earthquake of El Centro (USA, 1940), scaled to 25% of its maximum amplitude. Figure 9.7 shows the first twenty seconds of the measured time histories of the first floor relative displacement and the absolute acceleration for the structure with no control. Figure 9.8 shows the same histories, as well as the control force supplied by the cables, for one of the tests performed under predictive control. Values of $\lambda = 9$ and $\beta = 0$ have been assigned for this test. By comparing Figures 9.7 and 9.8, a very significant reduction in the controlled response is observed with a control effort that is compatible with the actuator limits.

The above experimental results have been obtained by designing the predictive control based on the model of (9.15). There are certainly mismatches between the model and the actual plant dynamics. However, the control has been efficient and gives an intuitive idea of the robustness of the predictive control scheme. In the following we will analyze this issue in more detail, using the model of (9.15) to perform numerical simulations of control tests and comparing the results with those obtained in the experiments. These comparisons can give us an additional insight into the robustness of the control algorithm.

Figure 9.9 shows the first ten-second time histories of the first floor relative displacement and the control force for both the experimental and numerical control tests in the case of $\lambda = 9$ and $\beta = 0$. The same time histories are presented in Figure 9.10 for a control test with $\lambda = 2$ and $\beta = 0.4$. While the correlation between experimental and numerical results is very good for the case illustrated in Figure 9.9, noticeable differences exist for that illustrated in Figure 9.10. These differences arise from the discrepancies between the true and the simulated systems. These discrepancies are mainly related to inaccuracies in the identification of structural parameters and the interaction between the actuator

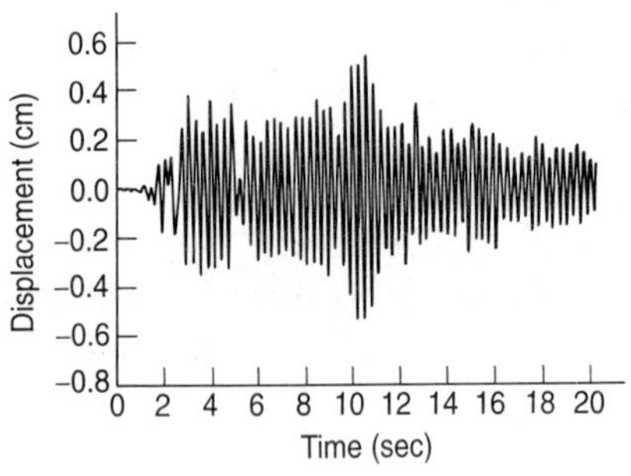

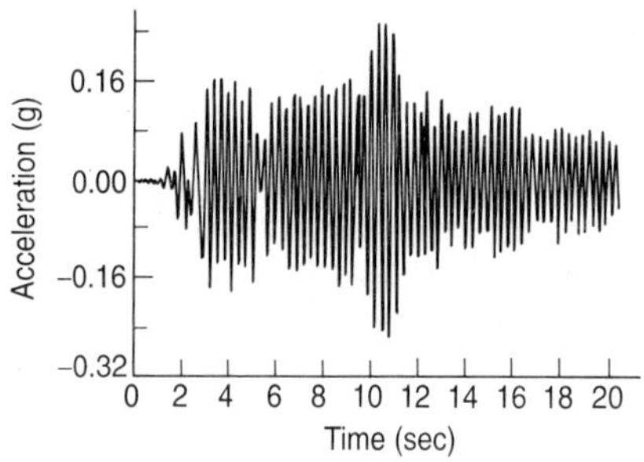

Fig. 9.7. Displacement and acceleration responses without control.

and the structure, as pointed out in [CRS88]. Another major discrepancy exists in the present implementation of predictive control related to the time delay. In this implementation a number of delays $r = 2$ have been assumed in the predictive model. Since a sampling period $T = 0.01$ seconds has been used, this implies the existence of a time delay of 0.02 seconds in the control loop. This value may cover the real lag introduced by the servovalve–actuator system as identified in [CRS88]. But there is an additional lag between the velocity signal and the control sequence since it takes 0.006 seconds for the analog differentiator to produce this signal from the measured displacement. Moreover, an extra lag of one sampling period is present in the experiments due to the real time operation of the digital computer to implement the control algorithm. The analysis performed in Chapter 3 (Section 3.5) shows that predictive control achieves stability and

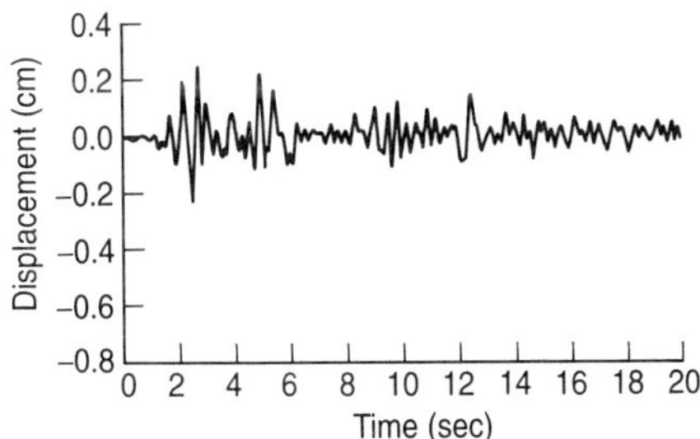

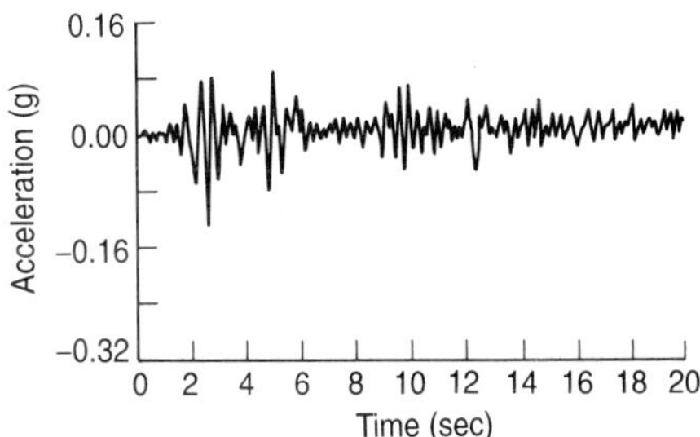

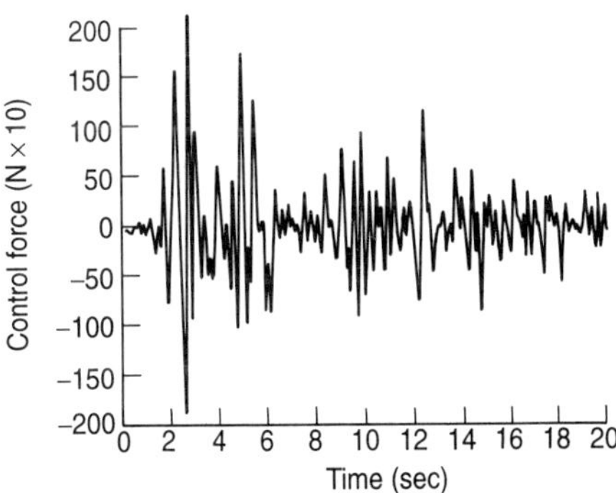

Fig. 9.8. Displacement and acceleration responses and control force under predictive control with $\lambda = 9$ and $\beta = 0$.

enhances robustness in the presence of inaccuracies in the model and errors in the time delay as the prediction horizon increases its length λ. While stability exists in the implementation for the two cases with $\lambda = 9$ and $\beta = 0$ (Figure 9.9) and $\lambda = 2$ and $\beta = 0.4$ (Figure 9.10), this comparison shows how robustness is significantly improved when the prediction horizon is extended to $\lambda = 9$.

9.5.2 Six-storey structure

Table 9.2 gives the frequency and damping of the six modes of vibration of the structure identified experimentally [RSLW89].

From these values, the following mass (in tons), damping (in kN s/m) and stiffness matrices (in kN/m) in the equations of motion of (9.9) were obtained:

$$M = \begin{pmatrix} 3.188 & 0.0 & 0.0 & 0.0 & 0.0 & 0.0 \\ 0 & 3.188 & 0.0 & 0.0 & 0.0 & 0.0 \\ 0.0 & 0.0 & 3.188 & 0.0 & 0.0 & 0.0 \\ 0.0 & 0.0 & 0.0 & 3.188 & 0.0 & 0.0 \\ 0.0 & 0.0 & 0.0 & 0.0 & 3.188 & 0.0 \\ 0.0 & 0.0 & 0.0 & 0.0 & 0.0 & 3.188 \end{pmatrix}$$

$$C = \begin{pmatrix} 6.508 & -0.417 & -1.393 & -0.744 & 2.038 & -1.209 \\ -0.417 & 3.460 & -2.190 & 0.198 & 0.417 & -0.090 \\ -1.393 & -2.190 & 4.343 & -0.637 & -0.677 & 0.679 \\ -0.744 & 0.198 & -0.637 & 2.686 & -1.198 & 0.549 \\ 2.038 & 0.417 & -0.677 & -1.198 & 3.917 & -1.466 \\ -1.209 & -0.090 & 0.679 & 0.549 & -1.466 & 2.243 \end{pmatrix}$$

$$K = \begin{pmatrix} 14332.6 & -3988.1 & -395.6 & -164.2 & -278.2 & -101.3 \\ -3988.1 & 12318.3 & -8154.0 & -93.4 & 247.6 & -81.5 \\ -395.6 & -8154.0 & 12869.1 & -4715.7 & 343.2 & -156.1 \\ -164.2 & -93.4 & -4715.7 & 9363.1 & -4742.7 & 102.0 \\ -278.2 & 247.6 & 343.2 & -4742.7 & 9635.8 & -5128.8 \\ -101.3 & -81.5 & -156.1 & 102.4 & -5128.8 & 5047.7 \end{pmatrix}$$

In the application of the predictive control law, the sampling period was chosen to be $T = 7.5$ ms, the time delay $r = 3$, the prediction horizon length $\lambda = 11$, the control weighting matrix $R = 0$ and the state weighting matrix Q, which is 12×12, with all its elements equal to 0 except the six first diagonal elements chosen equal to 1. This means that only the displacement of each floor in the performance criterion is weighted.

A series of tests was performed, subjecting the structure to the action of base accelerations corresponding to recorded real earthquakes. As a sample case, the results given here correspond to the earthquake Hachinoe (Japan). Figure 9.11 shows the displacement of the top floor for both the cases with no control and

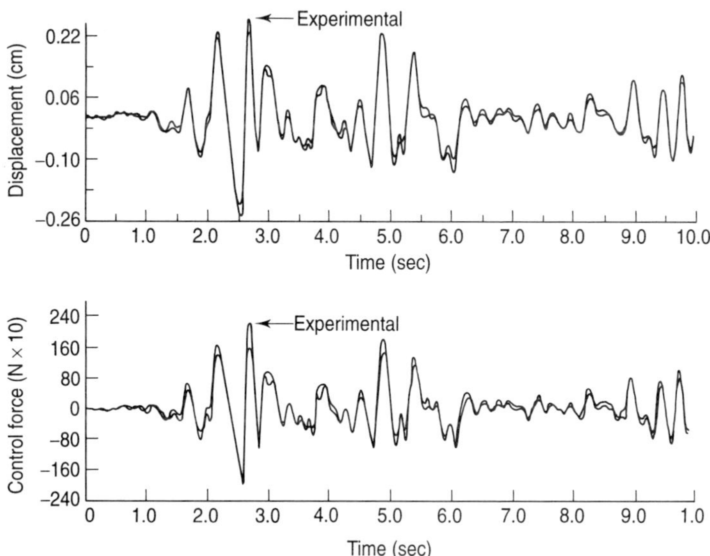

Fig. 9.9. Experimental and numerical results for $\lambda = 9$ and $\beta = 0$.

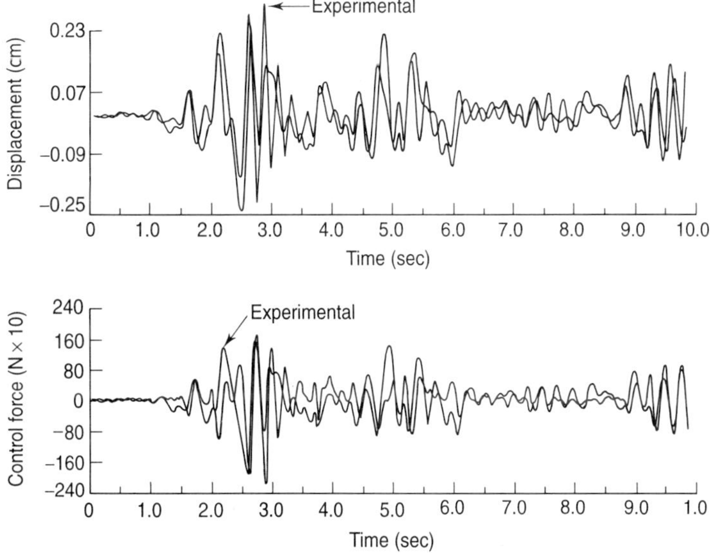

Fig. 9.10. Experimental and numerical results for $\lambda = 2$ and $\beta = 0.4$.

TABLE 9.2: Modal parameters of the six-storey structure

Vibration mode	Frequency (Hz)	Damping (%)
1	1.56	2.34
2	4.59	0.59
3	7.91	0.61
4	10.45	2.10
5	11.52	0.81
6	13.38	1.11

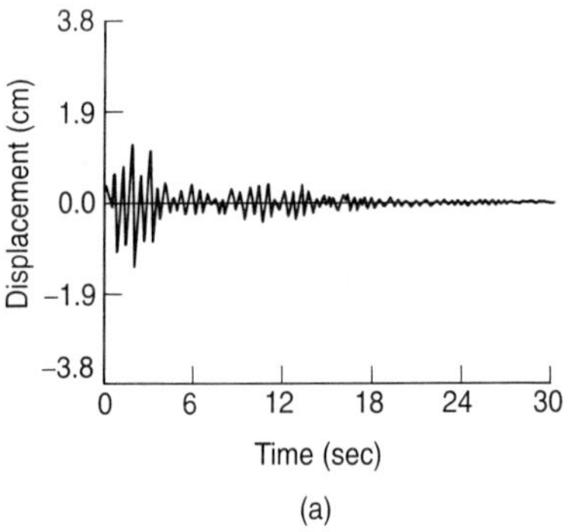

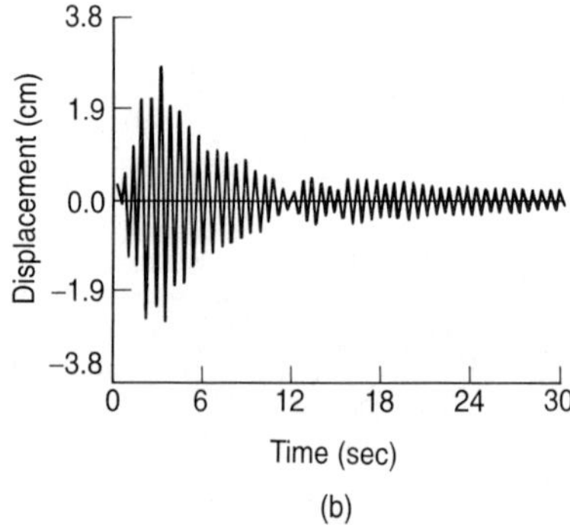

Fig. 9.11. Sixth-floor displacement with (a) and without (b) predictive control.

with predictive control supplied by the tendon controller placed between floors 2 and 3. By comparing both cases, we can see a significant reduction in the maximum peak response, as well as an increase in the damping mitigating the response.

9.6 CONCLUSIONS

This chapter has presented the first application of predictive control in the area of active vibration control of structures. The implementation was preceded by an experimental identification of a model of the system based on the equations of motion of the structure. Given this model, a control law derived from the extended strategy of predictive control was implemented without further parameter adaptation. The experimental results have shown the effectiveness of this control law, and comparisons of these results with numerical simulations have given a practical insight into the robustness as against the modelling errors. This robustness is enhanced by increasing the prediction horizon, as expected from the theoretical analysis given in Chapter 3.

Active control is a new and promising area of application of automatic control methodologies and systems in the context of civil, mechanical, marine, aerospace and other structures. The conclusion is that predictive control seems to be as suitable for this area as it has proved to be for other industrial areas discussed in the preceding chapters of this part of the book. The following part will illustrate how the capabilities of adaptive predictive control systems may be used to optimize the performance of industrial plants.

Part 3

THE OPTIMIZATION

Chapter 10

THE APCS OPTIMIZATION: APPLICATION TO THE CEMENT INDUSTRY

10.1 INTRODUCTION

The methodology, the theory and a set of practical applications of adaptive predictive control systems (APCS) have been presented in Parts I and II. In these applications APCS has proved to be a tool that is able to perform the precise control of the process variables; but this control may not be a sufficient condition to ensure that the plant is operating at a fully satisfactory level. Usually, the attainment of a desired performance in an industrial plant is associated with a form of *optimization* which involves a search of setpoint values that are appropriate for the critical process variables.

One approach is to derive setpoint values for an optimum operation based on theoretical models of the plant using programming techniques [DW88, RKS88, Lin93]. Another, more empirical, approach is based on an evolutionary operation [BL72] which may move the setpoint values dynamically throughout process operation in the search for an experimentally verified optimum performance. However, the practical use of this idea is severely limited by the need for precise control in the critical variables of complex industrial plants. As a control tool, APCS may be associated with any sensible optimization approach.

In this part of the book we propose a concept of optimization, related to the above-mentioned empirical approach, in the context of the application of adaptive predictive control, and we describe the so-called *APCS optimization systems* that have been developed specifically to transform this concept into reality in industrial plants. This is illustrated by describing the application of APCS optimization systems to the cement industry.

The cement industry has been chosen here to illustrate the concept of APCS optimization and its application because the difficulties involved in the control of processes such as cement kilns and mills are paramount within the industrial context. In fact they are very complex dynamic processes which deal with solids which are undergoing sinterization and milling, with numerous pure time delays, significant interaction between their variables, have a time varying nature and which are affected by unmeasurable disturbances that are unpredictable both in time and in intensity.

The first attempt at automation of these processes was based upon the use of conventional control techniques. Later, an attempt was made to replace the operator by so-called *expert systems* which are based on rules that attempt to emulate human behavior [HL89, HST87, Fli91, SSHS94]. Therefore, the best performance that can be expected from such expert systems is, by definition, that of good manual control.

This chapter defines the concept of APCS optimization, describes the optimization systems and introduces the application to the cement industry from a conceptual perspective. In the following two chapters, which complete this part of the book, we present the specific details of the application and the results obtained.

10.2 APCS OPTIMIZATION

The APCS optimization concept is based on the precise control of the critical process variables that can be achieved through the application of adaptive predictive control. In the first instance, this capability will allow the stabilization of these variables, that is, the minimization of their oscillations around their setpoints or the deviations caused by unmeasured disturbances that may act on the process. Additionally, once stabilization is obtained, the APCS optimization concept involves driving the setpoints of the process variables towards those operating points where some or all of the following conditions are met in real time:

- minimization of the consumption of resources (energy, raw materials and others);
- maximization of production in quantity and quality;
- regular and stable operation;
- maintenance of the required or specified safety and reliability conditions.

Thus, given the time varying nature of the industrial process and its operating context and objectives, the concept of APCS optimization is that of a permanent on-line search. This search may be conducted through a real time evaluation of the performance of the process operation, approaching the conditions stated above and moving the setpoint values in the direction in which this performance is increased.

APCS optimization systems have been developed to turn the above optimization concept into reality and to integrate the following subsystems in their operation:

- an adaptive predictive control system;
- a logic system for analog and digital signals;
- a system of data acquisition and analysis;
- a master system.

The adaptive predictive control system and the logic system for analog and digital signals guarantee the precise control required by the optimization and provide a general solution to all types of industrial control problem.

The data acquisition and analysis system allows the study of the process, its variables and its dynamic characteristics, by the generation of reports and graphic representations in real and historic time.

The master system is a rule-based system that is capable of searching for the optimum operating conditions in the way we have previously described, that is to say, evaluating the performance of the process operation periodically and moving the setpoint values in the direction in which this performance is increased, as will be illustrated in the following chapters. The optimum operating points may also vary over time. Thus the search conducted by the master system is permanent. The master system must also be capable of:

- reacting to orders entered in a simple manner by the operator to impose any kind of previously planned sequence of actions on the process;
- identifying the different contingencies that may arise during the process operation and reacting to these in an optimum predefined manner.

After APCS optimization systems are implemented, an initial phase of experimentation begins. In this initial phase the optimum points of operation, the possible transitions and the specific reactions to the different contingencies that may occur are established. In this way, the optimization of the process operation will finally be reached, as is illustrated conceptually in the examples given in the following sections.

10.3 APPLICATION TO CEMENT KILNS

10.3.1 Brief description of a kiln

Figure 10.1 presents the general scheme of a typical cement kiln. The mixture of raw materials that is fed into the process, usually called crude, is introduced in the preheater, where it reaches a temperature of approximately 900 °C before entering the kiln. The fall of the crude into the kiln inlet is produced by gravity and is opposed to the gases flowing towards the main fan. The crude heating is controlled by a cyclone system and is produced by heat absorption from the rising gases, in many cases with the aid of a burner installed in the preheater.

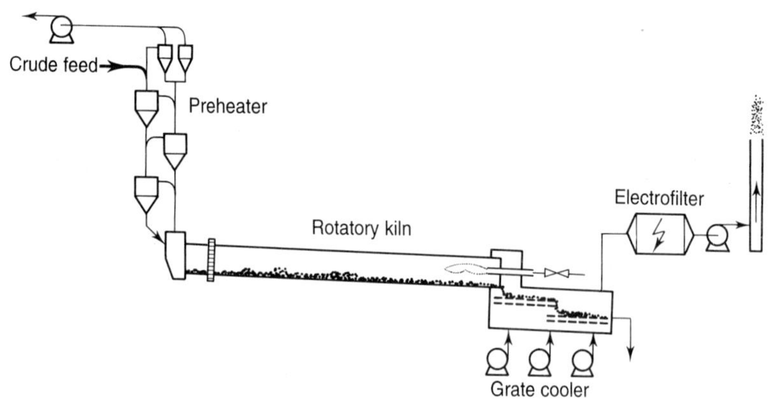

Fig. 10.1. General scheme of a cement kiln.

Once inside the kiln, the crude descends progressively by gravity due to the inclination of the kiln and a rotating movement towards the so-called sinterization area, where the reactions transforming the crude into clinker (the basic component of the cement) take place. The sinterization temperature, the maximum temperature for the kiln, is approximately 1400 °C and thus the clinker is incandescent when it falls onto the grate. The grate is a flat platform of porous tiles which, by horizontal movements, drives the clinker downstream. It consists of two separate parts, one situated below the other. Located underneath the grate is a unit of fans which spread cooling air through the clinker layer. Most of the air passing through the first grate is conducted towards the kiln and is called secondary air. Its function, besides cooling the clinker, is to ensure that the combustion air is heated by the heat remaining in the clinker that is produced, thus increasing efficiency. The grates, fans and all the mechanisms installed for cooling the clinker form the so-called cooler.

10.3.2 Typical kiln operation context

Stability of kiln operation is the main objective of the operator's performance. In this context, stability is understood as the maintenance of the critical variables of the process within certain bands in order to ensure the continuity of kiln operation and, as a result, the continuity of production.

In order to maintain this stability, the operator will try to retain a thermic level which will in general be higher than is strictly necessary for the production of clinker. Nevertheless, if the operator is asked to decrease the thermic level, the answer will usually be negative. This reaction is not surprising, due to the fact that operating at lower thermic levels often means an increase in the oscillations of the process variables, which could cause the kiln to cool. Consequently, all the possible energy savings achieved by working at a lower thermic level would be undone by the drawbacks that such a situation may involve.

In the same way, once the operator achieves satisfactory kiln operation, if an increase of crude feed is called for the answer will usually be negative again. This behaviour is not unreasonable because increasing the feed of the kiln could provoke more pronounced oscillations in the variables, which may force them out of their acceptable bands and render the operator unable to resume satisfactory control.

The preceding examples justify the pragmatic operator's criterion: to focus on achieving continued conservative kiln operation more than on minimizing the specific consumption of energy or on maximizing kiln production.

10.3.3 The new context of APCS optimization

The application of the APCS optimization systems to cement kilns introduces a new perspective. Under adaptive predictive control, the variables are centred in the assigned setpoints and typical dynamic oscillations are minimized. Consequently, the kiln behaves in a stable manner without depending on the operator's actions. The situation has changed drastically and what was a priority objective before (i.e. stability) now becomes a starting point which is obtained simply by the application of adaptive predictive control.

In this new context, if the possibility of driving the process operation to a lower thermic level is suggested to the operator, he or she may agree. The APCS optimization system will execute this while keeping the rest of the variables permanently centred around their setpoints. This successful kiln thermic level decrease in the presence of stability of the variables may encourage the operator to continue decreasing the thermic level until a minimum compatible with quality clinker production is achieved.

Clearly, this minimum thermic level can only be attained and maintained, without provoking cooling situations that a manual control would inevitably imply, by adaptive predictive control, due to its capacity to achieve precise control of the process variables around their setpoints under varying conditions. In order to do this, the adaptive predictive model uses the input/output information of more than one hour of process operation and the prediction horizon for thermic level evolution is extended over a similar period. To complement the optimization in the sense of energy savings, the oxygen level must also to be driven to a minimum that is compatible with acceptable carbon monoxide (CO) levels.

As described above, APCS can make the kiln work at a minimum thermic level, which means minimizing the specific energy consumption of the clinker produced and, ultimately, saving energy. Achieving this objective means producing the same amount as before but with less energy or, in other words, having extra energy which could be used to produce more clinker.

Indeed, with the kiln stabilized under APCS, with its variables centred at their setpoints and with the specific energy consumption minimized, if we request the operator to increase the crude feed (s)he will now probably agree as the kiln stability no longer depends on the operator but is assured by APCS. The first crude feed increase will be successful, with APCS maintaining the controlled variables at their setpoints and actuating over the manipulated variables, mainly the fuel. Subsequent attempts will follow, allowing kiln production to reach its physical limits; in other words, until maximization of production is achieved.

10.3.4 Role of the master system

As previously described, at the initial stage the optimum operation points of the kiln may be sought by the operator deciding himself the setpoints that APCS has to reach for each case. However, within the application of the APCS optimization systems, the master system has the role of searching automatically for the optimum operation points by varying the setpoints of the kiln critical variables progressively, at the same time as it monitors its overall operation, until the optimum operation points that are compatible with the physical limits are achieved.

The following chapter gives the complete details of the APCS optimization of cement kilns. It presents plots that illustrate the evolution of the variable NO_x, representative of the thermic level, and how it is driven through a smooth ramp to the minimum energy consumption point. In the same way, it illustrates how the oxygen level is conducted to a minimum compatible with an acceptable level of CO, and how the crude feed is increased gradually until it reaches the physical limits of production. Once these three optimum points are achieved the APCS optimization system stabilizes the process operation around them.

10.4 APPLICATION TO MILLS

10.4.1 Brief description of the process

Tube mills are rotating steel cylinders that grind materials into particles. Figure 10.2 represents the usual layout of the elements which perform the grinding in closed circuit.

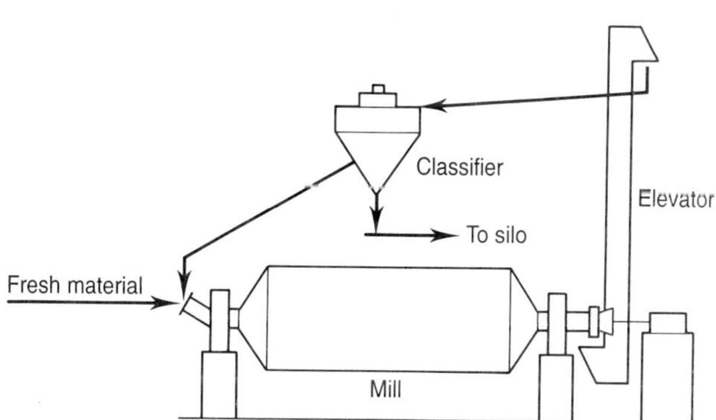

Fig. 10.2. General scheme of a cement tube mill.

Within the cylinder, the grinding bodies occupy approximately 1/3 of its volume: steel balls of different sizes with diameters between 10 mm and 100 mm. Materials to be ground enter the mill on one side and are dragged through the mill by a current of gases that sweeps through it. During this time, the materials collide with the steel balls, which results in their breakdown. The continuous rotating movement of the cylinder maintains the collisional activity of the balls. The cylinder interior is divided into chambers separated by walls which contain diverse sizes of balls. Balls are classified by size, so that the largest are found at the inlet chamber and the size decreases as the chambers are closer to the outlet. This makes the grinding effect more effective and the particles are minute upon exiting the mill. However, the size of every particle leaving the mill may not always be acceptable, and thus selection must be made in the classifier or separator. Upon exiting the mill, an elevator transports the materials to a height which allows introduction into the classifier. Inside it, particles are separated: particles of acceptable size are sent to silos, while those that are too big are returned to the mill for further grinding.

10.4.2 Typical mill operation context

The complexity of efficient control of ball mills with horizontal axes is widely recognized in the cement industry. Among other reasons, the dynamics of this process is complicated due to the return of oversized particles to the mill from the classifier.

In this type of process, if the mill load is relatively small, grinding will yield excellent results but will not be very efficient economically, since the energy used to rotate the mill yields low production. On the other hand, if the load is excessive, the mill could become obstructed, with subsequent losses in both production and energy. From this we can deduce that an optimum load exists for which grinding quality is satisfactory and production is maximized. Since a large part of the energy is used in moving the mill, the production maximization point corresponds with the minimization of energy consumption per ton ground.

The objective of good mill control is to control the process so that load degree is maintained at the optimum point and, as a result, production and energy efficiency is optimized. Nevertheless, two serious problems complicate such an objective. The first is derived from the oscillation trend of the load, which is inherent in the structure and dynamics of the grinding process. The second problem is related to the absence of a direct and reliable measure of the load degree. The load value is usually obtained by measuring sound levels from the mill with a special microphone, usually called a folaphone. However, this measure can change significantly with equivalent loads, depending on material quality, humidity and other circumstances.

If, in attempting to improve the mill's efficiency, the operator increases the feed to approach the optimum load, the oscillations characterizing this variable may produce an obstruction, which turns attempts of improvement into considerable losses. Consequently, the operator tends to manipulate the feed cautiously or simply to leave it at a sufficiently conservative level to avoid risking an obstruction caused by the oscillations. Clearly, this guarantees security and continuity of mill operation but is far from obtaining the highest possible efficiency.

10.4.3 APCS optimization

In this application, APCS, through the control of the folaphone signal, can eliminate the oscillations of the mill load degree, thus stabilizing this variable at its assigned setpoint. This result is achieved by manipulating the feed of the kiln.

Once the load degree is stabilized, the APCS optimization system will proceed to search for its optimum level, varying the setpoint assigned to the folaphone signal 'intelligently'. The action of the master system is required to execute this. It monitors all the measured process variables and, in particular, the critical ones such as the mill motor power and the elevator power. The master system, in considering the grinding efficiency derived periodically from such measures, moves the folaphone signal setpoint accordingly and, as a result, moves the mill load degree towards the optimum point of efficiency. It is important to emphasize that, despite the presence of bias in the measurement of the folaphone signal, the master system will find the best load setpoint step-by-step. Additionally, the master system, on monitoring the global process operation, will prevent any preobstruction situation arising. Therefore, in addition to achieving the objective of optimum efficiency, continuity of the mill operation is guaranteed.

Chapter 12 discusses the above described APCS optimization of cement mills in detail. It presents the control strategy used, the optimization criteria chosen and the results obtained by means of plots that illustrate the evolution of the different variables during the milling process.

Chapter 11

APCS OPTIMIZATION OF A CEMENT KILN

11.1 INTRODUCTION

As explained in Chapter 10, the application of APCS optimization systems to processes of complex dynamics, such as a cement kiln, consists of two clearly defined stages. In the first stage, the objective is to control the critical variables of the process with precision and, as a result, to stabilize its operation. In the second stage, these critical variables are driven towards their optimum points of operation, where the process continues to function in a stable and permanent manner.

This chapter describes the generic application of an APCS optimization system to a cement kiln, where the experimental results presented belong to the plants of HISALBA-Gador and Cementos del Cantábrico-Aboño (Spain). First, it considers the optimization criteria, centred basically on the minimization of energy consumption and, once these are obtained, on the maximization of production. Subsequently, it describes the general configuration of the system, the adaptive predictive control loops and the operation of the master system. Likewise, it presents and analyzes the results obtained in the stabilization of the kiln as well as in the application of the optimization strategies.

11.2 OPTIMIZATION CRITERIA

In general, the kiln operates at a higher thermic level than is strictly required for clinker production. Even though more energy under consumed under these conditions, greater stability is ensured in the operation, thus making the kiln's manual control easier for the operator. Nevertheless, once kiln stability is ensured by the operation of the APCS system, there is no reason to maintain such a high thermic level. Consequently, the energy optimization of the kiln will involve a gradual decrease in its thermic level from the usual operation value to the minimum required for production of clinker of the desired quality. This optimization will conclude with the stabilization of the kiln's operation at the required thermic level. Significant energy savings and an increase in clinker quality due to the regular functioning of the kiln will result from such an optimized operation.

A second criterion of energy optimization concerns oxygen excess in the gases exiting the kiln. Once this variable is stabilized, it should be driven to a minimum level that is compatible with an admissible concentration of CO. In this manner, the calories carried by the gases exiting the kiln will also be minimized.

Once the energy optimization is achieved, the objective will be to maximize production. In fact, under manual control, the stability of the critical variables is frequently precarious and a certain level of oscillation cannot usually be avoided. In these circumstances, an increase in production may carry the risk of destabilizing the process and, therefore, attempts to go further than the established manual production limits will not be permitted. Nevertheless, once energy stabilization and optimization is achieved, the same production requires lower energy consumption and, consequently, the process has an extra energy capacity. On the other hand, the process could absorb a reasonable increase in production since its stability will be maintained by the APCS system. In order to achieve the goal of maximizing production, we proceed to increase the feed of the kiln gradually as long as stability is preserved and until the system detects its physical limits of operation, which make further increases in feed not admissible.

11.3 APPLICATION OF THE APCS OPTIMIZATION SYSTEM

11.3.1 Configuration of inputs and outputs

In a typical application to a kiln with a grate cooler, the APCS optimization system receives the following signals:

- velocity of the kiln's rotation;
- crude feed rate;
- coal feed into the main burner;
- nitrogen oxide (NO_x) concentration in the exit gases;
- oxygen (O_2) concentration at the kiln entrance;
- pressure beneath the first grate of the cooler;
- pressure beneath the second grate of the cooler;
- flow of the five ventilators of the cooler;

- concentration of carbon monoxide in the exit gases;
- depression at the kiln's entrance;
- pressure at the kiln's exit;
- torque of the kiln;
- gas temperature in the main ventilator;
- pyrometer in the sinterization area;
- lime content in the clinker.

It is clear that a complex and changing interrelation exists between all these variables. Nevertheless, experience shows that it is possible to control the kiln appropriately through the manipulation of a reduced number of variables, namely:

- velocity of the kiln's rotation;
- crude feed;
- coal feed into the main burner;
- draught of the main ventilator;
- velocity of the two grates of the cooler;
- aspiration of the cooler's ventilators.

In the application of the APCS optimization system, the analog and digital logic system played an auxiliary role, while the adaptive predictive control system as well as the master system played the relevant roles described in the following subsection.

11.3.2 Adaptive predictive control

An important factor in achieving the objectives of the stabilization stage is that efficient control of a key variable eliminates its negative influence in the control of other variables. For instance, if the control on the draught aims to maintain the concentration of O_2 close to the desired setpoint, the appearance of high levels of CO and their repercussion over other variables is generally eliminated. Likewise, if the pressure under the grate is maintained constant, the interaction of its variations on other variables ceases to deteriorate their control.

Stabilization of the kiln was achieved by the application of adaptive predictive control to the following control loops:

Crude feed: The feed of crude to the kiln is measured by two flowmeters that send their signals to the adaptive predictive control system so that it can generate a control signal acting on the potentiometer which determines dosage.

Velocity of the kiln's rotation: The measure of velocity of the kiln's rotation is used by the system to work through a potentiometer on the regulator of thyristors of the kiln's motor. The setpoint of velocity is fixed manually or by the master system.

Thermic level: The concentration of NO_x is taken to be a reliable signal (under certain limits that are analyzed in the following section) in order to estimate the temperature variations in the sinterization area. Another input is considered in this loop: the rotation velocity, which is assumed to be a measurable disturbance and is synchronized with the variable of crude feed. As a matter of fact, a decrease in the rotation velocity is accompanied by a decrease in the crude feed and a longer period of residency of the material in the kiln. When this occurs, at the same flow rate of combustible, the temperature increases. The control action of the loop is the coal flow rate to the main burner (the precalcination is maintained constant). The varying time delay between the control action and its repercussion upon the temperature presents a control difficulty that challenges the stability and robustness of the control system.

Draught: The process signal used for the regulation of the draught is the concentration of O_2 at the kiln's entrance. This choice allows us to eliminate the influence of false air that is present in the preheater and in the gases circuit on the measured variable. The coal feed into the main burner, that is, the output of the preceding loop is considered as a measurable disturbance due to the fact that any variation in the coal flow rate will modify the draught necessary to maintain the optimum level of O_2. The control action on the draught will be carried out by modifying the opening of the main ventilator gate or the ventilator speed where possible.

Pressure beneath the grates of the cooler: The precise control of this pressure is important for the regular and stable operation of the kiln, given its influence on the flow and temperature of the secondary air. The average pressure beneath the first grate is measured and used by the system to manipulate the velocity of the motor which moves its tiles. The speed of the second grate is dependent on the velocity of the first grate, according to a fixed relation.

Flow of the cooler's ventilators: With the same objective of reducing energy consumption and stabilizing the kiln's operation, as well as regulating the pressure, it is convenient to control the flow of air passing through the grates in order to use only the necessary air to cool the clinker and maintain the required pressures.

11.3.3 Master system

The master system is composed of a set of so-called master programs. Each of these programs includes instruction sequences to be executed when certain conditions are verified. No programming language is needed for their creation, since they are configured to obey the basic logic of condition–action. Nevertheless, the function of the master programs should not be confused with the alarms and securities of the adaptive predictive control system which, for instance, operate when the absolute or incremental limits of a variable are reached.

In this application, the difference between the master programs related to the stabilization of the kiln and the programs related to its optimization must be distinguished. Figure 11.1 represents the interaction between the master programs and the different loops of the adaptive predictive control system.

The objective of the master programs related to the stabilization is the reaction in the face of special situations of the kiln or confusing tendencies of a variable. The following example will explain this reaction:

- The NO_x signal is unreliable in some circumstances: reduction conditions in the sinterization area (detected by an increase in CO) or an excess of dust (detected by a decrease in the pyrometer to the bottom of the scale) reduce the reliability of this variable. These conditions provoke a sensible reduction in the measurement of NO_x without altering the thermic level of the kiln. In this situation, the control loop would increase the coal flow rate unnecessarily. The opportune master programs are in charge of detecting the situation and changing the mode of the loop's operation.

The master programs related to the optimization generate smooth ramps, driving the setpoints of the different loops to their optimum values. Likewise, they monitor the different variables of the process in order to detect immediately the physical limits which may prevent to continue with said ramps. The master system generates descending ramps for the NO_x and the draught loop setpoints, and ascending ramps for the feed setpoint.

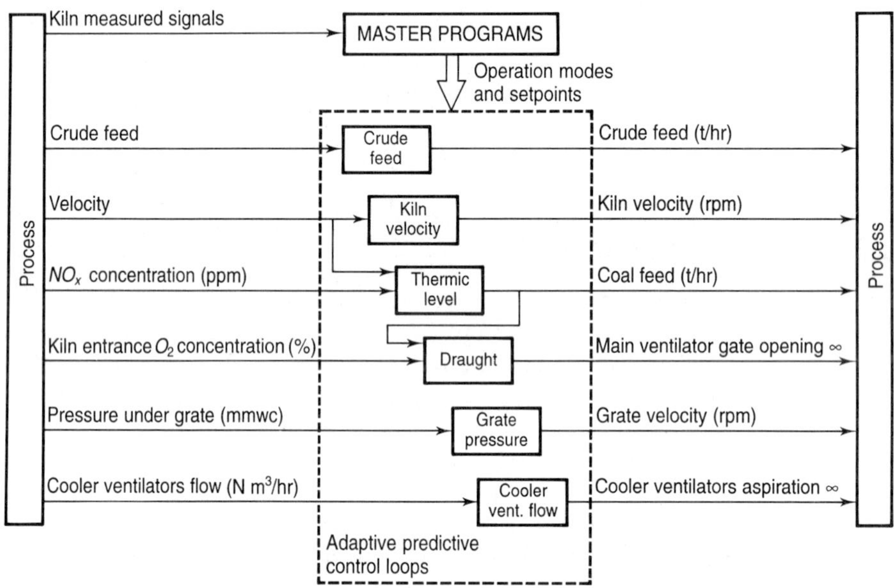

Fig. 11.1. General configuration of the APCS optimization system.

11.4 RESULTS

In the results obtained, we may distinguish those which purely concern the stabilization of the kiln from those which concern the optimization strategy. In all the critical variables of the kiln that will be considered in the following, previous PID automation attempts had always been disappointing.

11.4.1 Kiln stabilization

Cooler

The upper plot of Figure 11.2 represents the time evolution of the measured pressure (in mm of water column) under the first grate of the cooler. The lower plot represents the corresponding control action on the grate motor velocity. The part of the diagram to the left of the appearance of the straight line in the upper plot (setpoint line) corresponds to the manual operation in which the pressure is maintained in a wide band between 400 and 600 mm. It is clear that the operator cannot be continuously correcting the frequent variations of pressure caused by the irregular unloading of the kiln. Therefore, as long as there is no clear deviation from that band, the operator will not modify the grate velocity.

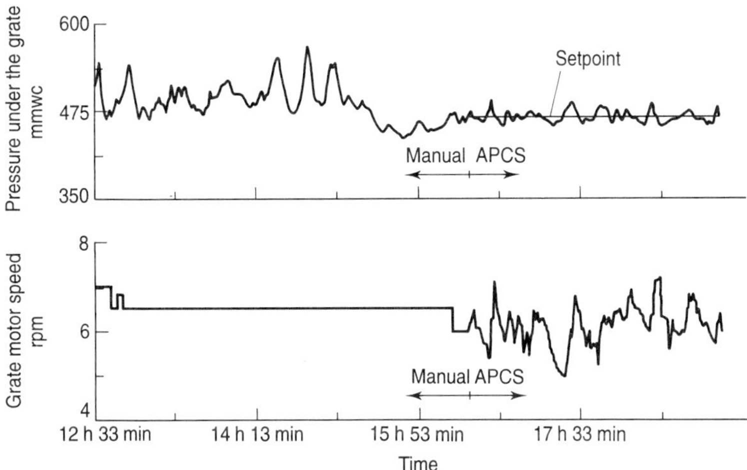

Fig. 11.2. Control of the pressure under the grate.

The appearance of the setpoint line marks the beginning of adaptive predictive control, characterized by a drastic descent of the pressure oscillations around its setpoint: the band is reduced to ±30 mm around the setpoint of 475 mm. To achieve this precision, it is necessary to vary the velocity of the motor that moves the tiles, as shown in the lower plot.

Figure 11.3 shows the results of Figure 11.2 between times 16:40:08 and 19:20:08 in more detail. We can see how different velocity values are necessary to maintain a constant pressure. A broader time scale, as shown in Figure 11.4, permits appreciation of the variations of pressure as well as the requirement of less abrupt velocity changes and the fine control performance achieved.

Figure 11.5 shows the response of the system in the face of a great disturbance such as an avalanche of clinker which unexpectedly fills the grate and makes the pressure rise. The system increases the grate velocity to moderate and then stop this pressure rise, and then drive it through a desired trajectory to the primitive setpoint value without oscillations. The complete pressure stabilization is achieved after approximately twelve minutes, while in manual operation an avalanche of this importance would lead to much longer instabilities.

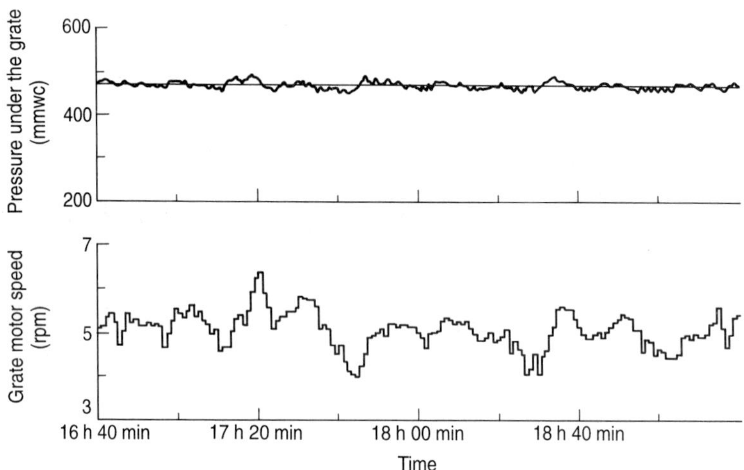

Fig. 11.3. Control of the pressure under the grate.

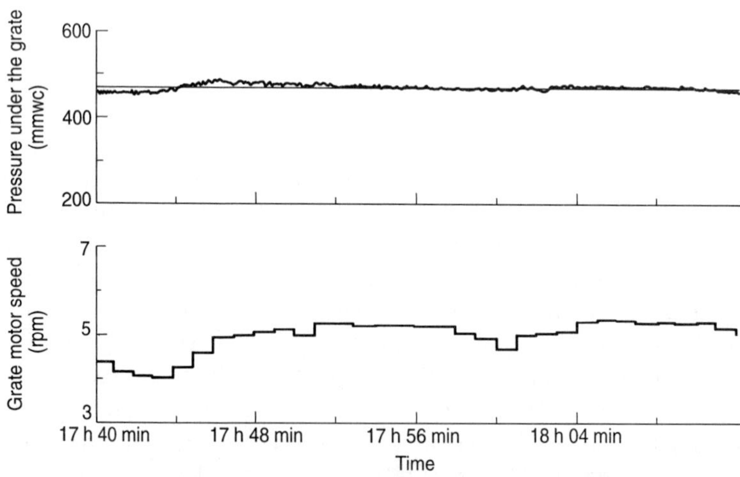

Fig. 11.4. Control of the pressure under the grate.

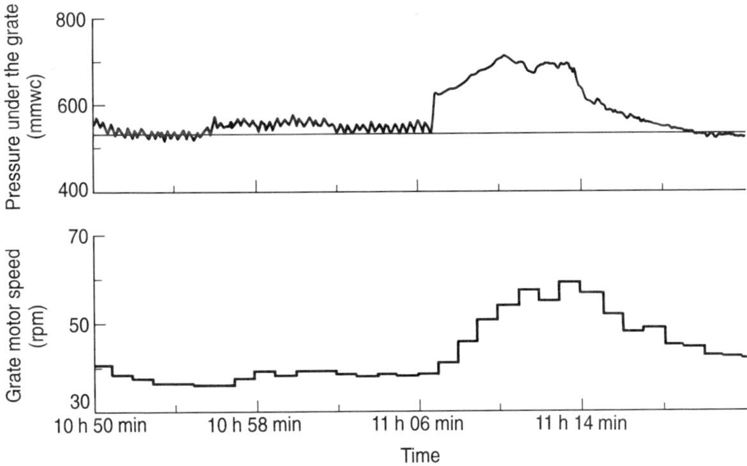

Fig. 11.5. Control of the pressure under the grate.

Kiln operation

In this section the operation of the kiln throughout a shift (eight hours) is analyzed. According to the usual control strategy, the velocity of the kiln will be at its highest value and the thermal level, if there are no short-term cooling tendencies, will be controlled by the flow rate of coal. The first plot of Figure 11.6 shows how the thermic level of the sinterization area, represented by the NO_x signal, is maintained close to the setpoint (straight line). The second plot represents the coal flow (in metric tonnes per hour) during this shift, which increases in order to correct the decreasing tendencies of the NO_x, finishing at 11.9 ton/hr after having started at 11.6 ton/hr. This increase is related to typical changes in the process.

Precise control of the draught during this shift can be seen in the third and fourth plots of Figure 11.6. While the third plot represents the evolution of the oxygen concentration at the entrance of the kiln, the fourth plot shows the corresponding control action on the main ventilator gate.

The evolution during the shift of other important variables of the kiln was also correct. In the first plot of Figure 11.7 it can be observed how the CO is maintained at low levels and without spikes; the second plot of this figure represents the evolution of the current consumed by the kiln's motor (proportional to the torque). The band of the signal is due to the variation of the torque in each turn of the kiln, since the distribution of the adhered material is not homogeneous.

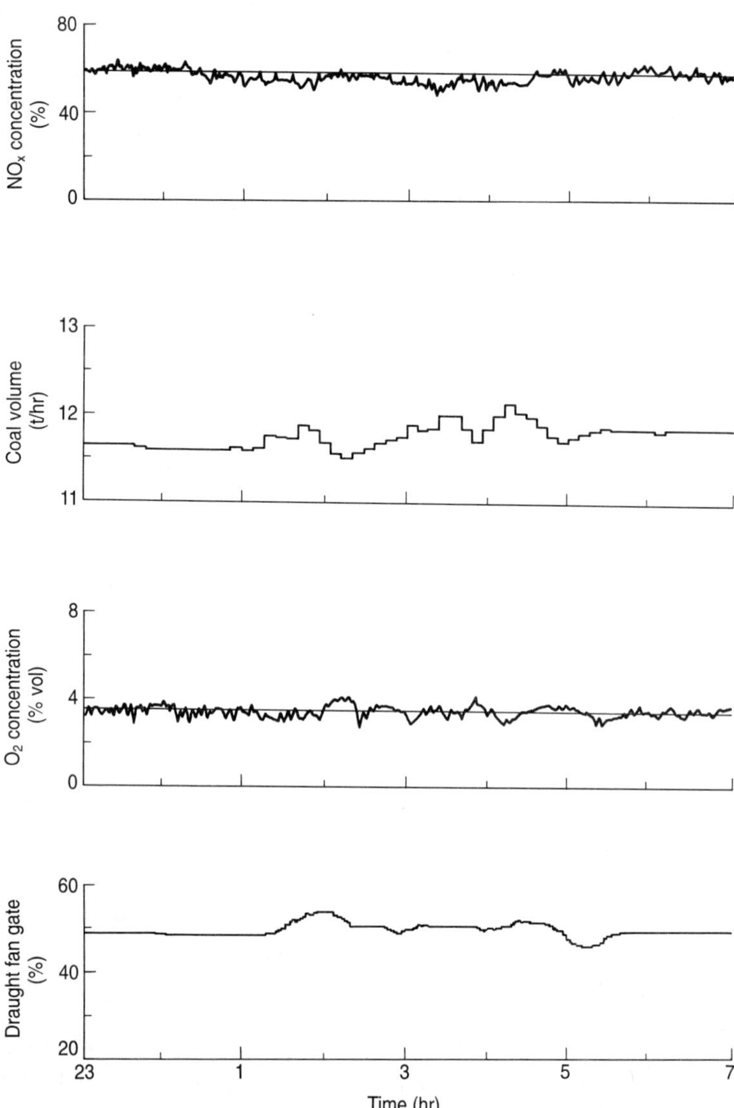

Fig. 11.6. Control of the kiln thermic level.

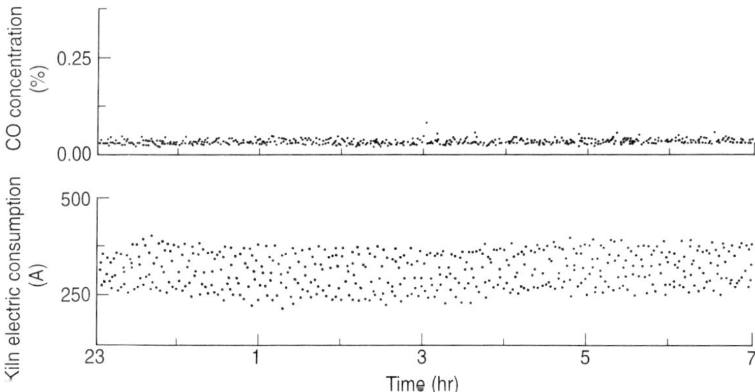

Fig. 11.7. Control of the kiln thermic level.

Correction of a precooling situation

A cooling tendency in the kiln may be produced by different circumstances, such as a fall of crust from the kiln walls, changes in the quality of coal and/or crude and other non-measurable variations in the process operation. This type of situation is not correctable by simply increasing the combustible flow. This action has a delayed effect and, therefore, is not capable of correcting a short-term tendency. In these conditions the master system changes the control strategy through reduction of the velocity of the kiln and the crude feed, both reductions having a much faster heating effect.

Figure 11.8 reflects an example of this situation. The first plot shows the NO_x concentration (in ppm) and its setpoint between 20 and 24 hr. A decreasing tendency of the NO_x, becoming more pronounced after 21 hr, is observed. The second plot shows the coal flow (in ton/hr) changes over the same period of time.

The control loop, as shown in this second plot, reacts in the first moment by adding more coal. Since the decreasing tendency of the NO_x becomes more pronounced, and once confirmed by the torque of the kiln, the master system, as previously mentioned, decides to reduce the velocity and the crude feed at 21:35 hr, as illustrated in the third and fourth plots of Figure 11.8.

This control strategy is maintained for approximately one hour. In the second plot of Figure 11.8 we observe how, after the descent in velocity, the coal flow decreases again. The multivariable correction stops when the NO_x is close to the setpoint, leaving the final stabilization up to the coal.

Figure 11.9 shows the interesting evolution of the pressure beneath the first

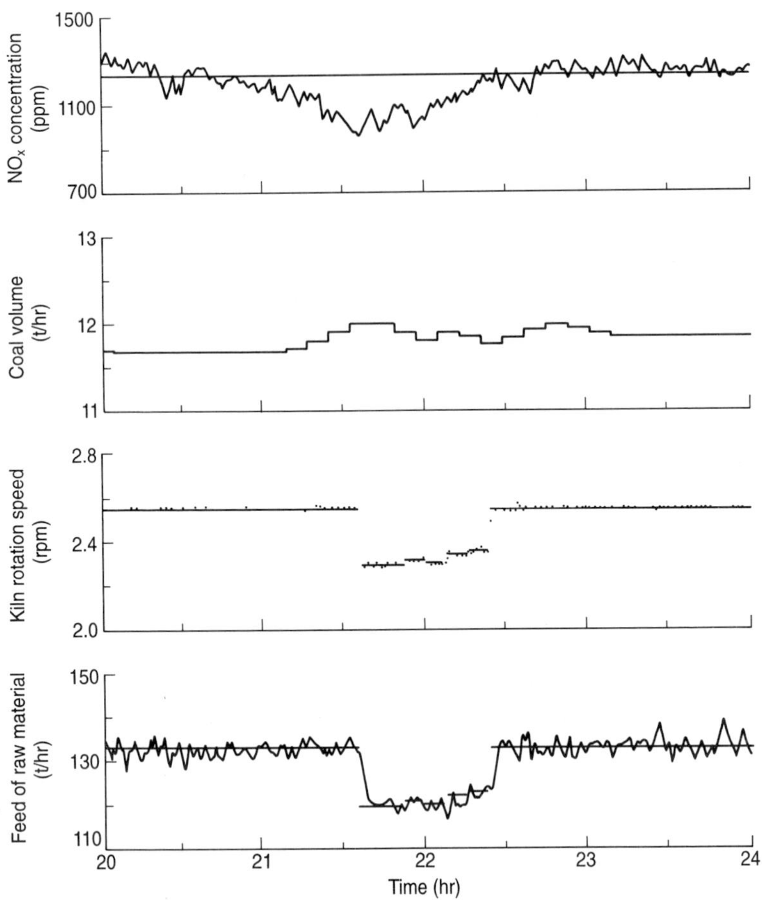

Fig. 11.8. Control of the kiln thermic level.

grate during the cooling tendency. The beginning of this situation, illustrated in the first plot of Figure 11.8, provokes a decreasing tendency in the pressure, caused by the smaller granulation of the clinker produced. This tendency, illustrated in the upper plot, is compensated for by the decrease of the grate velocity, as shown in the lower plot. The pressure tendency becomes decreasing again after 21:35, when the velocity of the kiln is reduced (see the third plot of Figure 11.8) due to the descent of the clinker into the grate. The inverse effect is observed when the kiln recovers its former thermic level and its velocity returns to its normal operation value.

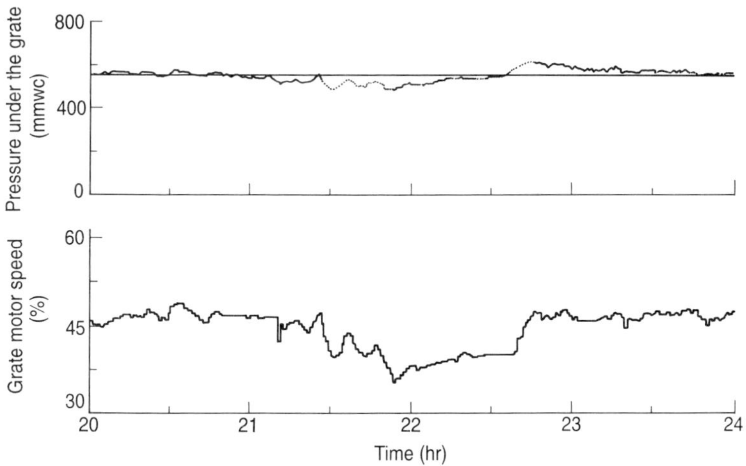

Fig. 11.9. Control of the pressure under the grate.

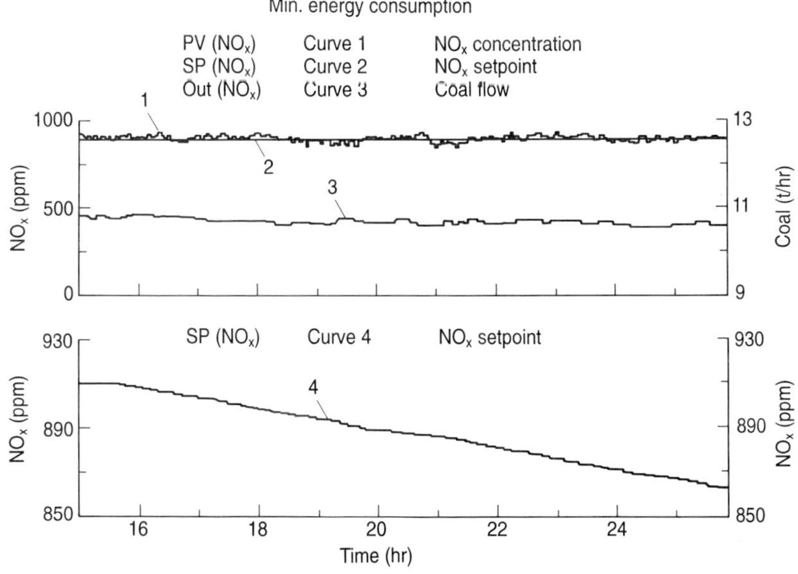

Fig. 11.10. Minimization of energy consumption.

11.4.2 Optimization strategy

The first plot of Figure 11.10 shows in its upper part the evolutions of the NO_x level and its setpoint in the presence of an optimization master program, which decreases the loop setpoint gradually from 910 ppm to 865 ppm in ten hours. The control actions on the coal flow can be seen under the curves shown in the first plot, the scale of which, in tons per hour, is given in the right-hand side of the plot. The NO_x evolution and the corresponding coal flow variations occur under a practically constant crude feed flow. The second plot of Figure 11.10 shows, on a different scale, the details of the NO_x setpoint ramp. The descent in the coal flow rate can be clearly observed, which obviously results in corresponding energy savings.

Figure 11.11 presents the application of the optimization strategy to the minimization of the oxygen level in the exit gases. The first plot of this diagram shows the evolutions of the oxygen concentration and its setpoint, while the second plot displays the corresponding control actions on the gate opening and the carbon monoxide level, the scale of which is given on the right-hand side of the plot. The setpoint changes to drive the oxygen level to its optimum value are fixed by the master program of draught optimization. The results presented in this diagram refer to the operation of the kiln of Cementos del Cantábrico in Aboño.

The optimization strategy presented in Figure 11.11 corresponds to the following logic: the O_2 concentration setpoint will gradually decrease while the average measure of the CO concentration remains lower than a certain limit and no sustained peaks are produced, in which case the O_2 setpoint would increase again. However, if the CO level oversteps the limit previously considered but does not achieve a second level of danger, the O_2 setpoint is maintained constant. The application of this optimization strategy, as shown in Figure 11.11 has led the oxygen concentration to decrease from a level close to 3% to a level close to 2% under an average constant coal flow in an interval of sixteen hours of operation.

Figure 11.12 presents the optimization strategy for the maximization of production. The first of the plots in this diagram shows, in its upper part, how the crude feed setpoint is conducted by a master program through a ramp from its initial value of 137 ton/hr to a value of approximately 148 ton/hr. This increase is produced in approximately forty-eight hours. This first plot presents, in its lower part, the evolution of the coal flow. The second of the plots in Figure 11.12 shows the evolutions of the NO_x level and its setpoint, which is maintained constant during this time period. The increase in the crude feed flow is produced at a constant velocity of the kiln. The results of this figure refer to the kiln of HISALBA in Gador.

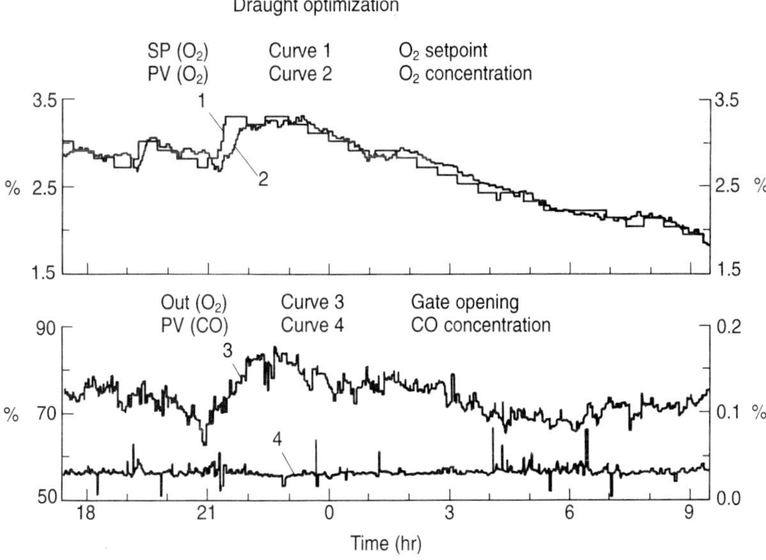

Fig. 11.11. Minimization of energy consumption.

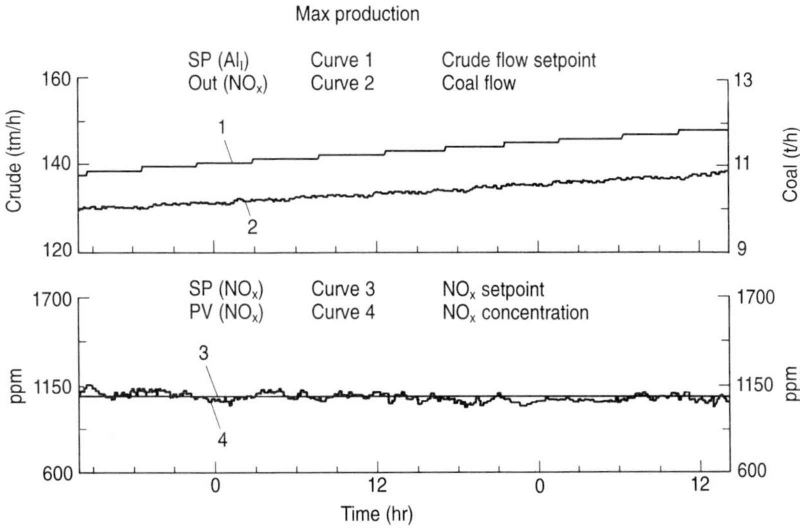

Fig. 11.12. Maximization of production.

11.5 CONCLUSIONS

The application of the APCS optimization system to a cement kiln pursues, in the first stage, the stabilization of the different variables of the process and, in the second stage, the application of the optimization strategy.

The results obtained, over more than four years of continuous use, have been fully satisfactory in all the cases. In this relatively short period of time, sixteen kilns are already operating with the APCS optimization systems. The optimization strategy drives the kiln to an operation point at the border of the physical limits of the process operation. As an example, the benefits reported after four years of operation of the HISALBA–Gador kiln may be summarized: 4.7% reduction in average calorific consumption, 10.9% increase in production and 30% reduction in refractory consumption. Another benefit to be considered, derived from the stability achieved in the kiln's operation, is the improvement in the availability of maintenance personnel.

Chapter 12

APCS OPTIMIZATION OF MILLING PROCESSES

12.1 INTRODUCTION

Within the cement production industry, the milling processes present important potential savings. Around 75% of the total electrical energy consumption of cement production is absorbed by the milling of raw materials and cement.

The rotating mills are machines of high consumption and low efficiency. Less than one-tenth of the electrical energy provided is used in breaking down the materials. Consequently, more than 90% of the energy consumed is wasted during this process and is basically dissipated in the form of heat, noise or vibration.

An increase in the production of the mill involves an increase in the energy necessary for the breakdown, which, as we have mentioned, is a small fraction of the total energy consumed by the motor. Furthermore, the total energy absorbed by the mill decreases as its filling degree is increased. This is due to the decrease in the torque of the mill caused by the reduction in the distance between the load gravity centre and the spin axis. Therefore, any increase in production represents a more than proportional reduction in the specific consumption (energy consumed per ton of ground material). The importance of maximizing mill production, with its corresponding energy savings, is evident. Nevertheless, the achievement of this objective is not simple in practice; among other things, due to the non-existence of a reliable direct measure of the mill load, to the time delays and to the variations of the process dynamics by external non-measurable factors.

In fact, in the manual regulation of the mill, the operator tends to feed very cautiously and to maintain the production at a sufficiently conservative level in order to prevent any obstruction. Such an obstruction could necessitate that the mill be stopped in order to solve the problem, leading to important increases in production costs, especially if the milling has to be executed during peak hours. The operator's behaviour guarantees security and continuity of mill operation but it is not an example of maximum efficiency. This is why the introduction of control and optimization systems that are capable of maximizing mill production without causing an overload has been in great demand. Nevertheless, attempts made to control these processes using traditional methodologies have generally failed until now.

The objective of optimization is to reach maximum mill production while maintaining the quality of the final product and the complete safety of the installation. To achieve this objective, the strategy followed by the APCS optimization system, as described in the following sections, is designed on two levels, each one centred upon a specific goal:

- fine and precise control of the mill load for a fixed production;

- maximization of the mill production.

Inadequate control and oscillations of the mill load are intolerable in proximity to the optimal point, since the oscillations of this variable could cause obstruction of the mill. For this reason, the attainment of the first objective is a priority and is decisive in order to approach the production maximum without risking overloading the mill.

In the following sections, this chapter describes the control and optimization strategies used by the APCS optimization system in order to maximize mill production. Subsequently, it shows the results obtained as a result of the application.

12.2 CONTROL STRATEGY

As previously mentioned, in the application of the APCS optimization system, the first step towards process optimization is the achievement of precise control and stabilization of the mill load. A secure approach to the production maximum, without risking overload, is not possible until this step is complete. One of the main difficulties involved in this process control problem is the low reliability of the direct signals measuring the mill load. An electronic microphone is generally situated opposite the camera in order to obtain a more pronounced sound gradient. It is assumed that the sound measured from the mill is inversely proportional to its load. This signal is not very reliable since it is highly influenced by the type of material ground (humidity, hardness and others), as well as by the wear degree on the balls by the wind and the accumulation of dust in the microphone.

Another signal that is traditionally used in the control strategy is the power consumed by the elevator of ground material. This magnitude is proportional to the mill load (outside the overload zone). This signal presents considerable alterations due to the different disturbances of the elevation process. This drawback is easily solved by the application of an adequate filter. More important is the time delay presented by this variable; a problem that is always a challenge for a control system.

The strategy used to control the mill load is described in Figure 12.1. A master loop controls the elevator power by changing the setpoint of a slave loop. This slave loop controls the filling degree, measured with the microphone, through actuation on the feed of fresh material.

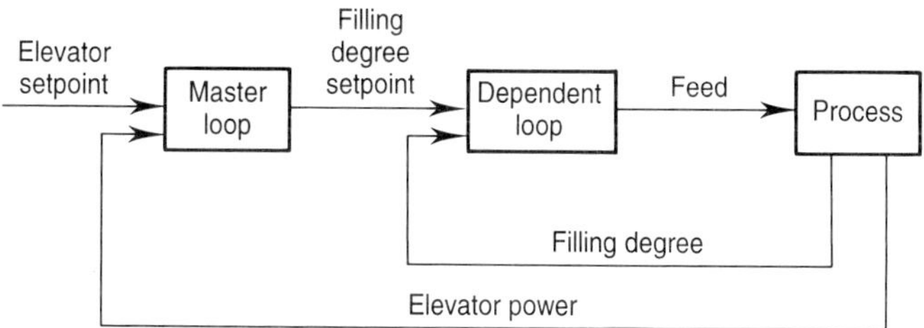

Fig. 12.1. APCS regulation scheme.

This strategy has not produced the desired results using conventional control methods. However, the adaptive predictive methodology, using such a strategy, has achieved clearly satisfactory results, as is shown in Section 12.4.

12.3 MAXIMIZATION OF PRODUCTION

When the optimum point for the mill load is surpassed, the milling performance decreases. This is due to a loss in the breaking capacity of the falling balls. Therefore, the proportion of ground material recirculated by the separator is increased, decreasing production and increasing the recirculation flow. In order to balance this increase in recirculation, the slave loop (filling degree) decreases the feed of fresh material to the mill. Further progression towards obstruction of the mill will lead to a decrease in the power consumed by the drive motor. As mentioned, this is caused by the decrease of the turning torque of the mill. Thereafter, the material output of the mill (elevator power) will fall and, if the situation continues, an obstruction will eventually occur.

In order to maximize production, a master program will search and maintain the mill's load at its optimum point, thus maximizing production. This program examines the long- and mid-term time trends of the fresh material feed and the power consumption of the mill motor continuously. Figure 12.2 represents the scheme of the APCS optimization described.

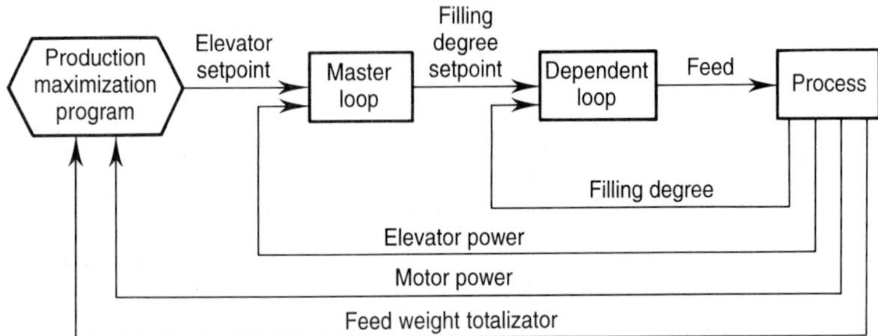

Fig. 12.2. APCS optimization scheme.

In accordance with the following criteria, the program modifies the elevator power setpoint (master loop), driving and maintaining the mill's load at the optimum point, which maximizes its production:

- While the trend of the fresh feed is not decreasing, the elevator setpoint is increased.
- When the trend of the feed decreases, and is accompanied by a fall in the motor power (nearing obstruction), the setpoint is decreased.

12.4 RESULTS

A typical manual control of the mill is presented in Figure 12.3, while Figure 12.4 shows the control under the APCS optimization system.

Where indicated by the arrow in the lower plot of Figure 12.3, approximately at time 4:40, we may observe a typical preobstruction situation. The filling degree increases drastically, while the motor power has previously been decreasing. The operator reacts to this situation by decreasing the mill feed for approximately forty minutes. Once the filling degree returns to a normal situation, the operator tends to re-establish the feed.

In the case of Figure 12.4 (APCS control), fast and constant actions on the feed permit the maintenance of the filling degree close to the setpoint values, thus reducing the risk of overload. The setpoint of the filling degree is varied by the master loop, that is to say, by the elevator loop. The setpoint of this last

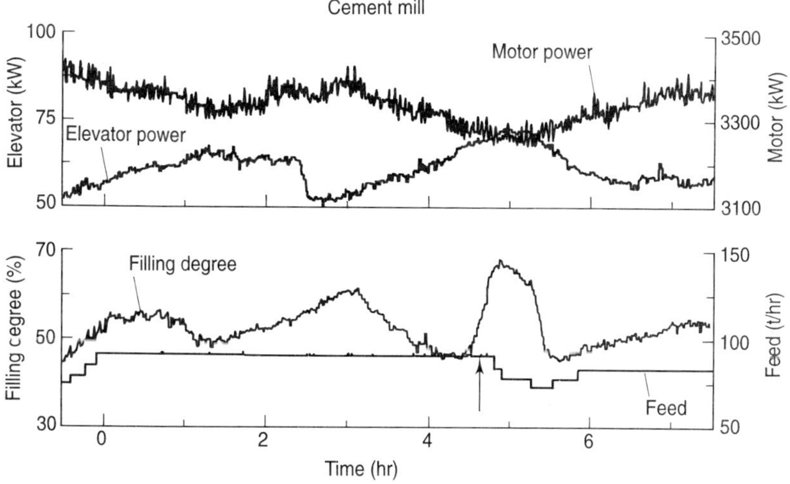

Fig. 12.3. Manual mill conduction.

always increases when the circumstances allow, according to the optimization strategy described in the preceding section. Where indicated by the arrow in the lower plot of Figure 12.4, approximately at time 4:20, we may observe a situation in which the system has automatically started to decrease the feed of fresh material, indicating an increase in the recirculated ground material as a result of a decrease in the milling efficiency, while the mill motor power was previously decreasing. As a result, the setpoint of the elevator power is decreased by the master program according to the criteria described in the preceding section. This kind of automatic operation of the APCS optimization system not only allows the mill to be driven towards its optimum operating points, but it can also, as shown clearly in this experiment, prevent an obstruction situation.

It is very important to note that, even in the presence of bias in the measurement of the filling degree by the microphone, this control strategy will be able to find automatically the appropriate levels for the setpoint of the filling degree, irrespective of the bias error since this variable is the control signal of the elevator power adaptive predictive control loop.

Comparing plots, the significant increase in production and elevator power achieved with the introduction of the APCS optimization system can clearly be seen. The decrease in motor power consumption in the optimized milling can also be seen. Figure 12.5 presents the fresh material flows for both cases and shows

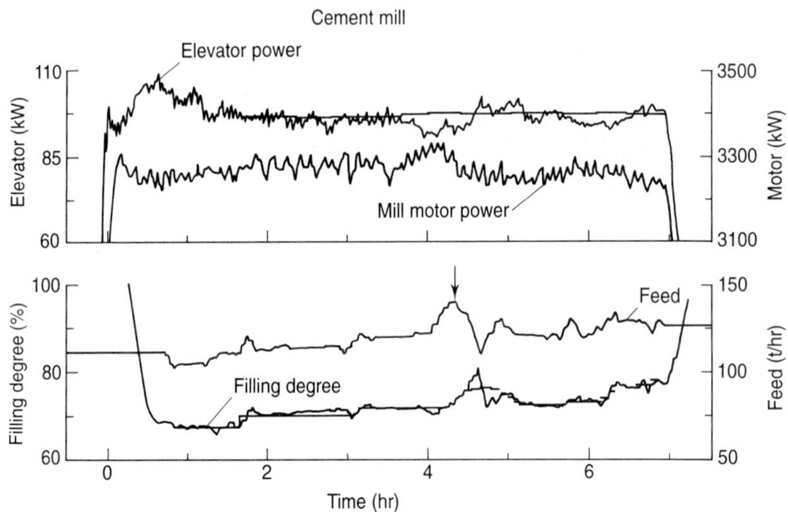

Fig. 12.4. APCS mill conduction.

a significant increase in fresh material fed in the optimized mill. The recovery in the event of an overload situation is much faster in the optimized case. In the latter case, production still increases, while the feed flow under manual control is decreased to avoid another preobstruction situation.

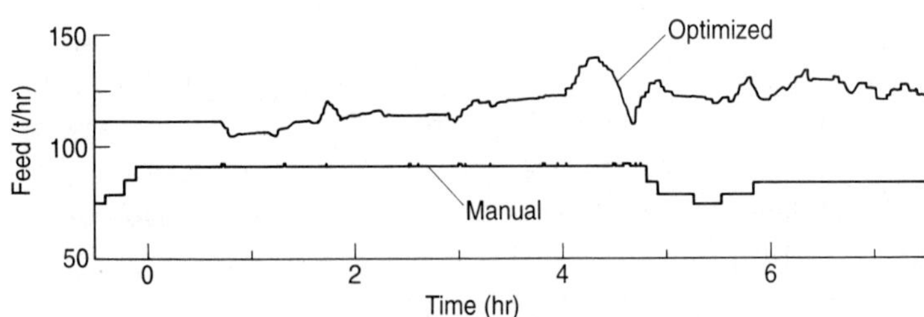

Fig. 12.5. Comparison between the manual and the APCS conduction of the mill.

It must be considered that the operator controls the mill in a rather conservative way, since (s)he does not know with precision when an obstruction situation may occur and (s)he tries to avoid it at all cost. It is very probable that the preobstruction situation considered in Figure 12.3, under manual control, would never have become a real obstruction even without decreasing the feed to the mill, since under APCS control the optimized operation has overcome, almost permanently, the filling degree that motivated the operator's reaction. However, it must be taken into account that the operator did not know which was the optimum operating point and, in this situation, a drastic increase in the filling degree, such as the one observed in Figure 12.3, fully justifies his or her behaviour.

It is clear that the APCS optimized operation increases the knowledge of the process. Therefore, in the same way that the lack of knowledge on the part of the operator of the optimal operating conditions clearly explains the spectacular increase in performance by the APCS optimization system, it is also true that when the operator observes the optimized operation of the mills, his or her own manual control performance increases significantly. Then the relative advantage of the optimized operation observed in the experimental results presented in this chapter is reduced.

Nevertheless, the improvements obtained in the application of the APCS optimization system to the milling process, alternating periods of automatic and manual operation, are summarized in the following points:

- Mean values of production increase and consumption decrease of 9.5% and 10.5%, respectively, with various types of clinker.
- Increase in the availability of the installation.
- APCS optimization system operation time is 86%.
- Risk of overload is reduced to a minimum.
- Self-tuning for different types of clinker.
- Simplicity in the process operation.
- Operators are released from control functions, which allows them to concentrate on supervision.

The consistency of these results, provided by the HISALBA plant in Gador is based upon nearly three years of operation of the APCS optimization system.

12.5 CONCLUSIONS

The application of the adaptive predictive control system (APCS) has allowed the optimization of the milling processes in an efficient and satisfactory manner. The following conclusions may be deduced from the results obtained, described in the preceding section:

- The APCS system is capable of optimizing cement mills by using a classic regulation strategy, generally abandoned due to deficiencies when applied with conventional control methodologies.

- The basis for this success lies in the methodology of adaptive predictive control and its ability to achieve and maintain fine and precise control of the variables of complex processes. This effective control allows one to reach mill loads and production values that would, with other controls, be impossible to reach without risking an obstruction.

- The solution obtained has a general character, allowing its direct application to all milling processes.

- Few signals are needed to attain optimization.

- The consumption per ton decreases, partly directly due to the increase in production, and also to the reduction in the motor power consumption, which is caused by the increase itself.

Finally, let us point out that the advantages of the optimization obtained by the use of adaptive predictive control in this application, as in many others in industry, not only rely in the higher 'ability' of the new control technology over human operators, but also in the specific knowledge of the process that it acquires, which is inaccessible (or unveiled) as long as the process remains under manual control.

Closure

THE STRENGTH OF THE CONCEPTS

The conclusions given in this book may also be considered as the conclusions of more than two decades of continuous research and development effort carried out by the authors and their collaborators. As pointed out in Chapter 1, these efforts have seen the birth of the concepts involved in adaptive predictive control and the mathematical methodology derived from them, the development of stability theory and the first applications to unit plants and industrial processes, the formulation of a new concept of optimization and the development of systems able to apply this concept systematically, and, finally, their industrial and commercial exploitation [GA90, TO90, DAM92, GBHU93, GCM94, MGK94, PPCC94, CBC95].

In parallel, and as indicated previously in this book, the research community has paid increasingly more attention to the basic concepts of predictive and adaptive predictive control. Many applications have been reported in academic, experimental and industrial contexts in a variety of areas [MM86, GL87, HS87, Cla88, LL89, PHAS89, RSS89, CAT91, CD91, DOD91, AFS92, LM92, BM93, CB93, GAB93, CDLE94, CKC94, DIP94, GR94, Has94, KS94, MBPB94, NFKH94, Ogo94, Ogu94, OHOT94, WCTG94, WMT94].

Most of these applications are based on a non-adaptive predictive control, in which the fixed-parameter predictive model must be obtained prior to the control application. The generic name *model based predictive control* has been popularized for this kind of approach. It is important to emphasize here that predictive control is only a subset of adaptive predictive control in which adaptation of the predictive model simply does not take place. In this case, the performance may be satisfactory as long as a certain linear time invariant process dynamic behaviour is maintained within the bounds allowed by the intrinsic robustness of predictive control. However, due to the time varying, non-linear nature of the industrial processes, this is not always the real case. Therefore, although local satisfactory results may be obtained, it necessarily lacks the generality and robustness of complete adaptive predictive control solutions.

The industrial implementations of adaptive predictive control already carried out have proved that it is an advanced solution able to answer satisfactorily the requirements described in the first chapter of this book and demanded by the nature of the industrial process and the optimization of its operation.

Adaptive predictive control is therefore the second methodology, after PID, that has been applied in a systematic and commercial way in the industrial field. In this sense, APCS may be considered as the successor of PID in a new era where digital computers have transformed many human activities and, particularly, the operation of industrial processes.

We would like to explain in the following how the success of APCS is essentially derived from the strength of its new concepts, from which the mathematical development is easily obtained, as we hope this book has shown.

As a matter of fact the classical PID control methodology was based on the *negative feedback principle*, according to which the control signal is generated from the difference between the setpoint and the measured process output, that is to say, *from the error already produced.* It was this basic principle that introduced the stability problem in the control scheme.

In an attempt to improve the performance of PID control, a whole class of control systems, named *model reference adaptive systems*, was developed before the introduction of predictive control. These systems attempted to adapt the parameters of the controller in order to maintain a desired control performance in the presence of changes in the process dynamic. However they remained attached to the negative feedback principle, and this was the limiting factor that, on introducing a problem of stability in the control scheme, frustrated their practical success. From the theoretical point of view, in spite of sophisticated mathematical treatment, the cause of the problem was seen as a set of 'very difficult' conditions to be fulfilled by the control scheme in order to guarantee stability. Obviously, these stability conditions were neither simple nor realistic and, consequently, were not practical. Thus when the new concept of predictive control was substituted for the traditional negative feedback principle, the inherent stability problem disappeared.

On the other hand, an alternative to the negative feedback principle had already been introduced in the 1960s by the so-called *optimal control theory.* This approach, which was first developed in continuous time, framed the control problem within a classical optimization context, requiring the solution of a complex Riccati equation. Most probably, it was this complexity which motivated the basic assumption that the process dynamics had to be known. As a consequence of this assumption, *identification* of the process dynamics became a main subject of research. However, due to the already mentioned nature of industrial processes, the problem of identification was indeed a very difficult problem to solve, even assuming a non-linear model. The problem became literally impossible to solve when trying to impose a linear structure for the model, which is required for the application of the control strategy. This is the main reason why there are virtually no reports of applications of optimal control in the industrial field.

When the concept of *identification* was substituted by that of *prediction with a time varying model* in the context of APCS, the problem changed from one that was impossible to solve to one that admitted a solution even with a reduced-order model. As a matter of fact, while identification requires the process and the model output sequences to match for any input sequence over an arbitrarily long time span, the problem of prediction only requires the extrapolation of the future evolution of the output variables in a short horizon, based on the previously measured input and output process variables, and this may be achieved by a time varying reduced-order model with an appropriate adaptation mechanism.

As a very brief summary of the evolution of the concepts, we may say that the principle of *negative feedback* and the objective of the *process identification* have been substituted by the new principle of *predictive control* and the objective of the *process output prediction with a time varying model.* These two new concepts have overcome the essential difficulties introduced in the control schemes by the previous ones, and have complemented each other to achieve a possible-to-solvc, casy-to-implcmcnt, practicable and efficient solution to industrial control problems. The new concepts were required to obtain the desired solution and they could not be substituted by the use of advanced mathematical tools in the framework of the old concepts.

Briefly, the solutions to the problems lie in the strength of the new concepts, which in this case may go beyond the purely technical context considered in this book.

Appendices

Appendix A

SOME BASIC CONCEPTS OF SYSTEM ANALYSIS

In this appendix we summarize some basic concepts of systems used in the book. These concern the mathematical modelling of systems, both using input/output and state space representations, and some essential features, such as stability, which define their dynamic behaviour. When dealing with the problem of modelling, the usual approach is based on the invokation of the fundamental laws describing the specific phenomena involved in the system dynamics and then translating these laws into mathematical equations. By nature, principles such as those of physics, chemistry, biology and many other disciplines, describe continuous dynamic changes involving the derivatives of input and output variables. Thus, the process of modelling frequently results in models that are based on differential equations. Another approach is based on the direct derivation of discrete time equations, as has been discussed in Chapter 1, where we are interested in considering the dynamic relation between inputs and outputs at sampling time instants only. The introduction of digital computers has motivated the present interest in this second approach. In fact, the case of a system under a digital computer-based control system, which we consider in this book, is typically described in discrete time. In this context, it may be useful to give the basic concepts of systems for both continuous and discrete time settings and outline some procedures for obtaining discrete time representations starting from continuous time models.

The material summarized in this appendix is standard, and extensive presentations can be found in many well-known textbooks such as [Oga67, Oga70, Dor80, Kai80, Che84, AW84, Gop84, Gop88, FPW90, Kuo91, Kuo92, PH93].

A.1 INPUT/OUTPUT LINEAR MODELS

Input/output models are derived by selecting the variables representing the actions on the system (inputs) and the measurable responses (outputs) and then obtaining the mathematical relations between these variables. It is well known that *linear systems* can describe the dynamic behaviour of a large class of processes as long as their inputs and outputs are defined as deviations from steady state values. In this section we outline some properties and tools that are useful for illustrating systems that are described by linear differential and difference equations.

A.1.1 Differential equations and s transfer functions

The dynamic behaviour of a wide class of systems with a single output $y(t)$ and a single input $u(t)$ can be described by the differential equation

$$a_m \frac{d^m y}{dt^m} + \ldots + a_1 \frac{dy}{dt} + a_o y = b_n \frac{d^n u}{dt^n} + \ldots + b_1 \frac{du}{dt} + b_o u \quad \text{(A.1)}$$

An important tool in system analysis is the *Laplace transform*, which, for a given time function $f(t)$, is defined in the form

$$\mathcal{L}\{f(t)\} = F(s) = \int_0^{\infty} e^{-st} f(t) dt \quad \text{(A.2)}$$

This transformation converts a time function into a function of a complex variable s and, among others, it exhibits the following properties:

(a) $\mathcal{L}\{\alpha_1 f_1(t) + \alpha_2 f_2(t)\} = \alpha_1 F_1(s) + \alpha_2 F_2(s)$ (A.3)
for any real numbers α_1, α_2 and functions f_1, f_2;

(b) $\mathcal{L}\{\frac{d^n f}{dt^n}\} = s^n F(s) - s^{n-1} f(0) - s^{n-2} \frac{df(0)}{dt} - \ldots - \frac{d^{n-1} f(0)}{dt^{n-1}}$; (A.4)

(c) $\lim_{t \to \infty} f(t) = \lim_{s \to 0} sF(s)$ (A.5)
provided $sF(s)$ is not infinite for any complex value s with positive real part. This is the so-called *final value theorem*.

By taking the Laplace transform of equation (A.1) and using properties (A.3) and (A.4) with zero initial conditions, we obtain

$$G(s) = \frac{Y(s)}{U(s)} = \frac{b_o + b_1 s + \ldots + b_n s^n}{a_o + a_1 s + \ldots + a_m s^m} \quad \text{(A.6)}$$

which is the so-called *transfer function* of the system.

A.1.2 Difference equations and z transfer functions

Consider that the input/output dynamics of systems is described only at sampling instants $k = 0, 1, 2, \ldots$, in such a way that $t = kT$, T being the sampling period.

In this context, a wide class of systems can be described by a difference equation of the form

$$y(k) = a_1 y(k-1) + \ldots + a_m y(k-m) + b_o u(k) + b_1 u(k-1) + \ldots + b_n u(k-n) \quad (A.7)$$

An important tool with which to analyze this class of systems is the Z *transform*, which, for a discrete time sequence $f(k)$, with $k = 0, 1, 2, \ldots$, is defined in the form

$$\mathcal{Z}\{f(k)\} = F(z) = \sum_{k=0}^{\infty} f(k) z^{-k} \quad (A.8)$$

This transformation converts a discrete time sequence into a function of a complex variable z and, among others, it exhibits the following properties:

(a) $\mathcal{Z}\{\alpha_1 f_1(k) + \alpha_2 f_2(k)\} = \alpha_1 F_1(z) + \alpha_2 F_2(z)$ (A.9)
for any real numbers α_1, α_2 and sequences f_1, f_2;

(b) $\mathcal{Z}\{f(k-n)\} = z^{-n} F(z)$; (A.10)

(c) $\lim_{k \to \infty} f(k) = \lim_{z \to 1} (z-1) F(z)$ (A.11)
provided that $(z-1)F(z)$ is not infinite for any complex value z with modulus greater than one. This is the so-called *final value theorem*.

By taking the Z transform of equation (A.7) and using properties (A.9) and (A.10), we obtain

$$H(z) = \frac{Y(z)}{U(z)} = \frac{b_o + b_1 z^{-1} + \ldots + b_n z^{-n}}{1 - a_1 z^{-1} - \ldots - a_m z^{-m}} \quad (A.12)$$

which is the so-called z *transfer function* of the discrete time system.

A.1.3 Stability, poles and zeros

Both transfer functions $G(s)$ and $H(z)$ are independent of the particular inputs and outputs and define the dynamic features of the system equivalently to equations (A.1) and (A.7) respectively.

Stability is the most basic dynamic feature of a system. For linear systems such as those described above, stability is not dependent on the particular input. Thus we may consider equations (A.1) and (A.7) with $u(t) = 0$ and $u(k) = 0$ respectively and analyze the output for given initial conditions. Stability concerns the ability of this output to converge to zero, as stated by the following definitions:

- Systems (A.1) and (A.7) are said to be *stable* when the unforced outputs remain bounded for any set of given non-zero initial conditions.

- Systems (A.1) and (A.7) are said to be *globally asymptotically stable* when, for any set of given non-zero initial conditions, the unforced outputs verify, respectively,

$$\lim_{t \to \infty} y(t) = 0 \quad \text{and} \quad \lim_{k \to \infty} y(k) = 0$$

- Systems (A.1) and (A.7) are said to be *unstable* when the outputs grow unbounded for any set of given non-zero initial conditions.

The above transfer functions $G(s)$ and $H(z)$ are written as quotients of polynomials:

$$G(s) = \frac{N(s)}{D(s)}; \qquad H(z) = \frac{N'(z)}{D'(z)} \tag{A.13}$$

The roots of $N(s)$ and $N'(z)$ are called the *zeros* and the roots of $D(s)$ and $D'(z)$ are called the *poles* of systems (A.1) and (A.7) respectively.

Poles are crucial in characterizing the system's response, and define the following *stability criteria*:

- System (A.1) is stable if and only if all poles have non-positive real parts and every pole with a zero real part is not repeated.

- System (A.1) is globally asymptotically stable if and only if all poles have negative real parts.

- System (A.1) is unstable in any case that differs from those considered in the two points above.

- System (A.7) is stable if and only if all poles have modulus not greater than one and if every pole with modulus one is not repeated.

- System (A.7) is globally asymptotically stable if and only if all poles have modulus less than one.

- System (A.7) is unstable in any case that differs from those given in the two immediate points above.

Although the influence of the zeros on the system's response is not so crucial as that of the poles, there is one aspect that may be interesting to note. Let us consider the *inverse* of a system, that is, the process obtained by reversing the inputs and outputs from the initial process. For the inverse systems we may also consider the above stability definitions. In particular, systems with an *unstable inverse* are those for which the input, required to produce an arbitrarily given bounded output, may be unbounded. According to the above criteria, the system of (A.1) has an unstable inverse if it has at least one zero with positive real part, and the system of (A.7) has an unstable inverse if at least one zero has modulus greater than one.

By its nature, the class of systems with an unstable inverse poses specific and difficult problems for control systems attempting to drive the outputs along trajectories. These problems are identified in Chapter 2 for the case of predictive control and are solved in Chapter 3 on introducing the extended strategy.

A.1.4 Time response

Once stability is analyzed, one way to characterize the dynamic behaviour of a system further is to obtain the time history of the output produced by given test input signals. Typical test inputs are impulse and step signals.

For continuous time systems, an impulse is formally represented by the Dirac δ function, whose response $g(t)$ is the inverse Laplace transform of the transfer function $G(s)$. By knowing the *impulse response* $g(t)$ the output $y(t)$ for any input $u(t)$ can be expressed for zero initial conditions by the so-called *convolution integral*

$$y(t) = \int_0^t g(\tau)u(t-\tau)d\tau \tag{A.14}$$

For discrete time systems, an impulse is represented by a sequence whose values are 0 except for the initial instant where the value is 1. The *impulse response* $h(k)$ is the inverse Z transform of the transfer function $H(z)$. By knowing $h(k)$, the output sequence $y(k)$ to any input sequence $u(k)$ can be expressed by the so-called *convolution summation*

$$y(k) = \sum_{i=0}^{k} h(i)u(k-i) \tag{A.15}$$

When $y(t)$ is the output of an asymptotically stable system (A.1) to a step input, it converges to a constant steady state value. By using the fact that the Laplace transform of a step equal to 1 is s^{-1} and recalling the *final value theorem* (A.5), the corresponding steady state value can be computed in the form

$$\lim_{t\to\infty} y(t) = \lim_{s\to 0} G(s) \tag{A.16}$$

which defines the so-called *static gain* of the system (A.1).

In a similar vein, for an asymptotically stable discrete time system (A.7), the output for a step input sequence 1 whose Z transform is $(z-1)^{-1}$ satisfies the following convergence property according to the discrete time final value theorem (A.11):

$$\lim_{k\to\infty} y(k) = \lim_{z\to 1} H(z) \tag{A.17}$$

which defines the so-called *static gain* of the system (A.7).

A.1.5 Discretization of continuous time models

Here we briefly consider the following problem: given a transfer function $G(s)$ of a continuous time system, find the equivalent discrete time transfer function $H(z)$.

Different approaches to this problem are treated in the literature. One approach consists of starting from the differential equation and deriving a difference equation through a *numerical integration* procedure. Another approach, referred to as the *hold equivalence*, is based on the scheme given in Figure A.1 and is briefly outlined here.

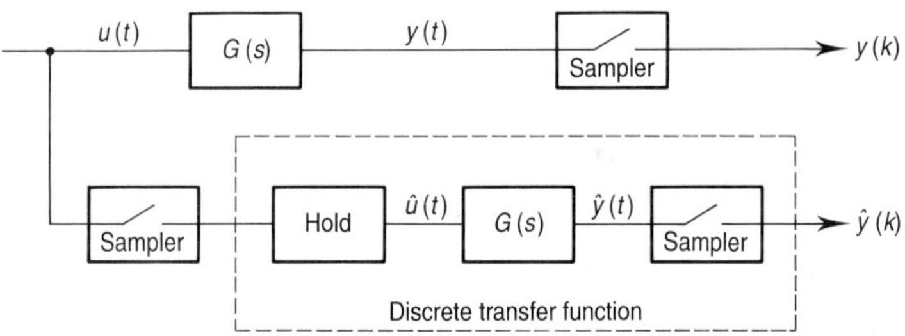

Fig. A.1.– Obtaining a discrete transfer function by hold equivalence.

The idea is as follows. Consider an input $u(t)$ producing an output $y(t)$ for the continuous system $G(s)$. Consider the sequence $y(k)$ consisting of the samples of $y(t)$. The objective is to find a discrete transfer function that, with an input sequence $u(k)$ which consists of samples of $u(t)$, has an output sequence $\hat{y}(k)$ that approximates $y(k)$. The discrete transfer function represents a discrete time system that contains the continuous $G(s)$ preceded by a hold and followed by a sampler. The hold has the role of constructing a continuous time signal $\hat{u}(t)$ from the discrete time sequence $u(k)$. This construction involves some kind of interpolation of the samples between sampling instants. Depending on the kind of interpolation, different forms of transfer function $H(z)$ can be derived.

The usual hold in digital control practice is the *zero-order* interpolation, which consists of producing a piecewise signal $\hat{u}(t)$ by keeping the values of $u(k)$ between consecutive sampling instants constant. In fact, the computer control loop considered as a basic prototype in Chapter 2 (schematized in Figure 2.1) includes D/A and A/D converters, whose operation is accurately described as a zero-order hold and a sampler respectively.

With a zero-order hold, the discrete transfer function is given by the following expression:

$$H(z) = (1 - z^{-1})\mathcal{Z}\{s^{-1}G(s)\} \tag{A.18}$$

In (A.18) $\mathcal{Z}\{s^{-1}G(s)\}$ denotes the operation of taking the Z transform to the sampled version of the inverse Laplace transform of $s^{-1}G(s)$. This operation can be tedious but is standardized in a number of software packages [Mat92].

It may be worth pointing out an interesting feature concerning the poles and the zeros of the discrete transfer function $H(z)$ obtained by discretizing the continuous one $G(s)$ as described above. It is known that the poles s_i of the continuous system are transformed into poles $z_i = \exp(s_i T)$ of the discrete system. This transformation maps the left half of the complex s-plane onto the disc of radius 1 in the complex z-plane. This implies that the stability properties of the discrete system are the same as those of the original continuous system. However, a simple transformation giving the zeros of the discrete system from those of the continuous one does not exist [AHS84]. A particularly significant consequence of this is that continuous systems with zeros in the left half plane, that is, with a stable inverse, may result in discrete time representations with zeros outside the unit disc, that is, with unstable inverses. This fact makes the existence of discrete time systems with unstable inverse unsurprising .

A.2 STATE SPACE MODELS

The models considered in the preceding section account for direct relations between inputs and outputs. These variables are external to the system and always have a physical meaning in terms of actions and responses. State models introduce a set of variables internal to the system whose evolution, described by first-order differential or difference equations, informs on the dynamic behaviour of the system induced by external inputs, and allows us to obtain the consequent outputs.

A.2.1 Concept of state and state equations

Consider a dynamic system described in the continuous time domain by a set of first-order differential equations and a set of algebraic output equations in the form

$$\dot{x}(t) = f[x(t), u(t), t] \tag{A.19}$$

$$y(t) = h[x(t), u(t), t] \tag{A.20}$$

where $x(t)$ is the *state vector* of dimension n, $u(t)$ is the r-dimensional *input vector* and $y(t)$ is the m-dimensional *output vector*, all of them at time instant t. f is a function defined to account for the system dynamics and having the mathematical properties required to ensure that a unique solution exists for equation (A.19). h is a function that uniquely determines the output.

The basic idea of the state vector is that it contains the minimum set of variables that completely determines the system's behaviour. In fact, if the state vector is known at any time t_o, then the state vector $x(t)$ and the output vector $y(t)$ are uniquely determined at any time $t > t_o$, provided the input vector u is known for all time between t_o and t.

In a similar vein, dynamic systems within a discrete time setting can be described by a state representation of the form

$$x(k+1) = f_d[x(k), u(k), k] \tag{A.21}$$

$$y(k) = h_d[x(h), u(k), k] \tag{A.22}$$

In general, the above functions f, h, f_d, h_d can be non-linear and involve time varying parameters. However, in many cases the dynamic behaviour of real processes around an equilibrium state (as defined in the following section) may be described by a linear model with constant parameters which, for continuous time systems, has the form

$$\dot{x}(t) = Fx(t) + Gu(t) \tag{A.23}$$

$$y(t) = Cx(t) + Du(t) \tag{A.24}$$

and, for discrete time systems,

$$x(k+1) = Ax(k) + Bu(k) \tag{A.25}$$

$$y(k) = Hx(k) + Lu(k) \tag{A.26}$$

A.2.2 Stability: definitions

Stability has been discussed previously for linear input/output models, being linked to the pole locations of the transfer functions. It is worth considering a wider perspective of stability for linear and non-linear systems in the state space setting. There are different approaches to studying the stability of a system, which usually split into two groups: (1) *equilibrium stability*, which deals with the maintenance of an equilibrium position when the system is free of external inputs, and (2) *external stability*, which deals with the boundedness of the response (state or output) when the system is forced by bounded outputs.

Let an unforced system be described by

$$\dot{x}(t) = f[x(t), t] \qquad\qquad x(t_o) = x_o \tag{A.27}$$

or, in discrete time,

$$x(k+1) = f_d[x(k), k] \qquad\qquad x(k_o) = x_o \tag{A.28}$$

x_e is said to be an *equilibrium state* of the system if, when x_e is the initial state of the system, it remains unchanged for any future time. For system (A.27), this means that

$$f[x_e, t] = 0 \qquad\qquad \text{for all} \quad t \geq t_o$$

and, for system (A.28),

$$x_e = f_d[x_e, k] \qquad\qquad \text{for all} \quad k \geq k_o$$

For the equilibrium state the following *stability definitions* are considered:

- An equilibrium state x_e (for system (A.27)) is said to be *stable* (in the sense of Lyapunov) if, for any t_o and any $\varepsilon > 0$, there exists a real number $\delta(\varepsilon, t_o)$ such that $\|x(t_o) - x_e\| < \delta$ implies that $\|x(t) - x_e\| < \varepsilon$ for all $t \geq t_o$.

- An equilibrium state x_e (for system (A.27)) is said to be *asymptotically stable* if it is stable and, if for any t_o, there exists $\delta(t_o) > 0$ such that $\|x(t_o) - x_e\| < \delta$ implies that $\|x(t) - x_e\| \to 0$ as $t \to \infty$.

- An equilibrium state is said to be *unstable* if it is neither stable nor asymptotically stable.

- An equilibrium state x_e (for system (A.27)) is said to be *globally asymptotically stable* at t_o if $x(t) \to x_e$ as $t \to \infty$ for any $x(t_o)$.

The practical meaning of Lyapunov stability is that the system state will remain close to equilibrium for all $t \geq t_o$ provided that the initial state is close enough to it. Asymptotic stability is stronger than Lyapunov stability, since it implies that, in addition to the equilibrium being stable, the state always converges to it when the initial condition is chosen sufficiently close. However, asymptotic stability has a local nature, since it requires the initial state to be sufficiently close to the equilibrium state, as defined by the value $\delta(t_o)$. When $\delta(t_o)$ is arbitrarily large, asymptotic stability becomes global. This means that it holds regardless of the initial state $x(t_o)$. In general, a system can have more than one equilibrium state. However, global asymptotic stability requires that there exist only one equilibrium state for the system. Thus, the concept of global asymptotic stability becomes a property inherent in the system.

Consider now the forced systems (A.19)–(A.20) and (A.21)–(A.22). The following *external stability definitions* can be stated:

- The system (A.19)–(A.20) is said to be *bounded input bounded state* stable if, for every bounded input $u(t)$ and for every initial state $x(t_o)$, the resulting state $x(t)$ is bounded.

- The system (A.19)–(A.20) is said to be *bounded input bounded output* stable if, for every bounded input $u(t)$, the resulting output $y(t)$ is bounded.

Similar stability definitions to those given above, considering k instead of t, can be stated for discrete time systems.

A.2.3 Stability: criteria for linear systems

Consider the linear systems

$$\dot{x}(t) = Fx(t) \qquad \text{or} \qquad x(k+1) = Ax(k)$$

The equilibrium states are the solutions of $0 = Fx$ or $x = Ax$. Thus, zero is always an equilibrium state for these linear systems and it is the only one unless F has a zero eigenvalue or A has a unity eigenvalue, in which case the systems have an infinity of equilibrium states. Consequently, zero is the only isolated equilibrium state for linear time invariant systems. For this reason, when dealing with linear time invariant systems we may refer to the *stability of the system* as synonymous with equilibrium stability. For this kind of system, stability criteria are drawn in terms of the eigenvalues of system matrices F and A respectively. The stability criteria for each case are the following:

- System (A.23) is stable if and only if all eigenvalues of F have non-positive real parts and if every eigenvalue of F with zero real part is not repeated.

- System (A.23) is globally asymptotically stable if and only if all eigenvalues of F have negative real parts.

- System (A.23) is unstable if F has any simple eigenvalue with real part or any repeated eigenvalue with zero or positive real part.

- System (A.25) is stable if and only if all simple eigenvalues of A have modulus ≤ 1 and all repeated eigenvalues have modulus < 1.

- System (A.25) is globally asymptotically stable if and only if all eigenvalues of A have modulus < 1.

- System (A.25) is unstable if A has any simple eigenvalue with modulus > 1 or any repeated eigenvalue with modulus ≥ 1.

- If the systems (A.23)–(A.24) and (A.25)–(A.26) are globally asymptotically stable, then they are bounded input bounded state and bounded output stable.

A.2.4 Discretization of continuous time state equation

Here we address the following problem: given a continuous time linear state equation of the form (A.23), obtain a discrete time state equation of the form (A.25).

By using the hold equivalence approach with a zero-order hold as in Section A.1.4, matrices A, B of (A.25) are obtained from matrices F, G of (A.22) by the following expressions:

$$A = \exp(TF); \qquad B = F^{-1}(A - I)G \tag{A.29}$$

T being the sampling period, I the identity matrix and $\exp(TF)$ the exponential matrix defined by the series

$$\exp(TF) = I + TF + \frac{T^2}{2!}F^2 + \frac{T^3}{3!}F^3 + \dots \tag{A.30}$$

This is a widely used procedure for discretizing linear time invariant state equations. The key point lies in the computation of the exponential matrix. There exists a variety of well-known procedures for this which are included in a number of software packages [Mat92].

Appendix B

OTHER PREDICTIVE MODELS AND PERFORMANCE INDICES

In Sections 3.3 and 3.4 we have presented two predictive control laws derived from the minimization of two performance indices using the difference equation (3.1) as the predictive model. Here we develop and discuss other control laws derived using different forms of predictive model and performance index. In Section B.1 we discuss the application of predictive control using difference equation models. In Section B.2 we deal with impulse and step response models and in Section B.3 state space models are considered. The basic tools on system analysis used here are outlined in Appendix A.

B.1 DIFFERENCE EQUATION MODELS

As has already been discussed in Chapter 1, difference equations may be used to describe the dynamic relationship between the inputs and outputs of physical processes. In Chapters 2 and 3 the particular difference equation models (2.4) and (3.1) have been used to formulate the basic and extended strategies of predictive control and to illustrate the concepts supporting them. Model (3.1) is the same as (2.4) but is used on a λ-step horizon. These models are a simplification of a more complete description, outlined in Chapter 1 which may include measurable disturbances and unmeasurable perturbations and which cope with multi-input/multi-output (MIMO) systems. In the following we will consider the multivariable case, the inclusion of disturbances and an incremental formulation for the difference equation models.

B.1.1 The multivariable case

The extension of the results of Chapter 3 for cases dealing with multivariable systems is really straightforward if we follow the derivations presented using a model such as (3.1) but in the matrix-vector form

$$\hat{Y}(k+j|k) = \sum_{i=1}^{\hat{n}} \hat{A}_i \hat{Y}(k+j-i|k) + \sum_{i=1}^{\hat{m}} \hat{B}_i \hat{U}(k+j-i|k) \qquad (j = 1, 2, \ldots, \lambda) \tag{B.1}$$

where

$$\hat{Y}(k+1-i|k) = Y(k+1-i) \qquad i = 1, \ldots, \hat{n}$$
$$\hat{U}(k+1-i|k) = U(k+1-i) \qquad i = 1, \ldots, \hat{m}$$

and a performance criterion such as (3.3) but in the form

$$J_k = \frac{1}{2}\sum_{j=1}^{\lambda}[\hat{Y}(k+j|k) - Y_r(k+j|k)]^T Q_j [\hat{Y}(k+j|k) - Y_r(k+j|k)] + \frac{1}{2}\sum_{j=0}^{\lambda-1} \hat{U}(k+j|k)^T R_j \hat{U}(k+j|k) \tag{B.2}$$

where Q_j and R_j are weighting matrices and Y_r is a reference trajectory such as that given in (3.4) but in the form

$$Y_r(k+j|k) = \sum_{i=1}^{p} A_{ri} Y_r(k+j-i|k) + \sum_{i=1}^{q} B_{ri} Y_{sp}(k+j-i) \qquad (j = 1, 2, \ldots, \lambda)$$

where

$$Y_r(k+1-i|k) = Y(k+1-i) \qquad i = 1, \ldots, p$$

and Y_{sp} is the setpoint. Control laws similar to (3.12) and (3.17) may be derived from the above in a matrix–vector form.

B.1.2 Model with disturbances

To consider the effect of disturbances, we can enlarge the model of (3.1) in the following form

$$\hat{y}(k+j|k) = \sum_{i=1}^{\hat{n}} \hat{a}_i \hat{y}(k+j-i|k) + \sum_{i=1}^{\hat{m}} \hat{b}_i \hat{u}(k+j-i|k) + \sum_{i=1}^{\hat{p}} \hat{c}_i \hat{w}(k+j-i|k) \qquad (j = 1, 2, \ldots, \lambda) \tag{B.3}$$

where

$$\begin{aligned} \hat{y}(k+1-i|k) &= y(k+1-i) & i &= 1,\dots,\hat{n} \\ \hat{u}(k+1-i|k) &= u(k+1-i) & i &= 1,\dots,\hat{m} \\ \hat{w}(k+1-i|k) &= w(k+1-i) & i &= 1,\dots,\hat{p} \end{aligned} \tag{B.4}$$

Note that (B.4) implies that the disturbance w is measurable at the present time k and that the presence of $\hat{w}(\cdot|k)$ in (B.3) implies that it is also measurable in advance over the prediction interval $[k, k+\lambda]$ or, at least, it can be estimated.

Using equations (B.3) and (B.4) instead of (3.1) and (3.2), the procedures of Sections 3.3 and 3.4 may be easily modified to extend the control laws of (3.12) and (3.17) including the extra terms associated with the sequence of present and past disturbances $w(k+1-i)$ $(i = 1,\dots,\hat{p})$ and with the sequence of disturbances on the prediction interval $\hat{w}(k+j|k)$ $(j = 1,\dots,\lambda-1)$.

In problems where disturbances are available in advance, this formulation allows the advantageous use of such information. As an example, this would be the case for active control of vehicle suspension where real time measurements of road roughness are possible [Hro91].

B.1.3 Advantages of an incremental formulation

First, we will extend the analysis made in Section 3.5 to illustrate some practical limitations of the control law of (3.17), specifically: (i) the permanent deviation of the process output from the setpoint where there exist unknown constant disturbances and (ii) the requirement for the predictive model to have the same gain as the process. Second, we will see how these limitations may be overcome through an incremental formulation.

Limitations of control law (3.17)

Consider the process described by a transfer function such as (3.18) but with an unknown disturbance p added to the control input. That is,

$$y(z) = \frac{B(z^{-1})}{A(z^{-1})}[u(z) + p(z)] \tag{B.5}$$

Solving for $u(z)$ in (3.20a) and substituting into (B.5) leads to

$$y(z) = \frac{B(z^{-1})}{\hat{\theta}'_{\lambda}(z^{-1})} y_r(z) + \frac{B(z^{-1})\hat{G}_{\lambda}(z^{-1})}{\hat{\theta}'_{\lambda}(z^{-1})} p(z) \tag{B.6}$$

$\hat{\theta}'_{\lambda}(z^{-1})$ being given in (3.26).

The substitution of (3.24a) into (B.6) leads to

$$y(z) = \frac{B(z^{-1})\Delta_\lambda(z^{-1})}{\hat{\theta}_\lambda(z^{-1})} y_{sp}(z) + \frac{B(z^{-1})\hat{G}_\lambda(z^{-1})}{\hat{\theta}_\lambda(z^{-1})} p(z) \tag{B.7}$$

where $\hat{\theta}_\lambda(z^{-1})$ is the characteristic polynomial defined in (3.29).

Assuming a constant setpoint y_{sp} and a constant disturbance $\bar{p}$, we can apply the discrete time final value theorem (Appendix A) to (B.7) in order to obtain the steady state value of the output:

$$\bar{y} = \lim_{k\to\infty} y(k) = \lim_{z\to 1} \left[\frac{B(z^{-1})\Delta_\lambda(z^{-1})}{\hat{\theta}_\lambda(z^{-1})} \bar{y}_{sp} + \frac{B(z^{-1})\hat{G}_\lambda(z^{-1})}{\hat{\theta}_\lambda(z^{-1})} \bar{p}\right] \tag{B.8}$$

Recalling properties (3.40)–(3.42) and using them in (B.8), we readily obtain

$$\bar{y} = \frac{G_s[1 - \Phi_\lambda(1)]}{[G_s - \hat{G}_s]\hat{E}_\lambda(1) + \hat{G}_s - \Phi_\lambda(1)G_s} [\bar{y}_{sp} + \hat{G}_\lambda(1)\bar{p}] \tag{B.9}$$

From (B.9) we note that the only cases when we have $\bar{y} = \bar{y}_{sp}$ are when both the process gain and the predictive model gain are equal ($G_s = \hat{G}_s$) and when there is no disturbance ($\bar{p} = 0$). For the case where $\bar{p} = 0$, (B.9) is equal to (3.43) and we have here the case analyzed in Section 3.5. In the case when $G_s = \hat{G}_s$, (B.9) reduces to

$$\bar{y} = \bar{y}_{sp} + \hat{G}_\lambda(1)\bar{p} \tag{B.10}$$

Since the second summation is non-zero, the steady state output differs from the setpoint, thus a permanent error exists that is proportional to the constant disturbance load $\bar{p}$. We remark that this deviation from the setpoint is present even in the ideal case having no modelling error.

Incremental formulation of predictive control

Let us revisit the model of (3.1) to relate the incremental values of the input process variables in the form

$$\Delta\hat{y}(k+j|k) = \sum_{i=1}^{\hat{n}} \hat{a}_i \Delta\hat{y}(k+j-i|k) + \sum_{i=1}^{\hat{m}} \hat{b}_i \Delta\hat{u}(k+j-i|k) \tag{B.11}$$
$$(j = 1, 2, \ldots, \lambda)$$

where

$$\begin{aligned}\Delta\hat{y}(k+j|k) &= \hat{y}(k+j|k) - \hat{y}(k+j-1|k) \\ \Delta\hat{u}(k+j|k) &= \hat{u}(k+j|k) - \hat{u}(k+j-1|k)\end{aligned} \tag{B.12}$$

and

$$\begin{aligned}&\Delta\hat{y}(k+1-i|k) = \Delta y(k+1-i) = y(k+1-i) - y(k-i) \\ &\qquad i = 1,\ldots,\hat{n} \\ &\Delta\hat{u}(k+1-i|k) = \Delta u(k+1-i) = u(k+1-i) - u(k-i) \\ &\qquad i = 1,\ldots,\hat{m}\end{aligned} \tag{B.13}$$

This incremental model may be combined with a cost function of the form

$$J_k = \frac{1}{2}\sum_{j=1}^{\lambda} Q_j[\hat{y}(k+j|k) - y_r(k+j|k)]^2 + \frac{1}{2}\sum_{j=0}^{\lambda-1} R_j \Delta\hat{u}(k+j|k)^2 \tag{B.14}$$

which penalizes the excessive excursions of the control signal considering the increments of the control variable.

The incremental formulation of the predictive model was proposed in [Mar76a] because it introduced fundamental practical advantages in the implementation of adaptive predictive control in a real context, as is discussed in Chapter 6. Later on, the derivation of the predictive control law using the performance index (B.14) was proposed in [Mar80] and further analyzed in [Rod82]. This performance index has also been used by other authors, particularly within the so-called generalized predictive control (GPC) [CMT87].

As an example, we now formulate a control law derived for a particular case of index (B.14) similar to the one presented in Section 3.4 but based on the incremental model of (B.11). This control law will be the incremental version of the one given in (3.17) and then, through an analysis similar to the one performed in the preceding subsection, the benefits of the incremental formulation of predictive control will be clearly pointed out.

Consider the performance criterion

$$J_k = [\hat{y}(k+\lambda|k) - y_r(k+\lambda|k)]^2 \tag{B.15}$$

along with the condition

$$\Delta\hat{u}(k+1|k) = \ldots = \Delta\hat{u}(k+\lambda-1|k) = 0 \tag{B.16}$$

This condition, as already discussed in Section 3.4, means that the control sequence will be constant during the prediction interval.

By using (B.11) recursively from the initial conditions (B.13), we may write

$$\Delta\hat{y}(k+j|k) = \sum_{i=1}^{\hat{n}} \hat{e}_i^{(j)} \Delta y(k+1-i) + \sum_{i=2}^{\hat{m}} \hat{g}_i^{(j)} \Delta u(k+1-i) + \hat{g}_1^{(j)} \Delta\hat{u}(k|k)$$
$$(j = 1, 2, \ldots, \lambda) \tag{B.17}$$

where $\hat{e}_i^{(j)}, \hat{g}_i^{(j)}$ are generated by the algorithms of (3.7).

By summing up the set of λ equations (B.17) and using (B.12), we obtain

$$\hat{y}(k+\lambda|k) - y(k) = \sum_{i=1}^{\hat{n}} \hat{\eta}_i^{(\lambda)} \Delta y(k+1-i) + \sum_{i=2}^{\hat{m}} \hat{\gamma}_i^{(\lambda)} \Delta u(k+1-i) + \hat{h}^{(\lambda)} \Delta\hat{u}(k|k) \tag{B.18}$$

where

$$\hat{\eta}_i^{(\lambda)} = \sum_{j=1}^{\lambda} \hat{e}_i^{(j)}; \qquad \hat{\gamma}_i^{(\lambda)} = \sum_{j=1}^{\lambda} \hat{g}_i^{(j)}; \qquad \hat{h}^{(\lambda)} = \sum_{j=1}^{\lambda} \hat{g}_1^{(j)} \tag{B.19}$$

To obtain the value of the incremental control, (B.18) is simply substituted into (B.15), the index J_k is cancelled and the following equation is solved:

$$\Delta u(k) = \Delta\hat{u}(k|k) = \frac{y_r(k+\lambda|k) - y(k) - \sum_{i=1}^{\hat{n}} \hat{\eta}_i^{(\lambda)} \Delta y(k+1-i) - \sum_{i=2}^{\hat{m}} \hat{\gamma}_i^{(\lambda)} \Delta u(k+1-i)}{\hat{h}^{(\lambda)}} \tag{B.20}$$

Then the control $u(k)$ may be obtained:

$$u(k) = \Delta u(k) + u(k-1) \tag{B.21}$$

Analysis of the incremental control law

Applying the Z transform to the control law (B.20)–(B.21), we may write

$$y_r(z) = y(z) + (1 - z^{-1})\hat{\Sigma}_\lambda(z^{-1})y(z) + (1 - z^{-1})\hat{\Gamma}_\lambda(z^{-1})u(z) \tag{B.22}$$

with the polynomials

$$\begin{aligned}\hat{\Sigma}_\lambda(z^{-1}) &= \hat{\eta}_1^{(\lambda)} + \hat{\eta}_2^{(\lambda)} z^{-1} + \ldots + \hat{\eta}_{\hat{n}}^{(\lambda)} z^{-\hat{n}+1} \\ \hat{\Gamma}_\lambda(z^{-1}) &= \hat{h}^{(\lambda)} + \hat{\gamma}_2^{(\lambda)} z^{-1} + \ldots + \hat{\gamma}_{\hat{m}}^{(\lambda)} z^{-\hat{m}+1}\end{aligned} \tag{B.23}$$

Solving for $u(z)$ in (B.22) and substituting into (B.5), the following equation is easily obtained:

$$y(z) = \frac{B(z^{-1})}{\hat{\psi}'_\lambda(z^{-1})} y_r(z) + \frac{B(z^{-1})\hat{\Gamma}_\lambda(z^{-1})(1-z^{-1})}{\hat{\psi}'_\lambda(z^{-1})} p(z) \tag{B.24}$$

where $\hat{\psi}'_\lambda(z^{-1})$ is the characteristic polynomial

$$\hat{\psi}'_\lambda(z^{-1}) = B(z^{-1})[1 + \hat{\Sigma}_\lambda(z^{-1})(1-z^{-1})] + A(z^{-1})\hat{\Gamma}_\lambda(z^{-1})(1-z^{-1}) \tag{B.25}$$

By substituting (3.24a) into (B.24) we obtain

$$y(z) = \frac{B(z^{-1})\Delta_\lambda(z^{-1})}{\hat{\psi}_\lambda(z^{-1})} y_{sp}(z) + \frac{B(z^{-1})\hat{\Gamma}_\lambda(z^{-1})(1-z^{-1})}{\hat{\psi}_\lambda(z^{-1})} p(z) \tag{B.26}$$

where $\hat{\psi}_\lambda(z^{-1})$ is the characteristic polynomial of the closed loop

$$\hat{\psi}_\lambda(z^{-1}) = \hat{\psi}'_\lambda(z^{-1}) - B(z^{-1})\Phi_\lambda(z^{-1}) \tag{B.27}$$

Using arguments similar to those used in Section 3.5, we can prove that the stability condition for polynomials $\hat{\psi}'_\lambda(z^{-1})$ and $\hat{\psi}_\lambda(z^{-1})$ is attained for a prediction length λ extended sufficiently. But here we are primarily interested in checking whether constant setpoints are reached without offsets in spite of constant disturbances.

Therefore, assuming again that the setpoint is equal to a constant $\bar{y}_{sp}$, and that the disturbance is also constant with a value $\bar{p}$, we can apply the discrete time final value theorem to (B.26) to obtain

$$\bar{y} = \lim_{k\to\infty} y(k) = \lim_{z\to 1}\left[\frac{B(z^{-1})\Delta_\lambda(z^{-1})}{\hat{\psi}_\lambda(z^{-1})}\bar{y}_{sp} + \frac{B(z^{-1})\hat{\Gamma}_\lambda(z^{-1})(1-z^{-1})}{\hat{\psi}_\lambda(z^{-1})}\bar{p}\right] \tag{B.28}$$

We can observe in (B.25) that $\psi'_\lambda(1) = B(1)$ and use it in (B.27) to write

$$\hat{\psi}_\lambda(1) = B(1)[1 - \Phi_\lambda(1)] \tag{B.29}$$

Selecting the reference trajectory (3.4) and λ in such a way that $\Phi_\lambda(1) \neq 1$, which is a simple natural condition, we see that (B.28) reduces to

$$\bar{y} = \frac{\Delta_\lambda(1)}{1 - \Phi_\lambda(1)} \bar{y}_{sp} \tag{B.30}$$

Choosing the reference trajectory (3.4) with a static gain equal to one, the quotient in (B.30) is equal to one as in (3.42). Then we finally have

$$\bar{y} = \bar{y}_{sp} \tag{B.31}$$

This analysis proves that the incremental control law (B.20)–(B.21) is able to drive the output to step setpoints in spite of constant step load disturbances. It also proves that the condition of having a predictive model with the same gain as the process model is not necessary in this case to ensure a zero offset. This issue, concerning robustness, is another significant advantage of the incremental formulation of predictive control.

It is interesting to remark on the presence of the term $1 - z^{-1}$ in the closed loop equation of (B.26) and the characteristic polynomials $\hat{\psi}'_\lambda(z^{-1})$ and $\hat{\psi}_\lambda(z^{-1})$. This term represents an integral action in the control loop which, ultimately, ensures the zero permanent error between the output and the setpoint.

B.2 IMPULSE AND STEP RESPONSE MODELS

As given in Appendix A, the impulse response model gives the output in the form

$$y(k) = \sum_{i=1}^{k} h_i u(k-i) \tag{B.32}$$

where the h_i are the values of the response of the process to a unit impulse input.

The step response model is of the form

$$y(k) = \sum_{i=1}^{k} s_i \Delta u(k-i) \tag{B.33}$$

where $\Delta u(k-i)$ is the increment of the control input

$$\Delta u(k-i) = u(k-i) - u(k-i-1)$$

The s_i are the sampled values of the response of the process to an unit step input.

Both types of model may be used as predictive models within the framework of the extended predictive control strategy presented in Chapter 3. As a matter of fact, impulse response models have been used extensively in the formulation of the so-called Identification and Command (IDCOM) [RRTP78] and Model Algorithmic Control (MAC) [RM82]. In the same context, step response models have been the vehicle for the formulation of Dynamic Matrix Control (DMC) [CR80].

To illustrate the use of this type of model within the predictive control setting, in the following we will formulate a control law similar to the one derived in Section 3.3 but now based on a predictive step model.

In order to do this, consider the prediction scenario illustrated in Figure 3.1 with the interval $[k, k+\lambda]$ defined at the sampling instant k. Assuming that we know the input applied to the process at instants prior to k, we may use equation (B.33) to write the output predicted for instant $k+j$ in the form

$$\hat{y}(k+j|k) = \sum_{i=1}^{j} \hat{s}_i \Delta\hat{u}(k+j-i|k) + \sum_{i=j+1}^{k+j} \hat{s}_i \Delta u(k+j-i) \tag{B.34}$$

for all $j = 1, 2, \ldots, \lambda$. Note that (B.34) is the same equation as (B.33) but assuming unknown inputs $\Delta\hat{u}(k|k), \ldots, \Delta\hat{u}(k+j-1|k)$ on the prediction horizon. Also, we distinguish $\hat{s}_i$ from s_i assuming estimated values of the process step response.

For stable systems, the step response tends to a steady state as $k \to \infty$. Thus we may assume that the coefficients $\hat{s}_i$ are constant for $i = N, \ldots, k+j$, N being an integer chosen to retain the most significant part of the transient step response and $N << k+j$. Therefore, we can truncate the second summation in (B.34) in the form

$$\sum_{i=j+1}^{k+j} \hat{s}_i \Delta u(k+j-i) \simeq \sum_{i=j+1}^{N-1} \hat{s}_i \Delta u(k+j-i) + \hat{s}_N \sum_{i=N}^{k+j} \Delta u(k+j-i) \tag{B.35}$$

Noting that the last summation in (B.35) is equal to the input $u(k+j-N)$ at $k+j-N$, we may write

$$\sum_{i=j+1}^{k+j} \hat{s}_i \Delta u(k+j-i) = y_{oj} \tag{B.36}$$

where

$$y_{oj} \simeq \sum_{i=j+1}^{N-1} \hat{s}_i \Delta u(k+j-i) + \hat{s}_N u(k+j-N) \tag{B.37}$$

Defining the $\lambda \times 1$ vectors

$$\begin{aligned}\hat{Y} &= [\hat{y}(k+1|k), \hat{y}(k+2|k), \dots, \hat{y}(k+\lambda|k)]^T \\ \Delta\hat{U} &= [\Delta\hat{u}(k|k), \Delta\hat{u}(k+1|k), \dots, \Delta\hat{u}(k+\lambda-1|k)]^T \\ Y_o &= [y_{o1}, y_{o2}, \dots, y_{o\lambda}]^T\end{aligned}$$

the set of λ equations of (B.34) can be written in the matrix-vector form

$$\hat{Y} = \hat{S}\,\Delta\hat{U} + Y_o \qquad (B.38)$$

where S is a $\lambda \times \lambda$ triangular matrix defined as

$$\hat{S} = \begin{pmatrix} \hat{s}_1 & 0 & 0 & \dots & 0 \\ \hat{s}_2 & \hat{s}_1 & 0 & \dots & 0 \\ \vdots & \vdots & \vdots & \ddots & \vdots \\ \hat{s}_\lambda & \hat{s}_{\lambda-1} & \hat{s}_{\lambda-2} & \dots & \hat{s}_1 \end{pmatrix} \qquad (B.39)$$

By way of an example, consider the evaluation of the process evolution on the prediction horizon defined by the performance index (B.14), which can be written in the form

$$J_k = \frac{1}{2}[\hat{S}\Delta\hat{U} + Y_o - Y_r]^T Q[\hat{S}\Delta\hat{U} + Y_o - Y_r] + \frac{1}{2}\Delta\hat{U}^T R\Delta\hat{U} \qquad (B.40)$$

Y_r, Q and R being defined as

$$\begin{aligned}Y_r &= [y_r(k+1|k), y_r(k+2|k), \dots, y_r(k+\lambda|k)]^T \\ Q &= \text{diag}[Q_1, Q_2, \dots, Q_\lambda] \\ R &= \text{diag}[R_0, R_1, \dots, R_{\lambda-1}]\end{aligned}$$

The minimization of J_k leads to

$$\Delta\hat{U} = (\hat{S}^T Q\hat{S} + R)^{-1}\hat{S}Q(Y_o - Y_r) \qquad (B.41)$$

From (B.41), the final control applied at each sampling instant k is

$$u(k) = \Delta\hat{u}(k|k) + u(k-1) = \hat{h}_o^T(Y_o - Y_r) + u(k-1) \qquad (B.42)$$

$\hat{h}_o$ being the first row of the matrix $(\hat{S}^T Q\hat{S} + R)^{-1}\hat{S}Q$.

A similar procedure would have been derived if the impulse response model (B.32) had been used.

B.3 STATE SPACE MODELS

Consider a multivariable system described by the following linear discrete time state model:

$$\begin{aligned} x(k+1) &= Ax(k) + Bu(k-r) + w(k) \\ y(k) &= Cx(k) + \Delta(k) \end{aligned} \tag{B.43}$$

Here x is the $n \times 1$ state vector, which, within the formulation considered in this section, is assumed to be available either by direct measurement or through an observer, u is the $p \times 1$ control vector and y is the $m \times 1$ output vector. A, B, C are matrices with appropriate dimensions. On comparing this model with the one described in Appendix A as a linear prototype case, we notice the presence of the $n \times 1$ vector w, which represents the disturbances, r, which is an integer representing the time delay in the number of sampling periods and the $m \times 1$ vector Δ, which represents the measurement noises corrupting the outputs.

The structure of model (B.43) can be useful in formulating predictive controllers for multivariable systems. In order to do this, we may start with the definition of a state predictive model of the form

$$\begin{aligned} &\hat{x}(k+j|k) = \hat{A}\hat{x}(k+j-1|k) + \hat{B}\hat{u}(k+j-1-\hat{r}|k) \\ &(j = 1, 2, \ldots, \lambda + \hat{r}) \end{aligned} \tag{B.44}$$

where $\hat{x}(k+j|k)$ denotes the state vector predicted at instant k for instant $k+j$ and $\hat{u}(\cdot|k)$ denotes the sequence of control vectors on the prediction interval. This model is redefined at each sampling instant k from the actual state vector and the controls previously applied, that is,

$$\hat{x}(k|k) = x(k); \qquad \hat{u}(k-j|k) = u(k-j) \qquad (j = 1, \ldots, \hat{r}) \tag{B.45}$$

Comparing (B.44) with (B.43), one may observe that we do not include the disturbance and measurement noise vectors since they are assumed to be unknown. It is straightforward to include measurable disturbances in the formulation that we will develop in the following, but, with the purpose of simplifying the presentation, we will not do so. In the predictive model we denote parameters as $\hat{A}, \hat{B}, \hat{r}$ to distinguish them from those of the system model of (B.43), assuming that we use estimates of them.

We may consider a linear quadratic performance index, similar to the one given in (3.3), and now written in the following form:

$$\begin{aligned} J_k = \frac{1}{2}\sum_{j=1}^{\hat{r}+\lambda}[\hat{x}(k+j|k) - x_r(k+j|k)]^T Q_j[\hat{x}(k+j|k) - x_r(k+j|k)] \\ +\frac{1}{2}\sum_{j=0}^{\lambda-1}\hat{u}(k+j|k)^T R_j\hat{u}(k+j|k) \end{aligned} \quad \text{(B.46)}$$

where $x_r(k+j|k)$ is a reference trajectory for the state vector which may be redefined at each sampling instant k from the current state vector and evolves towards the setpoint according to a chosen dynamics in a manner similar to the reference trajectory defined in (3.4). Q_j, R_j are symmetric weighting matrices.

In the following two subsections, we follow similar procedures to those described in Sections 3.3 and 3.4, respectively, in order to obtain two predictive control laws. In the first case the control law is obtained from the direct minimization of index (B.46). In the second case, we impose the condition for the control sequence on the prediction interval to be a constant, which leads to a computationally simplified control law.

B.3.1 Minimization of the cost function

By applying (B.44) recursively from the initial conditions of (B.45), we may write

$$\begin{aligned} \hat{x}(k+1|k) &= \hat{A}x(k) + \hat{B}u(k-\hat{r}) \\ \hat{x}(k+2|k) &= \hat{A}^2x(k) + \hat{A}\hat{B}u(k-\hat{r}) + \hat{B}u(k-\hat{r}+1) \\ &\cdots \\ \hat{x}(k+\hat{r}|k) &= \hat{A}^{\hat{r}}x(k) + \hat{A}^{\hat{r}-1}\hat{B}u(k-\hat{r}) + \ldots + \hat{B}u(k-1) \\ \hat{x}(k+\hat{r}+1|k) &= \hat{A}^{\hat{r}+1}x(k) + \hat{A}^{\hat{r}}\hat{B}u(k-\hat{r}) + \ldots + \hat{A}\hat{B}u(k-1) \\ &\quad + \hat{B}\hat{u}(k|k) \\ \hat{x}(k+\hat{r}+2|k) &= \hat{A}^{\hat{r}+2}x(k) + \hat{A}^{\hat{r}+1}\hat{B}u(k-\hat{r}) + \ldots + \hat{A}^2\hat{B}u(k-1) \\ &\quad + \hat{A}\hat{B}\hat{u}(k|k) + \hat{B}\hat{u}(k+1|k) \\ &\cdots \\ \hat{x}(k+\hat{r}+\lambda|k) &= \hat{A}^{\hat{r}+\lambda}x(k) + \hat{A}^{\hat{r}+\lambda-1}\hat{B}u(k-\hat{r}) + \ldots + \hat{A}^{\lambda}\hat{B}u(k-1) \\ &\quad + \hat{A}^{\lambda-1}\hat{B}\hat{u}(k|k) + \hat{A}^{\lambda-2}\hat{B}\hat{u}(k+1|k) + \ldots + \hat{B}\hat{u}(k+\lambda-1|k) \end{aligned} \quad \text{(B.47)}$$

Defining the $(\hat{r}+\lambda)n \times 1$ vectors

$$\begin{aligned} \hat{X} &= [\hat{x}(k+1|k)^T, \ldots, \hat{x}(k+\hat{r}|k)^T, \ldots, \hat{x}(k+\hat{r}+\lambda|k)^T]^T \\ X_r &= [x_r(k+1|k)^T, \ldots, x_r(k+\hat{r}|k)^T, \ldots, x_r(k+\hat{r}+\lambda|k)^T]^T \end{aligned}$$

and the $\hat{r}p \times 1$ and $\lambda p \times 1$ vectors

$$
\begin{aligned}
U_k &= [u(k-1)^T, u(k-2)^T, \ldots, u(k-\hat{r})^T]^T \\
\hat{U} &= [\hat{u}(k|k)^T, \hat{u}(k+1|k)^T, \ldots, \hat{u}(k+\lambda-1|k)^T]^T
\end{aligned}
$$

the index (B.46) can be written in the following form:

$$
J_k = \frac{1}{2}[\hat{X} - X_r]^T Q[\hat{X} - X_r] + \frac{1}{2}\hat{U}^T R\hat{U} \tag{B.48}
$$

where the weighting matrices Q, R are

$$
\begin{aligned}
Q &= \operatorname{diag}[Q_1, \ldots, Q_{\hat{r}}, \ldots, Q_{\hat{r}+\lambda}] \\
R &= \operatorname{diag}[R_0, \ldots, R_{\lambda-1}]
\end{aligned}
$$

The set of $\hat{r} + \lambda$ equations (B.47) can be packed in the form

$$
\hat{X} = Zx(k) + TU_k + N\hat{U} \tag{B.49}
$$

where Z, T, N are matrices (with dimensions $(\hat{r}+\lambda)n \times n$, $(\hat{r}+\lambda)n \times \hat{r}p$, $(\hat{r}+\lambda)n \times \lambda p$ respectively) defined as

$$
Z = \begin{pmatrix} \hat{A} \\ \hat{A}^2 \\ \vdots \\ \hat{A}^{\hat{r}} \\ \hat{A}^{\hat{r}+1} \\ \hat{A}^{\hat{r}+2} \\ \vdots \\ \hat{A}^{\hat{r}+\lambda} \end{pmatrix}; \quad
T = \begin{pmatrix}
0 & 0 & \cdots & 0 & \hat{B} \\
0 & 0 & \cdots & \hat{B} & \hat{A}\hat{B} \\
\vdots & \vdots & \cdots & \vdots & \vdots \\
\hat{B} & \hat{A}\hat{B} & \cdots & \hat{A}^{\hat{r}-2}\hat{B} & \hat{A}^{\hat{r}-1}\hat{B} \\
\hat{A}\hat{B} & \hat{A}^2\hat{B} & \cdots & \hat{A}^{\hat{r}-1}\hat{B} & \hat{A}^{\hat{r}}\hat{B} \\
\hat{A}^2\hat{B} & \hat{A}^3\hat{B} & \cdots & \hat{A}^{\hat{r}}\hat{B} & \hat{A}^{\hat{r}+1}\hat{B} \\
\vdots & \vdots & \cdots & \vdots & \vdots \\
\hat{A}^{\lambda}\hat{B} & \hat{A}^{\lambda-1}\hat{B} & \cdots & \hat{A}^{\hat{r}+\lambda-2}\hat{B} & \hat{A}^{\hat{r}+\lambda-1}\hat{B}
\end{pmatrix}
$$

$$N = \begin{pmatrix} 0 & 0 & 0 & \cdots & 0 & 0 \\ 0 & 0 & 0 & \cdots & 0 & 0 \\ \vdots & \vdots & \vdots & \cdots & \vdots & \vdots \\ 0 & 0 & 0 & \cdots & 0 & 0 \\ \hat{B} & 0 & 0 & \cdots & 0 & 0 \\ \hat{A}\hat{B} & \hat{B} & 0 & \cdots & 0 & 0 \\ \vdots & \vdots & \vdots & \cdots & \vdots & \vdots \\ \hat{A}^{\lambda-1}\hat{B} & \hat{A}^{\lambda-2}\hat{B} & \hat{A}^{\lambda-3}\hat{B} & \cdots & \hat{A}\hat{B} & \hat{B} \end{pmatrix}$$

Substituting (B.49) into (B.48) and imposing the condition on the gradient

$$\frac{\partial J_k}{\partial \hat{U}} = 0$$

we obtain the vector $\hat{U}$ minimizing J_k in the form

$$\hat{U} = -LZx(k) - LTU_k + LX_r \tag{B.50}$$

where

$$L = (N^T QN + R)^{-1} N^T Q \tag{B.51}$$

The control vector $u(k)$ applied to the process at time k is the first vector component of $\hat{U}$, that is,

$$u(k) = \hat{u}(k|k) = -D_1 x(k) - D_2 U_k + D_3 X_r \tag{B.52}$$

D_1 is a $p \times n$ feedback state gain matrix consisting of the first p rows of matrix LZ. D_2 is a $p \times \hat{r}p$ matrix including the first p rows of matrix LT and D_3 is a $p \times (\hat{r} + \lambda)n$ matrix having the first p rows of L.

B.3.2 Particular solution

The implementation of the control law of (B.52) can be computationally expensive, particularly if we are dealing with high-dimensional systems. In an attempt to reduce the calculations involved in the minimization of the index (B.46), we can impose a specified shape to the control sequence. As we discussed in Chapter 3 (Section 3.4), we can impose the control sequence to be constant over the prediction interval. That is

$$\hat{u}(k|k) = \hat{u}(k+1|k) = \ldots = \hat{u}(k+\lambda-1|k) \tag{B.53}$$

Also, with the purpose of reducing the formulation further, we can use a simplified version of the index (B.46) in the form

$$\begin{aligned} J_k = \frac{1}{2}[\hat{x}(k+\hat{r}+\lambda|k) - x_r(k+\hat{r}+\lambda|k)]^T Q' \\ \times [\hat{x}(k+\hat{r}+\lambda|k) - x_r(k+\hat{r}+\lambda|k)] + \frac{1}{2}\hat{u}(k|k)^T R' \hat{u}(k|k) \end{aligned} \tag{B.54}$$

Using (B.53) in the last prediction of (B.47), we may write

$$\hat{x}(k+\hat{r}+\lambda|k) = Z^* x(k) + T^* U_k + N^* \hat{u}(k|k) \tag{B.55}$$

where

$$\begin{aligned} Z^* &= \hat{A}^{\hat{r}+\lambda} \\ T^* &= [\hat{A}^{\lambda}\hat{B} \quad \hat{A}^{\lambda-1}\hat{B} \quad \ldots \quad \hat{A}^{\hat{r}+\lambda-2}\hat{B} \quad \hat{A}^{\hat{r}+\lambda-1}\hat{B}] \\ N^* &= \hat{A}^{\lambda-1}\hat{B} + \hat{A}^{\lambda-2}\hat{B} + \ldots + \hat{A}\hat{B} + \hat{B} \end{aligned} \tag{B.56}$$

Substituting (B.55) into (B.54) and imposing

$$\frac{\partial J_k}{\partial \hat{u}(k|k)} = 0$$

we obtain the control law

$$u(k) = \hat{u}(k|k) = -D_1^* x(k) - D_2^* U_k + D_3^* x_r(k+\hat{r}+\lambda|k) \tag{B.57}$$

where

$$\begin{aligned} D_1^* &= D_3^* Z^* \\ D_2^* &= D_3^* T^* \\ D_3^* &= ({N^*}^T Q' N^* + R')^{-1} {N^*}^T Q' \end{aligned} \tag{B.58}$$

As we mentioned before, under a state space formulation, predictive control can be well suited to multivariable applications. In Chapter 9, this is pointed out through the application of predictive control to the active vibration control of civil engineering structures within a state space formulation. Also, from a theoretical point of view, state predictive control may be appealing since it can exploit the advantages of the powerful framework of the state space representation to analyze such properties as stability or robustness. Some of these issues are discussed in references [RLM88, LR89].

Appendix C

INPUT/OUTPUT PROPERTIES OF STABLE LINEAR PROCESSES

C.1 INTRODUCTION

In this appendix we prove that discrete time, linear and stable processes with constant parameters and with a proper rational discrete transfer function satisfy property (5.3), stated in Chapter 5, which defines processes of a linear and stable nature. First, we will prove this result for single input/single output processes without pure time delay, and later we will consider the case of multivariable delayed processes. Proper rational transfer functions are those expressed as a quotient of polynomials with the degree of the numerator being less than or equal to the degree of the denominator, which corresponds with a difference equation with more terms associated with the output than with the input, as is usually found in the description of real processes.

C.2 SISO PROCESSES WITH NO TIME DELAY

Let us consider a discrete time, single input/single output linear process, with constant parameters, defined by the difference equation

$$\begin{aligned} y(k) &= a_1 y(k-1) + a_2 y(k-2) + \ldots + a_n y(k-n) + b_0 u(k) \\ &\quad + b_1 u(k-1) + b_2 u(k-2) + \ldots + b_m u(k-m) \qquad m \leq n \end{aligned} \tag{C.1}$$

Theorem C.1: If process (C.1) is stable (i.e. if its poles are within the unit circle), the following inequality is satisfied for any time instant $k > 0$:

$$\max_{0 \leq j \leq k} |u(j)| > \tau_1 \max_{0 \leq j \leq k} |y(j)| - \tau_2$$

where τ_1 and τ_2 are constants such that $0 < \tau_1 < +\infty$ and $0 \leq \tau_2 < +\infty$.

■

Proof: This proof considers two different cases depending on the process having distinct or repeated poles.

Process with distinct poles

The discrete transfer function of the system of (C.1) is

$$H(z) = \frac{b_0 + b_1 z^{-1} + \ldots + b_m z^{-m}}{1 - a_1 z^{-1} - a_2 z^{-2} - \ldots - a_n z^{-n}} \qquad m \leq n \tag{C.2}$$

Multiplying the numerator and denominator by z^n and factoring the denominator into its n distinct poles $p_1, \ldots, p_n$, (C.2) can be written in the form

$$H(z) = \frac{b_0 z^n + b_1 z^{n-1} + \ldots + b_m z^{n-m}}{(z - p_1)(z - p_2) \ldots (z - p_n)} \tag{C.3}$$

Through a partial fraction expansion, (C.3) can be written in the form

$$H(z) = c_0 + \frac{c_1}{z - p_1} + \frac{c_2}{z - p_2} + \ldots + \frac{c_n}{z - p_n} \tag{C.4}$$

where the coefficients $c_0, c_1, \ldots, c_n$ may be calculated as follows:

$$\begin{aligned} c_0 &= \lim_{z \to \infty} H(z) \\ c_i &= (z - p_i) H(z) \Big|_{z = p_i} \qquad i = 1, \ldots, n \end{aligned} \tag{C.5}$$

We can derive an equivalent state formulation of (C.4) using the parallel programming method [CM70] as illustrated in Figure C.1. In such a representation the state variables are defined by

$$x_i(z) = \frac{1}{z - p_i} u(z) \qquad i = 1, \ldots, n \tag{C.6a}$$

$$z x_i(z) = p_i x_i(z) + u(z) \qquad i = 1, \ldots, n \tag{C.6b}$$

and in the time domain:

$$x_i(k+1) = p_i x_i(k) + u(k) \qquad i = 1, \ldots, n \tag{C.7}$$

The process output $y(k)$ is the sum of the partial outputs $y_i(k)$:

$$y(k) = \sum_{i=0}^{n} y_i(k) \tag{C.8}$$

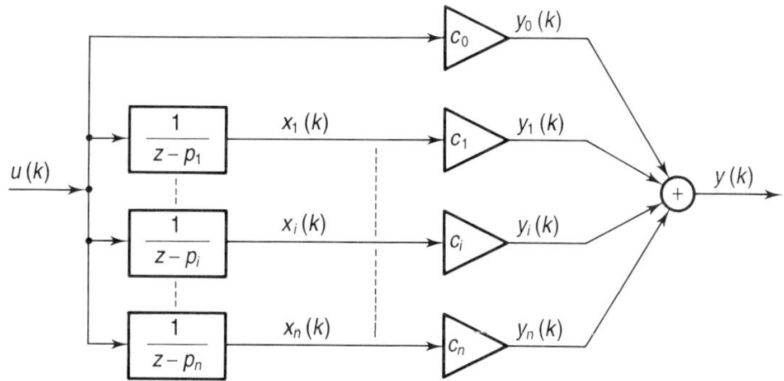

Fig. C.1. Equivalent state diagram for the process with distinct poles.

where

$$y_0(k) = c_0 u(k) \tag{C.9}$$

$$y_i(k) = c_i x_i(k) \qquad\qquad i = 1, \ldots, n \tag{C.10}$$

Clearly from (C.8)–(C.10):

$$y(k) = c_0 u(k) + \sum_{i=1}^{n} c_i x_i(k) \tag{C.11}$$

Let $k > 0$ be an arbitrarily chosen time instant and $\bar{k}\,(0 < \bar{k} \leq k)$ the time instant in which

$$|y(\bar{k})| = \max_{0 \leq j \leq k} |y(j)| \tag{C.12}$$

By recursive application of (C.7) from the initial instant 0, we may write

$$x_i(\bar{k}) = p_i^{\bar{k}} x_i(0) + \sum_{j=0}^{\bar{k}-1} p^{\bar{k}-1-j} u(j) \qquad\qquad i = 1, \ldots, n \tag{C.13}$$

Taking absolute values and using the Cauchy–Schwarz inequality:

$$|x_i(\bar{k})| \leq |p_i^{\bar{k}}||x_i(0)| + \sum_{j=0}^{\bar{k}-1} |p^{\bar{k}-1-j}||u(j)| \tag{C.14}$$

Since the process is stable, we have $|p_i| < 1$ and thus

$$|p_i^{\bar{k}}| < 1 \tag{C.15}$$

$$\lim_{\bar{k}\to\infty} \sum_{j=0}^{\bar{k}-1} |p_i^{\bar{k}-1-j}| = \frac{1}{1-|p_i|} \tag{C.16}$$

From (C.16) we may write

$$\sum_{j=0}^{\bar{k}-1} |p_i^{\bar{k}-1-j}| < \frac{1}{1-|p_i|} \tag{C.17}$$

and so

$$\sum_{j=0}^{\bar{k}-1} |p_i^{\bar{k}-1-j}||u(j)| < \frac{1}{1-|p_i|} \max_{0\leq j\leq \bar{k}-1} |u(j)| \tag{C.18}$$

From (C.11), (C.14), (C.15) and (C.18) we may write

$$\begin{aligned} |y(\bar{k})| &\leq |c_0||u(\bar{k})| + \sum_{i=1}^{n} |c_i||x_i(\bar{k})| \\ &< |c_0||u(\bar{k})| + \sum_{i=1}^{n} |c_i||x_i(0)| + \max_{0\leq j\leq \bar{k}-1} |u(j)| \sum_{i=1}^{n} \frac{|c_i|}{1-|p_i|} \end{aligned} \tag{C.19}$$

Now, using in (C.19) the fact that

$$\begin{aligned} |u(\bar{k})| &\leq \max_{0\leq j\leq k} |u(j)| \\ \max_{0\leq j\leq \bar{k}-1} |u(j)| &\leq \max_{0\leq j\leq k} |u(j)| \end{aligned} \tag{C.20}$$

we may write

$$|y(\bar{k})| < \sum_{i=1}^{n} |c_i||x_i(0)| + \max_{0\leq j\leq k} |u(j)| \left[|c_0| + \sum_{i=1}^{n} \frac{|c_i|}{1-|p_i|}\right] \tag{C.21}$$

Define τ_1 and τ_2 as

$$\begin{aligned}\tau_1 &= \frac{1}{|c_0| + \sum_{i=1}^{n} \frac{|c_i|}{1-|p_i|}} \\ \tau_2 &= \tau_1 \sum_{i=1}^{n} |c_i||x_i(0)|\end{aligned} \tag{C.22}$$

Then (C.21) may be written in the form

$$\max_{0\le j\le k} |u(j)| > \tau_1 |y(\bar{k})| - \tau_2 \tag{C.23}$$

The substitution of (C.12) into (C.23) completes the proof of the theorem for this case. □

Process with repeated poles

We will first consider the case in which the transfer function $H(z)$ has one pole for $z = p_n$ with multiplicity q, all other poles being distinct. In this case $H(z)$ may be written in the form

$$H(z) = \frac{b_o z^n + b_1 z^{n-1} + \ldots + b_m z^{n-m}}{(z-p_1)(z-p_2)\ldots(z-p_{n-q})(z-p_n)^q} \tag{C.24}$$

Likewise, it admits the following partial fraction expansion:

$$\begin{aligned}H(z) = c_0 &+ \frac{c_1}{z-p_1} + \ldots + \frac{c_{n-q}}{z-p_{n-q}} \\ &+ \frac{e_1}{z-p_n} + \frac{e_2}{(z-p_n)^2} + \ldots + \frac{e_q}{(z-p_n)^q}\end{aligned} \tag{C.25}$$

where the coefficients c_0 and c_i are calculated as given in (C.5) and the coefficients e_i are calculated as follows:

$$\begin{aligned}e_q &= (z-p_n)^q H(z)\Big|_{z=p_n} \\ e_{q-1} &= \frac{d}{dz}[(z-p_n)^q H(z)]\Big|_{z=p_n} \\ &\vdots \\ e_1 &= \frac{1}{(n-1)!}\frac{d^{q-1}}{dz^{q-1}}[(z-p_n)^q H(z)]\Big|_{z=p_n}\end{aligned} \tag{C.26}$$

In this case, from (C.25) and by means of the same parallel programming method considered in the preceding case, we can derive a state representation as illustrated in Figure C.2. The corresponding state variables are defined by

$$
\begin{aligned}
x_i(z) &= \frac{1}{z-p_i}u(z) \qquad\qquad i=1,\ldots,n-q \\
x_{n-q+1}(z) &= \frac{1}{z-p_n}u(z) \\
x_{n-q+2}(z) &= \frac{1}{z-p_n}x_{n-q+1}(z) \\
&\vdots \\
x_n(z) &= \frac{1}{z-p_n}x_{n-1}(z)
\end{aligned}
\tag{C.27}
$$

and in the time domain:

$$
\begin{aligned}
x_i(k+1) &= p_i x_i(k) + u(k) \qquad\qquad i=1,\ldots,n-q \\
x_{n-q+1}(k+1) &= p_n x_{n-q+1}(k) + u(k) \\
x_{n-q+2}(k+1) &= p_n x_{n-q+2}(k) + x_{n-q+1}(k) \\
&\vdots \\
x_n(k+1) &= p_n x_n(k) + x_{n-1}(k)
\end{aligned}
\tag{C.28}
$$

As for the preceding case, the process output can now be represented as the addition of n partial outputs that can be grouped as follows:

$$y(k) = y_s(k) + y_m(k) \tag{C.29}$$

with

$$y_s(k) = \sum_{i=0}^{n-q} y_i(k) \tag{C.30}$$

$$y_m(k) = \sum_{i=1}^{q} y_{n-q+i}(k) \tag{C.31}$$

where

$$y_0(k) = c_0 u(k) \tag{C.32}$$

$$y_i(k) = c_i x_i(k) \qquad\qquad i=1,\ldots,n-q \tag{C.33}$$

$$y_{n-q+i}(k) = e_i x_{n-q+i} \qquad\qquad i=1,\ldots,q \tag{C.34}$$

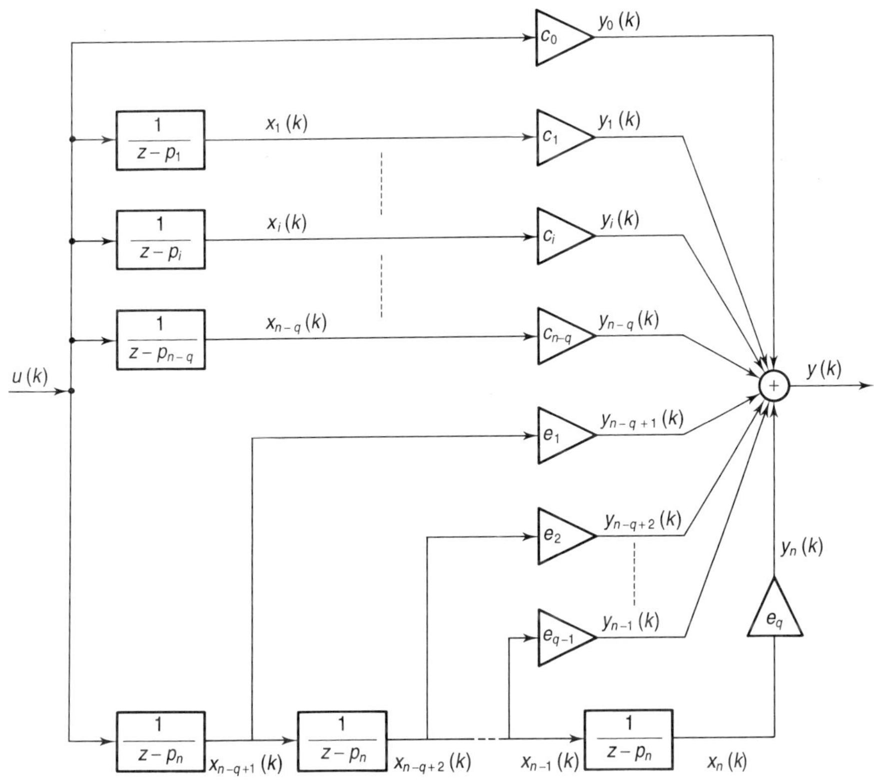

Fig. C.2. Equivalent state diagram for the process with a repeated pole.

Now, for instant $\bar{k}$ in which condition (C.12) is satisfied, we have

$$y(\bar{k}) = y_s(\bar{k}) + y_m(\bar{k})$$

$y_s(\bar{k})$, since it is the superposition of partial outputs associated with distinct poles, is bounded in a similar way as is (C.21):

$$|y_s(\bar{k})| < \sum_{i=1}^{n-q} |c_i||x_i(0)| + \max_{0 \le j \le k} |u(j)| \Big[|c_0| + \sum_{i=1}^{n-q} \frac{|c_i|}{1 - |p_i|} \Big] \tag{C.35}$$

Now we analyze how $y_m(\bar{k})$ is bounded, considering the contribution of the states $x_{n-q+1}(\bar{k}), \ldots, x_n(\bar{k})$.

By recursive application of (C.28) we may write

$$x_{n-q+1}(\bar{k}) = p_n^{\bar{k}} x_{n-q+1}(0) + \sum_{j=0}^{\bar{k}-1} p^{\bar{k}-1-j} u(j) \tag{C.36}$$

Since $|p_n| < 1$, we can use similar arguments to those contained in (C.15)–(C.20) to write

$$|x_{n-q+1}(\bar{k})| < |x_{n-q+1}(0)| + \frac{1}{1-|p_n|} \max_{0 \le j \le k} |u(j)| \tag{C.37}$$

Also the recursive application of (C.28) gives

$$x_{n-q+2}(\bar{k}) = p_n^{\bar{k}} x_{n-q+2}(0) + \sum_{l=0}^{\bar{k}-1} p^{\bar{k}-1-l} x_{n-q+1}(l) \tag{C.38}$$

Using the same arguments as used for (C.37), we may write

$$\begin{aligned} &|x_{n-q+1}(l)| < |x_{n-q+1}(0)| + \frac{1}{1-|p_n|} \max_{0 \le j \le k} |u(j)| \\ &l = 0, 1, \ldots, \bar{k}-1 \end{aligned} \tag{C.39}$$

Using (C.39) in (C.38), we may write

$$\begin{aligned} |x_{n-q+2}(\bar{k})| < |p_n^{\bar{k}}||x_{n-q+2}(0)| + |x_{n-q+1}(0)| \sum_{l=0}^{\bar{k}-1} |p_n^{\bar{k}-1-l}| \\ + \frac{1}{1-|p_n|} \max_{0 \le j \le k} |u(j)| \sum_{l=0}^{\bar{k}-1} |p_n^{\bar{k}-1-l}| \end{aligned} \tag{C.40}$$

Using (C.15) and (C.17) (with $i = n$) in (C.40) we have

$$\begin{aligned} |x_{n-q+2}(\bar{k})| < |x_{n-q+2}(0)| + \frac{1}{1-|p_n|} |x_{n-q+1}(0)| \\ + \frac{1}{[1-|p_n|]^2} \max_{0 \le j \le k} |u(j)| \end{aligned} \tag{C.41}$$

The reader can easily check that (C.39) and (C.41) are particular cases of the general inequality

$$|x_{n-q+i}(\bar{k})| < \sum_{j=1}^{i} \frac{1}{[1-|p_n|]^{i-j}} |x_{n-q+j}(0)| + \frac{1}{[1-|p_n|]^i} \max_{0 \le j \le k} |u(j)| \tag{C.42}$$

which holds for all $i = 1, \ldots, q$.

Using (C.34) and (C.42) in (C.31) we obtain

$$|y_m(\bar{k})| < \sum_{i=1}^{q} |e_i| \sum_{j=1}^{i} \frac{1}{[1-|p_n|]^{i-j}} |x_{n-q+j}(0)| + \max_{0 \leq j \leq k} |u(j)| \sum_{i=1}^{q} \frac{|e_i|}{[1-|p_n|]^i} \tag{C.43}$$

From (C.35) and (C.43) we may write

$$\begin{aligned} |y(\bar{k})| \leq |y_s(\bar{k})| + |y_m(\bar{k})| &< \sum_{i=1}^{n-q} |c_i||x_i(0)| \\ &+ \sum_{i=1}^{q} |e_i| \sum_{j=i}^{i} \frac{1}{[1-|p_n|]^{i-j}} |x_{n-q+j}(0)| \\ &+ \max_{0 \leq j \leq k} |u(j)| \left[|c_0| + \sum_{i=1}^{n-q} \frac{|c_i|}{1-|p_i|} + \sum_{i=1}^{q} \frac{|e_i|}{[1-|p_n|]^i} \right] \end{aligned} \tag{C.44}$$

Finally, the substitution of (C.12) into (C.44) gives

$$\max_{0 \leq j \leq k} |u(j)| > \tau_1 \max_{0 \leq j \leq k} |y(j)| - \tau_2 \tag{C.45}$$

where τ_1 and τ_2 are defined by

$$\tau_1 = \frac{1}{|c_0| + \sum_{i=1}^{n-q} \frac{|c_i|}{1-|p_i|} + \sum_{i=1}^{q} \frac{|e_i|}{[1-|p_n|]^i}} > 0 \tag{C.46}$$

$$\tau_2 = \tau_1 \left[\sum_{i=1}^{n-q} |c_i||x_i(0)| + \sum_{i=1}^{q} |e_i| \sum_{j=1}^{i} \frac{1}{[1-|p_n|]^{i-j}} |x_{n-q+j}(0)| \right] \geq 0 \tag{C.47}$$

The derivation of an expression such as (C.45) for the case of a process with several repeated poles is immediate by considering, for each pole, an additional partial output such as $y_m(k)$, and by deducing the corresponding boundedness condition as was done before for p_n. This concludes the proof of this theorem.

□

C.3 SISO PROCESSES WITH TIME DELAY

The single input/single output processes with delay r may be described by an equation such as:

$$\begin{aligned} y(k) = a_1 y(k-1) + a_2 y(k-2) + \ldots + a_n y(k-n) + b_0 u(k-r) \\ + b_1 u(k-r-1) + b_2 u(k-r-2) + \ldots + b_m u(k-r-m) \end{aligned} \tag{C.48}$$

By imposing the variable change $u'(k) = u(k-r)$, an equation relating $y(k)$ and $u'(k)$ and analogous to (C.1) is obtained. Consequently, if the process is stable, according to Theorem C.1, the following property can be derived:

$$\max_{0 \leq j \leq k} |u'(j)| > \tau_1 \max_{0 \leq j \leq k} |y(j)| - \tau_2 \tag{C.49}$$

where $0 < \tau_1 < +\infty$ and $0 \leq \tau_2 < +\infty$. Reversing the variable change and considering that expression (C.49) is valid for all k, we can write

$$\max_{0 \leq j \leq k} |u(j)| > \tau_1 \max_{0 \leq j \leq k} |y(j+r)| - \tau_2 \tag{C.50}$$

Thus, we conclude that property (5.3) is verified by linear stable processes with time delay.

C.4 MULTIVARIABLE PROCESSES

Linear and stable multivariable processes, with m outputs and n inputs, can always be decomposed into m processes with one output and n inputs. Each of these outputs may be considered as the sum of n partial outputs, which correspond to n SISO subprocesses with inputs that are the previously considered n inputs of the multivariable process. Each of these subprocesses verifies a condition of the type (C.50). From such conditions, an analogous condition for the norms of the input and output vectors of the multivariable process may be derived, that is,

$$\max_{0 \leq j \leq k} \|u(j)\| > \Gamma_1 \max_{0 \leq j \leq k} \|y(j+r)\| - \Gamma_2 \tag{C.51}$$

$$0 < \Gamma_1 < +\infty; \qquad 0 \leq \Gamma_2 < +\infty$$

$u(j)$ and $y(j)$ are, in this case, vectors composed of the initially considered n process inputs and m process outputs. r is defined by

$$r = \min \{r_{ij}\} \qquad i = 1, \ldots, m; \quad j = 1, \ldots, n \tag{C.52}$$

where the r_{ij} are the pure time delays of the corresponding SISO subprocesses considered previously.

Appendix D

APCS STABILITY FOR PROCESSES WITH AN UNSTABLE INVERSE

D.1 INTRODUCTION

This appendix presents a proof of APCS stability for linear and stable processes, the inverse of which may be unstable. This proof considers a particular case of the extended strategy of predictive control, without imposing Assumption 5.2 in respect of the physical realizability of the DDT and knowledge of the process time delays.

The proof of stability is formulated under the ideal case as defined in Chapter 4, Section 4.2.1, but the process is considered to be multivariable. We will see that the theory for this type of process is essentially no more complicated than for the single-input/single-output (SISO) case, being a natural extension. Therefore, the theoretical complexity of the results presented here has its origins in the nature of the proposed problem and not in the SISO or multivariable characteristics of the process.

First the APCS system is described, considering the process, the AP model and the selected driver block. Later the basic assumptions and some definitions of variables involved in the proof are presented, and the lemmas that will be instrumental in obtaining the final stability result are derived. This result, with the only restrictive assumption being that the steady state gain matrix of the process must be non-singular, demonstrates the existence of a prediction horizon for which global asymptotic stability is guaranteed. In this case, *global asymptotic stability* means that the process input/output vectors remain bounded and the control vector tends asymptotically to that which would be obtained if the process parameters were known and used in the predictive control system.

D.2 APCS DESCRIPTION

D.2.1 Process description

Let the process be described by the following discrete time multivariable model:

$$y(k) = \sum_{i=1}^{n} A_i y(k-i) + \sum_{i=1}^{m} B_i u(k-i) \tag{D.1}$$

The input and output vectors u and y are assumed to have the same dimension. A_i and B_i are unknown process parameter matrices of appropriate dimensions. Equation (D.1) may also be written in the form

$$y(k) = \theta_o \phi_o(k-1) + \theta_1 u(k-1) \tag{D.2}$$

$$y(k) = \theta \phi(k-1) \tag{D.3}$$

where

$$\begin{aligned} \theta_o &= [A_1, A_2, \ldots, A_n, B_2, B_3, \ldots, B_m] \\ \theta_1 &= B_1 \\ \phi_o(k-1)^T &= [y(k-1)^T, \ldots, y(k-n)^T, u(k-2)^T, \ldots, u(k-m)^T] \\ \theta &= [\theta_o, \theta_1] \\ \phi(k-1)^T &= [\phi_o(k-1)^T, u(k-1)^T] \end{aligned} \tag{D.4}$$

We do not assume that the matrix θ_1 is non-singular. We assume that the process time delays will be determined by those parameters in the B_i matrices that are equal to zero, that the process is stable and that its steady state gain matrix G is non-singular.

D.2.2 AP model description

The estimation of the process output by the adaptive predictive (AP) model at time instant k is given by

$$\hat{y}(k|k) = \sum_{i=1}^{n} \hat{A}_i(k) y(k-i) + \sum_{i=1}^{m} \hat{B}_i(k) u(k-i) \tag{D.5}$$

or

$$\hat{y}(k|k) = \hat{\theta}_o(k)\phi_o(k-1) + \hat{\theta}_1(k) u(k-1) = \hat{\theta}(k)\phi(k-1) \tag{D.6}$$

where $\hat{A}_i(k)$ and $\hat{B}_i(k)$ are estimates at time k of the corresponding process matrices, and, analogously $\hat{\theta}_o(k)$, $\hat{\theta}_1(k)$ and $\hat{\theta}(k)$ are estimates of $\hat{\theta}_o$, $\hat{\theta}_1$ and $\hat{\theta}$.

The *a posteriori estimation error* is defined by

$$e(k|k) = y(k) - \hat{y}(k|k) = y(k) - \hat{\theta}(k)\phi(k-1) \tag{D.7}$$

In a similar manner, the *a priori estimation error* is defined by

$$e(k|k-1) = y(k) - \hat{y}(k|k-1) = y(k) - \hat{\theta}(k-1)\phi(k-1) \tag{D.8}$$

The parameters of the AP model will be updated using the *a priori* estimation error by means of adaptive mechanisms such as those considered for the ideal case in Chapter 4.

The AP model is used to predict at time k the process output vector in the interval $[k+1, k+\lambda]$, $\hat{y}(k+j|k)$ $(j = 1, \ldots, \lambda)$, as a function of a process input sequence in the interval $[k, k+\lambda-1]$, $\hat{u}(k+j-1|k)$ $(j$=$1, \ldots, \lambda)$ as follows:

$$\hat{y}(k+j|k) = \sum_{i=1}^{n} \hat{A}_i(k)\hat{y}(k+j-i|k) + \sum_{i=1}^{m} \hat{B}_i(k)\hat{u}(k+j-i|k) \qquad (j = 1, \ldots, \lambda) \tag{D.9a}$$

where λ is a positive integer and

$$\hat{y}(k+1-i|k) = y(k+1-i) \qquad (i = 1, \ldots, n) \tag{D.9b}$$

$$\hat{u}(k+1-i|k) = u(k+1-i) \qquad (i = 1, \ldots, m) \tag{D.9c}$$

D.2.3 Driver block design

Let us consider a particular driver block design, already considered in Chapter 3 (Section 3.4), which is based on the performance criterion defined by the two conditions stated below.

Condition D.1: The value of the projected desired trajectory (PDT) at time $k+\lambda$, $y_d(k+\lambda|k)$, is explicitly known or is determined by the designer at time k and bounded, that is, $\|y_d(k+\lambda|k)\| \leq \sigma^2 < +\infty \ \forall k \geq 0$.

□

Condition D.2: The process input sequence $\hat{u}(k+j-1|k)$ $(j$=$1, \ldots, \lambda)$ is constant in the prediction interval $[k, k+\lambda-1]$, that is, $\hat{u}(k|k) = \hat{u}(k+j-1|k)$ $\forall j = 1, \ldots, \lambda)$.

□

Under Condition D.2 the prediction at time k of the process output for time $k+\lambda$, $\hat{y}(k+\lambda|k)$, obtained by the AP model of (D.9), may also be computed by

$$\hat{y}(k+\lambda|k) = \sum_{i=1}^{n} \hat{A}_{i\lambda}(k)y(k+1-i)+\sum_{i=2}^{m} \hat{B}_{i\lambda}(k)u(k+1-i)+\hat{B}_{1\lambda}(k)\hat{u}(k|k) \quad \text{(D.10)}$$

where the matrices $\hat{A}_{i\lambda}$ and $\hat{B}_{i\lambda}$ may be obtained from the AP model matrices $\hat{A}_i$ and $\hat{B}_i$ in (D.9) by means of a recursive algorithm similar to the one considered in Chapter 3 (Section 3.3) for the single-input/single-output case [Rod82].

Equation (D.10) may also be written in the form

$$\hat{y}(k+\lambda|k) = \hat{\theta}_{o\lambda}(k)\phi_o(k) + \hat{\theta}_{1\lambda}(k)\hat{u}(k|k) \quad \text{(D.11)}$$

where

$$\begin{aligned} \hat{\theta}_{o\lambda}(k) &= [\hat{A}_{1\lambda}, \ldots, \hat{A}_{n\lambda}, \hat{B}_{2\lambda}, \ldots, \hat{B}_{m\lambda}] \\ \hat{\theta}_{1\lambda} &= \hat{B}_{1\lambda} \end{aligned} \quad \text{(D.12)}$$

Considering Condition D.1 and using (D.11), the control vector $u(k)$ may be computed as follow:

$$u(k) = \hat{u}(k|k) = \hat{\theta}_{1\lambda}(k)^{-1}[y_d(k+\lambda|k) - \hat{\theta}_{o\lambda}(k)\phi_o(k)] \quad \text{(D.13)}$$

As discussed later, the non-singularity of $\hat{\theta}_{1\lambda}$ may be prevented with an appropriate choice of λ.

D.3 STABILITY RESULTS

D.3.1 Basic assumptions

The assumptions considered in this appendix in order to prove the global asymptotic stability of APCS are presented below.

Assumption D.1: The process described by (D.1) is stable and its steady state gain matrix G is non-singular.

□

Assumption D.2: If, from any instant k, a finite sequence of λ control vectors is applied to the process such that:

$$u(k) = u(k+j) \qquad \forall j = 0, 1, \ldots, \lambda - 1$$

then the adaptation mechanism of APCS verifies:

(a) $\lim_{k\to\infty} \|e(k+j|k+j)\| = 0 \qquad \forall j = 0, 1, \ldots, \lambda.$

(b) $\lim_{k\to\infty} \|\hat{\theta}(k+\lambda) - \hat{\theta}(k)\| = 0.$

(c) $\|\hat{\theta}(k+j)\| \leq \theta_m < +\infty \qquad \forall j = 0, 1, \ldots, \lambda.$

(d) $\det \hat{\theta}_{1\lambda}(k) \neq 0.$

□

The properties of this assumption will be generally satisfied by adaptation mechanisms such as those presented for the ideal case in Chapter 4 of this book. In particular the non-singularity of $\hat{\theta}_{1\lambda}(k)$ may be ensured as discussed later in the comments on Lemma D.2.

D.3.2 Definitions

Definition D.1: $\hat{y}_p(k+j|k)$ $(j = 1, \ldots, \lambda)$ is the predicted output sequence that would have been obtained at time k as a function of a predicted input sequence $\hat{u}(k+j-1|k)$ if the process parameter matrices A_i and B_i had been known and used in the AP model prediction (D.9).

□

Therefore, $\hat{y}_p(k+j|k)$ $(j = 1, \ldots, \lambda)$ would be given by

$$\hat{y}_p(k+j|k) = \sum_{i=1}^{n} A_i \hat{y}_p(k+j-i|k) + \sum_{i=1}^{m} B_i \hat{u}(k+j-i|k) \qquad \text{(D.14}a\text{)}$$

where

$$\hat{y}_p(k+1-i|k) = y(k+1-i) \qquad (i = 1, \ldots, n) \qquad \text{(D.14}b\text{)}$$

$$\hat{u}(k+1-i|k) = u(k+1-i) \qquad (i = 1, \ldots, m) \qquad \text{(D.14}c\text{)}$$

Since in this appendix we consider the ideal case, it is clear that $\hat{y}_p(k+j|k)$ $(j = 1, \ldots, \lambda)$ is equal to the process output sequence that we would obtain if the predicted input sequence were applied to the process, that is, $y_p(k+j|k) = y(k+j)$ if $\hat{u}(k+j-1|k) = u(k+j-1)$ $(j = 1, \ldots, \lambda)$.

According to Condition D.2, the value of $\hat{y}_p(k+\lambda|k)$ could be computed by

$$\hat{y}_p(k+\lambda|k) = \theta_{o\lambda}\phi_o(k) + \theta_{1\lambda}u(k) \qquad \text{(D.15)}$$

where $\theta_{o\lambda}$ and $\theta_{1\lambda}$ could have been obtained by means of recursive algorithms from the process matrices A_i and B_i in a similar way to that considered for the derivation of $\hat{\theta}_{o\lambda}(k)$ and $\hat{\theta}_{1\lambda}(k)$ from $\hat{A}_i(k)$ and $\hat{B}_i(k)$ in (D.11).

Definition D.2: $\hat{u}_p(k|k)$ is the control vector that would have been applied to the process if process matrices A_i and B_i had been known and used in the AP model under the driver block defined by Conditions D.1 and D.2. □

Therefore, $\hat{u}_p(k|k)$ could be obtained directly from (D.15) as follow:

$$\hat{u}_p(k|k) = \theta_{1\lambda}^{-1}[y_d(k+\lambda|k) - \theta_{o\lambda}\phi_o(k)] \tag{D.16}$$

Here λ is chosen in such a way that $\det \theta_{1\lambda} \neq 0$. This choice is always possible as is shown in Lemma D.2.

Definition D.3: $\hat{\phi}(k+j|k)$ is the prediction of the I/O vector for time $k+j$ ($j \geq 0$) obtained at time k using the AP model equation (D.9), that is,

$$\begin{aligned}\hat{\phi}(k+j|k)^T = [\hat{y}(k+j|k)^T, \ldots, \hat{y}(k+j-n+1|k)^T, \hat{u}(k+j-1|k)^T, \ldots \\ \ldots, \hat{u}(k+j-m+1|k)^T, \hat{u}(k+j|k)^T] \qquad j \geq 0\end{aligned} \tag{D.17a}$$

where

$$\hat{y}(k+j-i|k) = y(k+j-i) \qquad \forall j-i \leq 0 \qquad (i = 1, \ldots, n-1) \tag{D.17b}$$

$$\hat{u}(k+j-i|k) = u(k+j-i) \qquad \forall j-i \leq 0 \qquad (i = 1, \ldots, m-1) \tag{D.17c}$$

$$\hat{\phi}(k|k) = \phi(k) \tag{D.17d}$$

□

D.3.3 Basic lemmas

The following lemmas are required in the proof of APCS stability.

Lemma D.1: Under Assumption D.1:

(a) $\lim_{\lambda \to \infty} \theta_{1\lambda} = G$;

(b) $\lim_{\lambda \to \infty} \|\theta_{o\lambda}\| = 0$.

□

Proof: This result can easily be derived as an extension to multivariable processes of the analysis performed in Chapter 3 (Section 3.5.1) [Rod82].

■

Lemma D.2: Under Condition D.1 and Assumption D.1, there exists a finite $\lambda_1 > 0$ such that the following hold for any $\lambda \geq \lambda_1$:

(a) $\det \theta_{1\lambda} \neq 0$

(b) $\hat{y}_p(k+\lambda|k) - y_d(k+\lambda|k) = \theta_{1\lambda}[u(k) - \hat{u}_p(k|k)]$ (D.18)

□

Proof: Since $\det G \neq 0$, property (a) of this lemma immediately follows from property (a) of Lemma D.1. Equation (D.16) may be written in the form:

$$y_d(k+\lambda|k) = \theta_{o\lambda}\phi_o(k) + \theta_{1\lambda}\hat{u}_p(k|k) \qquad \forall\lambda > \lambda_1 \tag{D.19}$$

Subtracting (D.19) from (D.15), (D.18) is obtaining, thus completing the proof of this Lemma D.2.

■

Lemma D.2 has ensured the non-singularity of $\theta_{1\lambda}$ under an appropriate choice of λ. Although the singularity of the corresponding $\hat{\theta}_{1\lambda}(k)$ is possible in principle, there is a provision in APCS adaptation algorithms that ensures its non-singularity, as stated in Assumption D.2(d). This provision can be proved by extending the proof presented in [MSF84] for the case in which $\lambda = 1$.

Lemma D.3: Under Assumption D.2(c) and Condition D.2, we have

$$\|\hat{\phi}(k+j|k)\| \leq \rho_j\|\phi(k)\|, \quad 0 < \rho_j < \rho_{m\lambda} < +\infty, \quad 0 \leq j \leq \lambda - 1 \tag{D.20}$$

□

Proof: This is a standard result which simply states that, under a constant input sequence, the norm of the predicted I/O vector cannot grow faster than exponentially to $+\infty$, since the parameters of the AP model are bounded.

■

Lemma D.4: Under the process description (D.1), the control law (D.13) and Assumption D.2(d), the sequence $\{\|\phi(k)\|\}$ will be unbounded only if there exists a subsequence $\{k_s\}$ of $\{k\}$ such that

(a) $\lim_{k_s \to \infty} \|\phi(k_s)\| = +\infty$;

(b) $\|u(k_s)\| > \Lambda_1\|\phi(k_s)\| - \Lambda_2$ with $0 < \Lambda_1 < +\infty$ and $0 \leq \Lambda_2 < +\infty$.

□

Proof: Assumption D.2d ensures that the control obtained from (D.13) verifies $\|u(k)\| < +\infty \ \forall k \geq 0$. Since process (D.1) is stable and linear, it satisfies property (C.51) of Appendix C. Using this property with similar arguments as in Lemma 5.1 of Chapter 5, the above conditions (a) and (b) are derived.

■

Lemma D.5: Assume that, along with the sequence $\{k\}$, we have

$$\begin{aligned} &\hat{u}(k+\lambda-1-i|k) = u(k+\lambda-1-i) \\ &\lambda \geq 1 \qquad\qquad i = 0, 1, \ldots, \lambda - 1 \end{aligned} \tag{D.21}$$

Then,

$$y(k+\lambda)-\hat{y}(k+\lambda|k) = \sum_{i=0}^{\lambda-1} \Pi_{\lambda-i}^{\lambda}(k)[e(k+\lambda-i|k+\lambda-i) \\ + \Delta\hat{\theta}(k+\lambda-i,k)\hat{\phi}(k+\lambda-1-i|k)] \tag{D.22}$$

where

$$\Delta\hat{\theta}(k+\lambda-i,k) = \hat{\theta}(k+\lambda-i) - \hat{\theta}(k) \tag{D.23}$$

and $\Pi_{\lambda-i}^{\lambda}(k)$ are matrices of appropriate dimension defined by

$$\begin{aligned} \Pi_{\lambda}^{\lambda}(k) &= I \\ \Pi_{\lambda-i}^{\lambda}(k) &= \sum_{j=1}^{i} \hat{A}_j(k+\lambda)\Pi_{\lambda-i}^{\lambda-j}(k) \qquad \forall i = 1,\ldots,\lambda-1 \\ \hat{A}_j(k+\lambda) &= 0 \qquad \forall j > n \end{aligned} \tag{D.24}$$

□

Proof: We will first prove that the lemma is true for $\lambda = 1$, and then will verify that, if it is true for all $\lambda \leq \lambda_1 - 1$ $(\lambda_1 > 1)$, it is also true for $\lambda = \lambda_1$.

Using (D.6), (D.8) and (D.17d) we obtain

$$\hat{y}(k+1|k+1) - \hat{y}(k+1|k) = \hat{\theta}(k+1)\phi(k) - \hat{\theta}(k)\phi(k) = \Delta\hat{\theta}(k+1,k)\hat{\phi}(k|k) \tag{D.25}$$

Using (D.7) we may write

$$y(k+1) - \hat{y}(k+1|k+1) = e(k+1|k+1) \tag{D.26}$$

Adding (D.25) and (D.26) gives

$$y(k+1) - \hat{y}(k+1|k) = e(k+1|k+1) + \Delta\hat{\theta}(k+1,k)\hat{\phi}(k|k) \tag{D.27}$$

which proves this lemma for $\lambda = 1$.

Let us now assume that the lemma is true for all $\lambda \leq \lambda_1 - 1$ and define $\epsilon(k+\lambda|k)$ as follows:

$$\epsilon(k+\lambda|k) = \sum_{i=0}^{\lambda-1} \Pi_{\lambda-i}^{\lambda}(k)[e(k+\lambda-i|k+\lambda-i) \\ + \Delta\hat{\theta}(k+\lambda-i,k)\hat{\phi}(k+\lambda-1-i|k)] \qquad 1 \leq \lambda \leq \lambda_1 - 1 \tag{D.28}$$

Since the lemma is true for all $\lambda \leq \lambda_1 - 1$, using (D.22) and (D.28) we have

$$y(k+\lambda) = \hat{y}(k+\lambda|k) + \epsilon(k+\lambda|k) \qquad 1 \leq \lambda \leq \lambda_1 - 1 \tag{D.29}$$

Combining (D.5) and (D.29) gives

$$\begin{aligned}\hat{y}(k+\lambda_1|k+\lambda_1) = &\sum_{j=1}^{n} \hat{A}_j(k+\lambda_1)[\hat{y}(k+\lambda_1-j|k) + \epsilon(k+\lambda_1-j|k)] \\ &+ \sum_{j=1}^{m} \hat{B}_j(k+\lambda_1)u(k+\lambda_1-j)\end{aligned} \tag{D.30}$$

where

$$\hat{y}(k+\lambda_1-j|k) = y(k+\lambda_1-j) \quad \text{and} \quad \epsilon(k+\lambda_1-j|k) = 0 \qquad \forall j \geq \lambda_1$$

Since this lemma assumes that condition (D.21) holds, we may use (D.9a) to obtain

$$\hat{y}(k+\lambda_1|k) = \sum_{j=1}^{n} \hat{A}_j(k)\hat{y}(k+\lambda_1-j|k) + \sum_{j=1}^{m} \hat{B}_j(k)u(k+\lambda_1-j) \tag{D.31}$$

where

$$\hat{y}(k+\lambda_1-j|k) = y(k+\lambda_1-j) \qquad \forall j \geq \lambda_1$$

Using (D.17) and (D.23) with condition (D.21), and subtracting (D.31) from (D.30) yields

$$\begin{aligned}\hat{y}(k+\lambda_1|k+\lambda_1) - \hat{y}(k+\lambda_1|k) = &\Delta\hat{\theta}(k+\lambda_1,k)\hat{\phi}(k+\lambda_1-1|k) \\ &+ \sum_{j=1}^{n} \hat{A}_j(k+\lambda_1)\epsilon(k+\lambda_1-j|k)\end{aligned} \tag{D.32}$$

where

$$\epsilon(k+\lambda_1-j|k) = 0 \qquad \forall j \geq \lambda_1$$

We may use equation (D.7) to write

$$y(k+\lambda_1) - \hat{y}(k+\lambda_1|k+\lambda_1) = e(k+\lambda_1|k+\lambda_1) \tag{D.33}$$

Let us define $\psi(k+\lambda_1-h|k)$ (for $h \geq 0$) as follows:

$$\begin{aligned}\psi(k+\lambda_1-h|k) = &e(k+\lambda_1-h|k+\lambda_1-h) \\ &+ \Delta\hat{\theta}(k+\lambda_1-h,k)\hat{\phi}(k+\lambda_1-h-1|k)\end{aligned} \tag{D.34}$$

Adding (D.32) and (D.33) and considering (D.28) and (D.34), we have

$$y(k+\lambda_1) - \hat{y}(k+\lambda_1|k) = \psi(k+\lambda_1|k) + \sum_{j=1}^{n} \hat{A}_j(k+\lambda_1)\Big[\sum_{i=0}^{\lambda_1-j-1} \Pi_{\lambda_1-j-i}^{\lambda_1-j}(k)\psi(k+\lambda_1-j-i|k)\Big] \tag{D.35}$$

where

$$\Pi_{\lambda_1-j-i}^{\lambda_1-j}(k) = \psi(k+\lambda_1-j-i|k) = 0 \qquad \forall i+j \geq \lambda_1$$

Equation (D.35) may also be written in the form

$$y(k+\lambda_1) - \hat{y}(k+\lambda_1|k) = \psi(k+\lambda_1|k) + \sum_{j=1}^{n} \hat{A}_j(k+\lambda_1) \sum_{s=j}^{\lambda_1-1} \Pi_{\lambda_1-s}^{\lambda_1-j}(k)\psi(k+\lambda_1-s|k) \tag{D.36}$$

where

$$\Pi_{\lambda_1-s}^{\lambda_1-j}(k) = \psi(k+\lambda_1-s|k) = 0 \qquad \forall s \geq \lambda_1$$

Hence,

$$y(k+\lambda_1) - \hat{y}(k+\lambda_1|k) = \psi(k+\lambda_1|k) + \sum_{s=1}^{\lambda_1-1} \sum_{j=1}^{s} \hat{A}_j(k+\lambda_1)\, \Pi_{\lambda_1-s}^{\lambda_1-j}(k)\psi(k+\lambda_1-s|k) \tag{D.37}$$

where

$$\hat{A}_j(k+\lambda_1) = 0 \qquad \forall j > n$$

Taking (D.24) and (D.34) into account, it can be easily verified that (D.37) is equivalent to (D.22). Therefore, if the lemma is true for all $\lambda \leq \lambda_1 - 1$, it is also true for $\lambda = \lambda_1$, which completes its proof.

■

D.3.4 APCS global asymptotic stability

The result of APCS global asymptotic stability is stated by the following theorem.

Theorem D.1: Under Assumptions D.1 and D.2 and the driver block design defined by Conditions D.1 and D.2, there exists $\lambda_s > \lambda_1 > 0$ such that, $\forall \lambda > \lambda_s$, adaptive predictive control of the process described by (D.1) is globally asymptotically stable in the sense that:

(a) The process input/output signals are bounded for all k, that is,

$$\|\phi(k)\| < \Omega < +\infty$$

(b) The control vector tends asymptotically to that obtained if the actual process parameters were known and used in the predictive control law, that is,

$$\lim_{k\to\infty} [u(k) - \hat{u}_p(k|k)] = 0$$

□

Proof: To prove this theorem, we will derive, within the prediction horizon, a relationship between variables already defined in the preceding sections. Then the stability result will be derived by proving the contradiction existing between this relationship and an unbounded I/O vector sequence.

At time k, the predicted process input sequence $\hat{u}(k+i|k) = u(k)$ $(i = 0, 1, \ldots, \lambda - 1)$ is computed, according to the driver block design Conditions D.1 and D.2, from the value of the PDT at time $k+\lambda$, $y_d(k+\lambda|k)$, which by definition is equal to $\hat{y}(k+\lambda|k)$. The same predicted input sequence would produce a predicted output sequence $\hat{y}_p(k+\lambda|k)$, whose value at $k+\lambda$, $\hat{y}_p(k+\lambda|k)$ may be related to $y_d(k+\lambda|k)$, using Lemma D.5, by means of the following equation:

$$\begin{aligned}\hat{y}_p(k+\lambda|k) - y_d(k+\lambda|k) = \sum_{i=0}^{\lambda-1} \Pi^{\lambda}_{\lambda-i}(k)[e^*(k+\lambda-i|k+\lambda-i) \\ + \Delta\hat{\theta}^*(k+\lambda-i,k)\hat{\phi}(k+\lambda-i|k)]\end{aligned} \tag{D.38}$$

where

$$\begin{aligned}&\Pi^{\lambda}_{\lambda}(k) = I \\ &\Pi^{\lambda}_{\lambda-i}(k) = \sum_{j=1}^{i} \hat{A}^*_j(k+\lambda)\Pi^{\lambda-j}_{\lambda-i}(k) \qquad \forall i = 1, \ldots, \lambda-1 \\ &\hat{A}^*_j(k+\lambda) = 0 \qquad \forall j > n\end{aligned} \tag{D.39}$$

$e^*(k+\lambda-i|k+\lambda-i), \Delta\hat{\theta}^*(k+\lambda-i,k)$ and $\hat{A}^*_j(k+\lambda)$ are the corresponding *a posteriori* estimation errors, increments of the AP model parameter matrix $\hat{\theta}$ and the updated AP model matrices $\hat{A}_j$ that would be generated by the APCS adaptive system in the prediction interval $[k+1, k+\lambda]$ if the predicted process input sequence were actually to be applied. Clearly, in this case $\hat{y}_p(k+\lambda|k)$ would be equal to $y(k+\lambda)$.

Now, considering $\lambda \geq \lambda_1$ and using Lemma D.2 and equation (D.38), we obtain

$$\theta_{1\lambda}[u(k)-\hat{u}_p(k|k)] = \sum_{i=0}^{\lambda-1} \Pi_{\lambda-i}^{\lambda}(k)[e^*(k+\lambda-i|k+\lambda-i) + \Delta\hat{\theta}^*(k+\lambda-i,k)\hat{\phi}(k+\lambda-i|k)] \tag{D.40}$$

Using (D.16), the left-hand side of (D.40) may be written as

$$\theta_{1\lambda}[u(k)-\hat{u}_p(k|k)] = \theta_{1\lambda}u(k) - y_d(k+\lambda|k) - \theta_{o\lambda}\phi_o(k) \tag{D.41}$$

Combining (D.40) and (D.41) gives

$$u(k) = \theta_{1\lambda}^{-1}\Big[\sum_{i=0}^{\lambda-1} \Pi_{\lambda-i}^{\lambda}(k)[e^*(k+\lambda-i|k+\lambda-i) + \Delta\hat{\theta}^*(k+\lambda-i,k)\hat{\phi}(k+\lambda-i|k)] + y_d(k+\lambda|k) + \theta_{o\lambda}\phi_o(k)\Big] \tag{D.42}$$

Using the triangular and Cauchy–Schwartz inequalities,

$$\begin{aligned}\|u(k)\| \le{}& \|\theta_{1\lambda}^{-1}\| \sum_{i=0}^{\lambda-1} \|\Pi_{\lambda-i}^{\lambda}(k)\|\,\|e^*(k+\lambda-i|k+\lambda-i)\| \\ &+ \|\theta_{1\lambda}^{-1}\| \sum_{i=0}^{\lambda-1} \|\Pi_{\lambda-i}^{\lambda}(k)\|\,\|\Delta\hat{\theta}^*(k+\lambda-i,k)\|\,\|\hat{\phi}(k+\lambda-i|k)\| \\ &+ \|\theta_{1\lambda}^{-1}\|\|y_d(k+\lambda|k)\| + \|\theta_{1\lambda}^{-1}\|\,\|\theta_{o\lambda}\|\,\|\phi_o(k)\|\end{aligned} \tag{D.43}$$

Using Lemma D.3 and (D.43) yields

$$\begin{aligned}\|u(k)\| \le{}& \|\theta_{1\lambda}^{-1}\| \sum_{i=0}^{\lambda-1} \|\Pi_{\lambda-i}^{\lambda}(k)\|\,\|e^*(k+\lambda-i|k+\lambda-i)\| \\ &+ \rho_{m\lambda}\|\theta_{1\lambda}^{-1}\|\|\phi(k)\| \sum_{i=0}^{\lambda-1} \|\Pi_{\lambda-i}^{\lambda}(k)\|\,\|\Delta\hat{\theta}^*(k+\lambda-i,k)\| \\ &+ \|\theta_{1\lambda}^{-1}\|\|y_d(k+\lambda|k)\| + \|\theta_{1\lambda}^{-1}\|\,\|\theta_{o\lambda}\|\,\|\phi_o(k)\|\end{aligned} \tag{D.44}$$

If we now assume that $\{\|\phi(k)\|\}$ is unbounded, there will exist a subsequence $\{k_s\}$ of $\{k\}$ such that properties (a) and (b) of Lemma D.4 will be satisfied. Using property (b) of Lemma D.4 and (D.44), along with $\{k_s\}$, we have

$$
\begin{aligned}
\Lambda_1 \|\phi(k_s)\| \leq \|\theta_{1\lambda}^{-1}\| \sum_{i=0}^{\lambda-1} \|\Pi_{\lambda-i}^{\lambda}(k_s)\| \, \|e^*(k_s + \lambda - i | k_s + \lambda - i)\| \\
+ \rho_{m\lambda} \|\theta_{1\lambda}^{-1}\| \|\phi(k_s)\| \sum_{i=0}^{\lambda-1} \|\Pi_{\lambda-i}^{\lambda}(k_s)\| \, \|\Delta\hat{\theta}^*(k_s + \lambda - i, k_s)\| \\
+ \|\theta_{1\lambda}^{-1}\| \|y_d(k_s + \lambda | k_s)\| + \|\theta_{1\lambda}^{-1}\| \, \|\theta_{o\lambda}\| \, \|\phi_o(k_s)\| + \Lambda_2
\end{aligned}
\tag{D.45}
$$

with $0 < \Lambda_1 < +\infty$, $0 < \rho_{m\lambda} < +\infty$ and $0 \leq \Lambda_2 < +\infty$.

According to Assumption D.2, the terms $\|e^*(k_s + \lambda - i | k_s + \lambda - i)\|$ and $\|\Delta\hat{\theta}^*(k_s + \lambda - i, k_s)\|$ in (D.45) tend to zero as $k_s \to \infty$. The terms $\|\theta_{1\lambda}^{-1}\|$ and $\|y_d(k_s + \lambda | k_s)\|$ are bounded, and the terms $\|\Pi_{\lambda-i}^{\lambda}(k_s)\|$ are also bounded since the matrices $\Pi_{\lambda-i}^{\lambda}(k_s)$ are computed by (D.39) from the AP model matrices $\hat{A}_j^*(k_s + \lambda)$, which are bounded as stated by Assumption D.2(c). Given that, according to property (a) of Lemma D.4, $\|\phi(k_s)\| \to +\infty$ as $k_s \to \infty$, inequality (D.45) can only be satisfied if $\|\theta_{1\lambda}^{-1}\| \, \|\theta_{o\lambda}\| \geq \Lambda_1$, since $\|\phi_o(k_s)\| \leq \|\phi(k_s)\|$. However, according to property (b) of Lemma D.1, $\|\theta_{o\lambda}\| \to 0$ as $\lambda \to \infty$. Therefore, there exists a certain $\lambda_s > \lambda_1 > 0$ such that, for all $\lambda \geq \lambda_s$, $\|\theta_{1\lambda}^{-1}\| \, \|\theta_{o\lambda}\| < \Lambda_1$. Consequently, inequality (D.45) cannot be satisfied for all $\lambda \geq \lambda_s$, which implies that a subsequence $\{k_s\}$ of the type considered in Lemma D.4 cannot exist, and hence $\{\|\phi(k)\|\}$ cannot be unbounded. This proves property (a) of this theorem.

Using Lemma D.3 and the previously proven property (a), it is easy to derive the fact that the vectors $\hat{\phi}(k + \lambda - 1 - i | k)$ on the right-hand side of (D.40) are bounded. Since the terms $\|e^*(k + \lambda - i | k + \lambda - i)\|$ and $\|\Delta\hat{\theta}^*(k + \lambda - i, k)\|$ in the same equation tend to zero as $k \to \infty$, the right-hand side of (D.40) tends to zero as $k \to \infty$, which implies the same property for the left-hand side. This completes the proof of this Theorem D.1.

■

D.4 CONCLUSIONS

This appendix has proved global asymptotic APCS stability for a class of multivariable, stable processes with an inverse that may be unstable, using a particular case of the extended strategy of predictive control. The proof has been approached in the context of the ideal case, without using the condition of the DDT being physically realizable or knowledge of the process time delays. Using the same basic ideas of the proof here presented, the reader can extend it to the different real cases considered in Chapters 4 and 5.

References

[ADDG77] M. Athans, D. Castañon, K.P. Dunn, C.S. Greene, W.H. Lee, N. Sandell and A.S. Willsky, 'The stochastic control of the F-8 aircraft using a multiple model adaptive control method. Part I: equilibrium flight', *IEEE Transactions on Automatic Control*, Vol. AC–22(5), pp. 768–780, 1977.

[AE71] K.J. Aström and P. Eykhoff, 'System identification, a survey', *Automatica*, Vol. 7, pp. 123–167, 1971.

[AF66] M. Athans and P.L. Falb, *Optimal Control*, McGraw Hill, New York, USA, 1966.

[AFMY92] H. Abiru, M. Fujishiro, T. Matsumoto, S. Yamazaki and N. Nagata, 'Tuned active dampers installed in the Yokohama Landmark Tower', *First International Conference on Motion and Vibration Control*, Yokohama, Japan, Vol. 1, pp. 110–115, 1992.

[AFS92] A. Afshari, A. Fausse and S. Sorasi, 'A predictive control scheme for automatic management of domestic gas heating', *Proc. of IGRC Conference*, Orlando, Florida, USA, pp. 381–391, 1992.

[AHS84] K.J. Aström, P. Hagander and J. Sternby, 'Zeros of sampled systems', *Automatica*, Vol. 20(1), pp. 31–38, 1984.

[AK77] G. Alag and H. Kaufman, 'An implementable digital adaptive flight controller designed using stabilized single-stage algorithms', *IEEE Transactions on Automatic Control*, Vol. AC–22(5), pp. 780–788, 1977.

[Ast70] K.J. Aström, *Introduction to Stochastic Control Theory*, Academic Press, New York, USA, 1970.

[AW73] K.J. Aström and B. Wittenmark, 'On self-tuning regulators', *Automatica*, Vol. 9, pp. 185–199, 1973.

[AW84] K.J. Aström and B. Wittenmark, *Computer Controlled Systems*, Prentice Hall, Englewood Cliffs, NJ, USA, 1984.

[BGW90] R.R. Bitmead, M. Gevers and V. Wertz, *Adaptive Optimal Control. The Thinking Man's GPC*, Prentice Hall, Englewood Cliffs, NJ, USA, 1990.

[BH69] A.E. Bryson and Y.C. Ho, *Applied Optimal Control*, Ginn and Company, Waltham, MA, USA, 1969.

[BL72] P. Benedek and A. Laszlo, *Les Bases Scientifiques du Genie Chimique* (in French), Dunod, Paris, France, 1972.

[BM93] P. Bendotti and M. M'Saad, 'A skid-to-turn missile autopilot design: the generalized predictive adaptive control approach', *International Journal of Adaptive Control and Signal Processing*, Vol. 7(1), pp. 13–32, 1993.

[Bro85] W.L. Brogan, *Modern Control Theory*, Prentice Hall, Englewood Cliffs, NJ, USA, 1985.

[BRRM95] A. H. Barbat, J. Rodellar, E. P. Ryan and N. Molinares, 'Active control of nonlinear base isolated structures', *Journal of Engineering Mechanics*, ASCE, Vol. 121(6), pp. 676–684, 1995.

[Bry77] A.E. Bryson, 'Mini–issue on NASA's advanced control program for the F-8 DFBW aircraft', *IEEE Transactions on Automatic Control*, Vol. AC–22(5), p. 752, 1977.

[CAT91] G.N. Charos, Y. Arkun and R.A. Taylor, 'Model predictive control of an industrial lime kiln', *TAPPI Journal*, Vol 74(2), pp. 203–211.

[CB93] E.F. Camacho and M. Berenguel, 'Application of generalized predictive control to a solar power plant', *Proc. of the World Congress on Advances in Model Based Predictive Control*, Oxford, UK, Vol. 2, pp. 182–188, 1993.

[CB95] E.F. Camacho and C. Bordons, *Model Predictive Control in the Process Industry*, Springer-Verlag, Berlin, Germany, 1995.

[CBC95] A. Cabanillas, A. Bahillo and J. Cerezo, 'Adaptive predictive control in a bubbling Fluidized bed boiler', *Proc. of the 3rd European Conference on Industrial Furnaces and Boilers*, Lisboa, Portugal, 1995.

[CD91] J.M. Caldwell and J.G. Dearwater, 'Model predictive control applied to FCC units', *Proc. of the Conference on Chemical Process Control (CPCIV)*, Padre Island, Texas, USA, 1991.

[CDLE94] J.M. Compas, P. Decarreau, G. Lanquetin, J.L. Estival, N. Fulget, R. Martin and J. Richalet, 'Industrial applications of predictive functional control to rolling mill, fast robot, river dam', *Proc. of the 3rd IEEE Conference on Control Applications*, Glasgow, UK, Vol. 3, pp. 1643–1655, 1994.

[CG75] D.W. Clarke and P.J. Gawthrop, 'On self-tuning controllers', *Proceedings of the IEE*, Vol. 122, pp. 929–934, 1975.

[Che84] C.T. Chen, *Linear System Theory and Design*, Holt, Rinehart and Winston, USA, 1984.

[CKC94] C.M. Chow, A.G. Kuznetsov and D.W. Clarke, 'Application of multivariable generalized predictive control to the benchmark paper machine', *Proc. of Control System'94*, Stockolm, Sweden 1994.

[Cla88] D.W. Clarke, 'Application of generalized predictive control to industrial processes', *IEEE Control Systems Magazine*, Vol. 8, pp. 49–55, 1988.

[CLSR89] L.L. Chung, R.C. Lin, A.M. Reinhorn and T.T. Soong, 'Experimental study of active control of MDOF seismic structures', *Journal of Engineering Mechanics*, ASCE, Vol. 115(12), pp. 1609–1627, 1989.

[CM70] J.A. Cadzow and H.R. Martens, *Discrete Time and Computer Control Systems*, Prentice Hall, Englewood Cliffs, NJ, USA, 1970.

[CM87] D.W. Clarke and C. Mohtadi, 'Properties of generalized predictive control', *Proc. of the 10th IFAC World Congress*, Munich, Germany, Vol. 10, pp. 63–74, 1987.

[CMSF88] W.R. Cluett, J.M. Martín Sánchez, S.L. Shah and D.G. Fisher, 'Stable discrete-time adaptive control in the presence of unmodeled dynamics', *IEEE Transactions on Automatic Control*, Vol. AC–33, pp. 410–414, 1988.

[CMT87] D.W. Clarke, C. Mohtadi and P.S. Tuffs, 'Generalized predictive control. Parts I and II', *Automatica*, pp. 137–160, 1987.

[CP75] R.W. Clough and J. Penzien, *Dynamics of Structures*, McGraw Hill, New York, USA, 1975.

[CR80] C.R. Cutler and B.L. Ramaker, 'Dynamic matrix control; a computer control algorithm', *Proc. Joint Automatic Control Conference*, San Francisco, USA, 1980.

[CRS88] L.L. Chung, A.M. Reinhorn and T.T. Soong, 'Experiments on active control of structures', *Journal of Engineering Mechanics*, ASCE, Vol. 114(2), pp. 241–256, 1988.

[CS82] C.C. Chen and L. Shaw, 'On receding horizon control', *Automatica*, Vol. 18, pp. 349–352, 1982.

[CSW83] G.W.M.Coppus, S.L. Shah and R.K. Wood, 'Robust multivariable control of a binary distillation column', *Proceedings of the IEE (D)*, Vol. 130, pp. 201–208.

[DAM92] M. Douas, G. Alvarez and J.M. Martín Sánchez, 'Adaptive predictive control of the first Spanish installation for the official approval of nuclear equipment', *Proc. of the Workshop on Industrial Applications of Model Based Predictive Control*, Cambridge, UK, 1992.

[DDDW77] J.C. Deckert, M.N. Desai, J.J. Deyst and A.S. Willsky, *IEEE Transactions on Automatic Control*, Vol. AC–22(5), pp. 795–803, 1977.

[DIP94], J. Daoudi, E. Irving and N. Pons, 'A multivariable predictive control with internal model of a laminated windshield bending furnace', *Proc. of the 3rd IEEE Conference on Control Applications*, Glasgow, UK, Vol. 3, pp. 1897–1902, 1994.

[DM77] H.J. Dunn and R.C. Montgomery, 'A moving window parameter adaptive control system for the F8–DFBW aircraft', *IEEE Transactions on Automatic Control*, Vol. AC–22(5), pp. 788–795, 1977.

[DMZ89] G. Dumont, J.M. Martín Sánchez and Ch.C. Zervos, 'Comparison of an auto–tuned PID regulator and an adaptive predictive control system on an industrial bleach plant, *Automatica*, Vol. 25(1), pp. 33–40, 1989.

[DOD91] R. Dittmar, Z. Ogonowski and K. Damert, 'Predictive control of a non-linear open loop unstable polymerization reactor', *Chemical Engineering Science*, Vol. 46(10), 1991.

[Dor80] R.C. Dorf, *Modern Control Systems*, Addison–Wesley, Reading, MA, USA, 1980.

[DV85] R.M. De Keyser and A.R. Van Cauwenberghe, 'Extended prediction self-adaptive control', *Proc. 7th IFAC Symposium on Identification and Control*, York, UK, 1985.

[DVD85] R.M. De Keyser, P.G. VandeVelde and F.A. Dumoriter, 'A comparative study of self-adaptive long-range predictive control methods', *Proc. 7th IFAC Symposium on Identification and Control*, York, UK, pp. 1317–1322, 1985.

[DW88] M.L. Darby and D.C. White, 'On line optimization of complex process units', *Chemical Engineering Progress*, 1988.

[Edg76] T.F. Edgard, 'Status of design methods for multivariable control', *AIChE Symposium Series*, 72, No. 159.

[Ell77] J.R. Elliot, 'NASA's advanced control law program for the F-8 digital fly–by–wire aircraft', *IEEE Transactions on Automatic Control*, Vol. AC–22, pp 753–757, 1977.

[Fli91] T. Flintham, 'Making the rules', *IEE Review*, Vol. 37(6), pp. 235–239, 1991.

[Fos73] A.S. Foss, 'Critique of chemical process control theory', *AIChE Journal*, Vol. 19, pp. 209–214, 1973.

[FPW90] G. Franklin, F.D. Powell and M.L. Workman, *Digital Control of Dynamic Systems*, Addison–Wesley, Reading, MA, USA, 1990.

[FSF93] M. Q. Feng, M. Shinozuka and S. Fujii, 'Friction-controllable sliding isolated system', *Journal of Engineering Mechanics*, ASCE, Vol. 119(9), pp. 1845–1864, 1993.

[GA90] D. García Alcazar and G. Alvarez, 'SCAP optimization and its industrial application' (in Spanish), *Química e Industria*, Vol. 36(4), pp. 337–342, 1990.

[GAB93] C. Georgescu, A. Afshari and G. Bornard, 'A model based adaptive predictor fault detection method applied to building heating, ventilating and air conditioning', *Proc. of ToolDiag'93*, Toulouse, France, pp. 1009–1018, 1993.

[GBHU93] J. González Delgado, J. Blanco, M. Hermoso and M. Uceda, 'Automation in olive oil factories' (in Spanish), *Agricultura*, Vol. 736, pp. 930–933, 1993.

[GCM94] S. González Díez, C. Corzo and J.M. Martín Sánchez, 'SCAP optimization of cement production', *World Cement*, Vol. 25(7), pp. 38–42, 1994.

[GCMR91] S. González, C. Corzo, J.M. Martín Sánchez, A.M. Ruiz and A. López, *Optimization in the Cement Production*, SCAP Europa, Madrid, Spain, 1991.

[GL87] D. Gustafson and W.M. Lebow, 'Model predictive control of injection molding machines', *Proc. of the 26th Conference on Decision and Control*, Los Angeles, California, USA, pp. 2017–2026, 1987.

[GM82] C.E. García and M. Morari, 'Internal model control. Part 1: A unifying review and some new results', *Industrial Engineering and Chemical Process Design and Development*, Vol. 21, pp. 308–323, 1982.

[GM90] E. Goles and S. Martínez, *Neural and Automata Networks*, Kluwer Academic, Dordrecht, The Netherlands, 1990.

[GMMZ84] C. Greco, G. Menga, E. Mosca and G. Zappa, 'Performance improvements on self-tuning controllers by multistep horizons. The MUSMAR approach', *Automatica*, Vol. 20, pp. 681–699, 1984.

[Gop84] M. Gopal, *Modern Control System Theory*, Wiley, New Delhi, India, 1984.

[Gop88] M. Gopal, *Digital Control Engineering*, Wiley, New Delhi, India, 1988.

[GPM89] C.E. García, D.M. Prett and M. Morari, 'Model predictive control: Theory and practice; a survey', *Automatica*, Vol. 25, pp. 335–348, 1989.

[GR94] M.S. Gelormino and N.L. Ricker, 'Model predictive control of a combined sewer system', *International Journal of Control*, Vol. 59(3), pp. 793–816, 1994.

[GRC80] G.C. Goodwin, P.J. Ramadge and P.E. Caines, 'Discrete time multivariable adaptive control', *IEEE Transactions on Automatic Control*, Vol. AC–25, pp. 449–456, 1980.

[Gri93] M.J. Grimble, 'Long range predictive optimal control law with guaranteed stability for process control applications', *Journal of Dynamic Systems, Measurement and Control*, ASME, Vol. 115, pp. 600–610, 1993.

[GWK87] R. Gorez, V. Wertz and Z. Kuan-Yi, 'On a generalized predictive control algorithm', *Systems and Control Letters*, Vol. 9, pp. 369–377, 1987.

[Has94] A. Hasegawa, 'Development of a strip temperature control system with adaptive generalized predictive control', *Proc. of the 3rd IEEE Conference on Control Applications*, Glasgow, UK, Vol. 3, pp. 1525–1531, 1994.

[HKP91] J. Hertz, A. Krogh and R.G. Palmer, *Introduction to Theory of Neural Computation*, Addison-Wesley, Redwood City, CA, USA, 1991.

[HL89] M. Hislop and A. Lorimer, 'An expert system for kiln control', *Second NCB International Seminar on Cement and Building Material*, India, Vol. 4, pp. IV.73–IV.78, 1989.

[Hro91] D. Hrovat, 'Optimal suspension performance for 2-D vehicle models', *Journal of Sound and Vibration*, Vol. 146(1), pp. 93–110, 1991.

[HS87] W.E. Houston and F. Schork, 'Adaptive predictive control of a semibatch polymerization reactor', *Polymer Process Engineering*, Vol. 5, p. 119, 1987.

[HST87] D.W. Haspel, C.J. Southan and R.A. Taylor, 'The benefits of kiln optimization using Linkman and high level control strategies', *World Cement*, Vol. 18(6), 1987.

[HSW89] K. Hornik, M. Stinchcombe and H. White, 'Multilayer feedforward networks are universal approximators', *Neural Networks*, Vol. 2, pp. 359–366, 1989.

[Iem94] H. Iemura, 'Active and hybrid control development in Japan. Experiments and implementation', *Passive and Active Structural Vibration Control in Civil Engineering*, T.T. Soong and M.C. Constantinou (eds.), International Center for Mechanical Sciences, CISM Courses and Lectures No. 345, Springer Verlag, Wien, Austria, pp. 355–371, 1994.

[IFF86] E. Irving, C.M. Falinower and C. Fonte, 'Adaptive generalized predictive control with multiple reference model', *Proc. 2nd IFAC Workshop on Adaptive Systems in Control and Signal Processing*, Lund, Sweden, 1986.

[IK90] J. Inaudi and J.M. Kelly, 'Active isolation', *Proc. of US National Workshop on Structural Control Research*, Los Angeles, USA, pp. 125–130, 1990.

[JF93] T.A. Johansen and B.A. Foss, 'Constructing NARMAX models using ARMAX models', *International Journal of Control*, Vol. 58, pp. 1125–1153, 1993.

[KA86] G. Kreisselmeier and B.D.O. Anderson, 'Robust model reference adaptive control', *IEEE Transactions on Automatic Control*, Vol. AC–31, pp. 127–133, 1986.

[Kai80] T. Kailath, *Linear Systems*, Prentice Hall, Englewood Cliffs, NJ, USA, 1991.

[Kal60] R.E. Kalman, 'A new approach to linear filtering and prediction problem', *Journal of Basic Engineering*, ASME, Vol. 82D, pp. 35–45, 1960.

[KF85] R.L. Kosut and B. Friedlander, 'Robust adaptive control: conditions for global stability', *IEEE Transactions on Automatic Control*, Vol. AC–30, pp. 610–624, 1985.

[KG88] S.S. Keerthi and E.G. Gilbert, 'Optimal infinite-horizon feedback laws for a general class of constrained discrete time systems: stability and moving horizon approximations', *Journal of Optimization Theory and Applications*, Vol. 57, pp. 265–293, 1988.

[KJ84] R.L. Kosut and C.R. Johnson, 'An input-output view of robustness in adaptive control', *Automatica*, Vol. 20, pp. 569–581, 1984.

[KLS87] J. M. Kelly, G. Leitmann and A. Soldatos, 'Robust control of base isolated structures under earthquake excitation', *Journal of Optimization Theory and Applications*, Vol. 53, pp. 159–181, 1987.

[KM92] T. Kobori and S. Kamagata, 'Active variable stiffness system', *Proc. of the US – Italy – Japan Workshop on Structural Control and Intelligent Systems*, University of Southern California Publication No. CE–9210, Los Angeles, USA, pp. 140–153, 1992.

[KN86], F. Kozin and H.G. Natke, 'System identification techniques', *Structural Safety*, Vol. 3(3–4), pp. 269–316, 1986.

[KP77] W.H. Kwon and A.E. Pearson, 'A modified quadratic cost problem and feedback stabilization of a linear system', *IEEE Transactions on Automatic Control*, Vol. AC–22, pp. 838–842, 1977.

[KP78] W.H. Kwon and A.E. Pearson, 'On feedback stabilization of time varying discrete linear systems', *IEEE Transactions on Automatic Control*, Vol. AC–23, pp. 479–481, 1978.

[KS72] H. Kwakernaak and R. Sivan, *Linear Optimal Control Systems*, Wiley, New York, USA, 1972.

[KS94] M. Kawahara and Y. Shimada, 'A predictive control method to operate the water gate of a dam', *Computer Methods in Applied Mechanics and Engineering*, Vol. 112, pp. 199–218, 1994.

[KST76] A. Kestenbaum, R. Shinnar and F.E. Than, 'Design concepts for process control', *Industrial Engineering and Chemical Process Design and Development*, Vol. 15, pp. 2–13, 1976.

[KU88] K. Krämer and H. Unbehauen, 'Survey to adaptive long-range predictive control', *Proc. 12th IMACS World Congress on Scientific Computation*, Paris, France, Vol. 1, pp. 358–363, 1988.

[Kuo91] B.C. Kuo, *Automatic Control Systems*, Prentice Hall, Englewood Cliffs, NJ, USA, 1991.

[Kuo92] B.C. Kuo, *Digital Control Systems*, 6th edition, Saunders College Publ., USA, 1992.

[Lan74] D. Landau, 'A survey of model reference adaptive techniques. Theory and application'. *Automatica*, Vol. 10, pp. 356–379, 1974.

[LARR94] F. López Almansa, R. Andrade, J. Rodellar and A.M. Reinhorn, 'Modal predictive control of structures. Part I and Part II', *Journal of Engineering Mechanics*, ASCE, Vol. 120(8), pp. 1743–1772, 1994.

[Lin93] T.D.V. Lin, 'FCCU advanced control and optimization', *Hydrocarbon Processing*, 1993.

[Lju87] L. Ljung, *System Identification: Theory for the User*, Prentice Hall, Englewood Cliffs, NJ, USA, 1987.

[LL83] K.S. Lee and W. Lee, 'Extended discrete time multivariable adaptive control using long term predictor', *International Journal of Control*, Vol. 38, pp. 495-514, 1983.

[LL89] K.W. Lim and K.V. Ling, 'Generalized predictive control of a heat exchanger', *IEEE Control Systems Magazine*, Vol. 9, pp. 9–12, 1989.

[LM92] D.A. Linkens and M. Mahfouf, 'Generalized predictive control with feedforward (GPCF) for multivariable anaesthesia', *International Journal of Control*, Vol. 56(5), pp. 1039–1057, 1992.

[LR89] F. López Almansa and J. Rodellar, 'Feasibility and robustness of predictive control of building structures', *Earthquake Engineering and Structural Dynamics*, Vol. 19, pp. 157–171, 1989.

[LW76] W. Lee and V.W. Weekmann, 'Advanced control practice in the chemical process industry: a view from industry', *AIChE Journal*, Vol. 22, pp. 27–38, 1976.

[Mar74] J.M. Martín Sánchez, 'Contribution to model reference adaptive systems from hyperstability theory' (in Spanish), Doctoral dissertation, Universidad Politécnica de Catalunya, Barcelona, Spain, 1974.

[Mar76a] J.M. Martín Sánchez, 'Adaptive predictive control system', *USA Patent* No. 4,197,576, 1976.

[Mar76b] J.M. Martín Sánchez, 'A new solution to adaptive control', *Proceedings of the IEEE*, Vol. 64, pp. 1209–1218, 1976.

[Mar77a] J.M. Martín Sánchez, *Modern Control Theory: Adaptive Predictive Method. Theory and realizations* (in Spanish), Juan March Foundation, University Series, Madrid, Spain, 1977.

[Mar77b] J.M. Martín Sánchez, 'Reply to comments on a new solution to adaptive control', *Proceedings of the IEEE*, Vol. 65, pp. 587–588, 1987.

[Mar80] J.M. Martín Sánchez, 'Adaptive predictive control system' (CIP), *European Patent*, No. 0037579, 1980.

[Mar84] J.M. Martín Sánchez, 'A globally stable APCS in the presence of bounded unmeasured noises and disturbances', *IEEE Transactions on Automatic Control*, Vol. AC–29, pp. 461–464, 1984.

[Mar86] J.M. Martín Sánchez, 'Adaptive control for time variant processes', *International Journal of Control*, Vol. 44, pp. 315–329, 1986.

[Mat92] Mathworks, *The Student Edition of MATLAB*, Prentice Hall, Englewood Cliffs, NJ, USA, 1992.

[MBPB94] T. Manayathara, J. Bentsman, G. Pellegrinetti and R. Blauwkamp, 'Application of stabilizing predictive control to boilers', *Proc. of the 3rd IEEE Conference on Control Applications*, Glasgow, UK, Vol. 3, pp. 1637–1642, 1994.

[MD85] J.M. Martín Sánchez and G. Dumont, 'Industrial comparison of an autotuned PID regulator and an adaptive predictive control system (APCS)', *IFAC Workshop on Adaptive Control of Chemical Processes*, Frankfurt, Germany, October 1985, pp. 72–78.

[Mei90] L. Meirovitch, *Dynamics and Control of Structures*, Wiley, New York, USA, 1990.

[Men73] J.M. Mendel, *Discrete Techniques of Parameter Estimation. The Equation Error Formulation*, Marcel Dekker, New York, USA, 1973.

[MFS81] J.M. Martín Sánchez, D.G. Fisher and S.L. Shah, 'Application of a multivariable adaptive predictive control system', *Proc. Workshop on Applications of Adaptive Control*, Yale University, USA, pp. 138–148, 1981.

[MG84] M. Morari and C.E. García, 'Internal model control', *Proc. American Control Conference*, San Diego, California, USA, pp. 661–667, 1984.

[MGK94] J.M. Martín Sánchez, M. Gurumurthy and S. Krishnaswamy, 'Adaptive predictive control system for process optimization of cement industry", *Proc. of the Fourth International Seminar for Cement and Building Materials*, New Delhi, India, Vol. 2, pp. V33–V42, 1994.

[MGPL93] M. Morari, C.E. García, D.M. Prett and J.H. Lee, *Model Predictive Control*, Prentice Hall, Englewood Cliffs, NJ, USA, 1993.

[MM86] D. Marqués and M. Morari, 'Model predictive control of gas pipeline networks', *Proc. of the 1986 American Control Conference*, pp. 349–354, 1986.

[MM90] D.Q. Mayne and H. Michalska, 'Receding horizon control of nonlinear systems', *IEEE Transactions on Automatic Control*, Vol. AC–35, pp. 814–824, 1990.

[MM91] D.Q. Mayne and H. Michalska, 'Robust receding horizon', *Proc. 30th IEEE Conference on Decision and Control*, Brighton, UK, pp. 64–69, 1991.

[MMDC88] R.K. Miller, S.F. Masri, T.J. Dehghanyar and T.K. Caughey, 'Active vibration control of large civil engineering structures', *Journal of Engineering Mechanics*, ASCE, Vol. 114(9), pp. 1542–1570, 1988.

[MMS88] P.R. Maurath, D.A. Mellichamp and D.E. Seborg, 'Predictive controller design for single-input single-output (SISO) systems', *Industrial Engineering Chemical Research*, Vol. 27, pp. 976–963, 1988.

[Mon74] R.V. Monopoli, 'Model reference adaptive control with an augmented error signal', *IEEE Transactions on Automatic Control*, Vol. AC–19, pp. 474–484, 1974.

[Mos94] E. Mosca, *Optimal, Predictive and Adaptive Control*, Prentice Hall, Englewood Cliffs, NJ, USA, 1994.

[MRER82] R.K. Mehra, J. Rouhani, J. Eterno, J. Richalet and A. Rault, 'Model algorithmic control: review and recent developments', *Engineering Foundation Conference on Chemical Process Control*, Georgia, pp. 287–310, 1982.

[MS84] J.M. Martín Sánchez and S.L. Shah, 'Multivariable adaptive predictive control of a binary distillation column', *Automatica*, Vol. 20, pp. 607–620, 1984.

[MSF84] J.M. Martín Sánchez, S.L. Shah and D.G. Fisher, 'A stable adaptive predictive control system', *International Journal of Control*, Vol. 39, pp. 215–234, 1984.

[MZL89] E. Mosca, G. Zappa and J.M. Lemos, 'Robustness of multipredictor adaptive regulators: MUSMAR', *Automatica*, Vol. 25, pp. 521–529, 1989.

[MZM84] E. Mosca, G. Zappa and C. Manfredi, 'Multistep horizon self-tuning control: the MUSMAR approach', *Preprints of the 9th IFAC World Congress*, Budapest, Hungary, pp. 935–939, 1984.

[NFKH94] D. Neumerkel, J. Franz, L. Krüger and A. Hidiroglu, 'Real time application of neural model predictive control for an induction servo drive', *Proc. of the 3rd IEEE Conference on Control Applications*, Glasgow, UK, Vol. 1, pp. 433–438, 1994.

[NK74] K.S. Narendra and P. Kundra, 'Stable adaptive schemes for system identification and control. Parts I and II', *IEEE Transactions on Systems, Man and Cybernetics*, Vol. SMC–4, pp. 542–560, 1974.

[NP90] K.S. Narendra and K. Parthasarathy, 'Identification and control of dynamical systems using neural networks', *IEEE Transactions on Neural Networks*, Vol. 1(1), pp. 4–27, 1990.

[NV78] K.S. Narendra and L.S. Valavani, 'Stable adaptive controller design. Direct control', *IEEE Transactions on Automatic Control*, Vol. AC–23, pp. 570–583, 1978.

[Oga67] K. Ogata, *State Space Analysis of Control Systems*, Prentice Hall, Englewood Cliffs, NJ, USA, 1967.

[Oga70] K. Ogata, *Modern Control Engineering*, Prentice Hall, Englewood Cliffs, NJ, USA, 1970.

[Ogo94] Z. Ogonowski, 'Application of model based predictive control to acrylin itryl polymerization reactor', *Proc. of the 3rd IEEE Conference on Control Applications*, Glasgow, UK, Vol. 3, pp. 1903–1908, 1994.

[Ogu94] B.A. Ogunnaike, 'On line modelling and predictive control of an industrial terpolymerization reactor', *International Journal of Control*, Vol. 59(3), pp. 711–729, 1994.

[OHOT94] M. Ohshima, I. Hashimoto, H. Ohno, M. Takeda, T. Yoneyama and F. Gotoh, 'Multirate multivariable predictive control and its application to a polymerization reactor', *International Journal of Control*, Vol. 59(3), pp. 731–742, 1994.

[OPL85] R. Ortega, L. Praly and I.D Landau, 'Robustness of discrete time direct adaptive controllers', *IEEE Transactions on Automatic Control*, Vol. AC–30, pp. 1179–1187, 1985.

[PA73] V. Peterka and K.J. Aström, 'Control of multivariable systems with unknown but constant parameters', *3rd IFAC Symposium on Identification and System Parameter Estimation*, The Hague, 1973.

[Pet84] V. Peterka, 'Predictor based self-tuning control', *Automatica*, Vol. 20, pp. 39–50, 1984.

[Pet90] V. Peterka, 'Predictive and LQG optimal control: Equivalences, differences and improvements', *Control of Uncertain Systems*, D. Hinrichsen and B. M. Mårtenson (eds.), Birkäuser, Boston, USA, pp. 221–244, 1990.

[PG80] D.M. Prett and R.D. Gillette, 'Optimization and constrained multivariable control of a catalytic cracking unit', *Proc. Joint Automatic Control Conference*, San Francisco, USA, 1980.

[PH93] P.C. Parks and V. Hahn, *Stability Theory*, Prentice Hall, Englewood Cliffs, NJ, USA, 1993.

[PHAS89] T. Peterson, E. Hernández, Y. Arkun and F.J. Schork, 'Nonlinear predictive control of a semi batch polymerization reactor by an extended DMC', *Proc. of the 1989 American Control Conference*, Pittsburg, USA, pp. 1534–1539, 1989.

[PPCC94] L. Pérez, F.J. Pérez, J. Cerezo J. Catediano and J.M. Martín Sánchez, 'Adaptive predictive control in a thermal power station', *Proc. of the 3rd IEEE Conference on Control Applications*, Glasgow, UK, Vol. 1, pp. 747–752, 1994.

[Pra83] L. Praly, 'Robustness of model reference adaptive control', *Proc. 3rd Yale Workshop on Adaptive Control*, Yale University, USA, pp. 224–226, 1983.

[Pra84] L. Praly, 'Robust model reference adaptive controllers. Part 1', *Proc. 23rd IEEE Conference on Decision and Control*, pp. 1009–1014, 1984.

[RBM87] J. Rodellar, A.H. Barbat and J.M. Martín Sánchez, 'Predictive control of structures', *Journal of Engineering Mechanics*, ASCE, Vol. 113(6), pp. 797–812, 1987.

[RCSR89] J. Rodellar, L.L. Chung, T.T. Soong and A.M. Reinhorn, 'Experimental digital control of structures', *Journal of Engineering Mechanics*, ASCE, Vol. 115(6), pp. 1245–1261, 1989.

[RF58] J.R. Ragazzini and G.F. Franklin, *Sample-Data Control Systems*, McGraw Hill, New York, USA, 1958.

[RGM89] J. Rodellar, M. Gómez and P. Martín Vide, 'Stable predictive control of open-channel flow', *Journal of Irrigation and Drainage Engineering*, ASCE, Vol. 115(4), pp. 701–713, 1989.

[RKS88] P. Rosendorf, M. Kubicek and J. Schongut, 'On line optimization of a rectification column', *Computers and Chemical Engineering*, Vol. 12(2/3), pp. 109–203, 1988.

[RLM88] J. Rodellar, F. López Almansa and J.M. Martín Sánchez, 'Stable predictive control in the presence of modelling errors', *Proc. 12th IMACS World Congress on Scientific Computation*, Paris, France, Vol. 1, pp. 384–386, 1988.

[RM82] R. Rouhani and R.K. Mehra, 'Model algorithmic control: basic theoretic perspectives', *Automatica*, Vol 18, pp. 401–414, 1982.

[RM83] J. Rodellar and J. M. Martín Sánchez, 'Adaptive predictive control of processes with unknown and variable time delay', *Methods and Applications of Measurement and Control*, S. G. Tzafestas and M. H. Hamza (eds.), Acta Press, Calgary, Canada, pp. 489–492, 1983.

[Rod82] J. Rodellar, 'Optimal design of the driver block in the adaptive predictive control system' (in Spanish), Doctoral dissertation, Universidad de Barcelona, Spain, 1982.

[Roo80] J. Roorda, 'Experiments in feedback control of structures', *Structural Control*, (H.H.E. Leipholz, ed.), North Holland, Amsterdam, pp. 629–661, 1980.

[Ros70] H.H. Rosenbrock, *State Space and Multivariable Theory*, Wiley, New York, USA, 1970.

[RRTP78] J.A. Richalet, J.L. Rault, J.L. Testud and J. Papon, 'Model predictive heuristic control: applications to an industrial process', *Automatica*, Vol. 14, pp. 413–428, 1978.

[RS76] E.J. Rijnsdorp and D.E. Seborg, 'A survey on experimental applications of multivariable control to process control problems', *AIChE Symposium Series*, 72, No. 159.

[RSLW89] A.M. Reinhorn, T.T. Soong, R.C. Lin, Y.P. Wang, Y. Fukao, H. Abe and M. Nakai, '1:4 scale model studies of active tendon systems and active mass dampers for aseismic protection', Technical Report NCEER–89–0026, National Center for Earthquake Engineering Research, Buffalo, USA, 1989.

[RSS89] N.L. Ricker, T. Subrahmanian and T. Sim, 'Case studies of model predictive control in pulp and paper production', *Proc. of the 1988 IFAC Workshop on Model Based Process Control*, Oxford, UK, pp. 13–22, 1989.

[Saf89] E. Safak, 'Adaptive modelling, identification and control of dynamic structural systems. Part I and Part II', *Journal of Engineering Mechanics*, ASCE, Vol. 115(11), pp. 2386–2426, 1989.

[SBM91] J. Saint–Donat, N. Bhat and T.J. McAvoy, 'Neural net based model predictive control', *International Journal of Control*, Vol. 54, pp. 1453–1468, 1991.

[SC94] T.T. Soong and M.C. Constantinou (eds.), *Passive and Active Structural Vibration Control in Civil Engineering*, International Center for Mechanical Sciences, CISM Courses and Lectures No. 345, Springer Verlag, Wien, Austria, 1994.

[SH77] G. Stein and G.L. Hartmann, 'Adaptive control laws for F-8 flight test', *IEEE Transactions on Automatic Control*, Vol. AC–22(5), pp. 758–767, 1977.

[Shi88] F.G. Shinskey, *Process Control Systems Application, Design and Tuning*, 3rd edition, McGraw Hill, New York, USA, 1988.

[Sim76] U.F. Simonsmeier, 'Nonlinear binary distillation column models', M.Sc. Thesis, Department of Chemical Engineering, University of Alberta, Canada, 1978.

[Soe92] R. Soeterboek, *Predictive Control. A Unified Approach*, Prentice Hall, Englewood Cliffs, NJ, USA, 1992.

[Soo90] T.T. Soong, *Active Structural Control*, Longman Scientific and Technical, London, England, 1990.

[SRWL91] T.T. Soong, A.M. Reinhorn, Y.P. Wang and R.C. Lin, 'Full scale implementation of active control. Part I: Design and Simulation', *Journal of Structural Engineering*, ASCE, Vol. 117(11), pp. 3516–3536, 1991.

[SS81] T.T. Soong and G.T. Skinner, 'Experimental study of active structural control', *Journal of Engineering Mechanics*, ASCE, Vol. 107(6), pp. 1057–1068, 1981.

[SSHS94] J.R. Shekhar, M. Sudhir, P.H. Kumar, J. Sorojini and M.L. Prasad, 'Expert system for kiln optimization', *Fourth NCB International Seminar on Cement and Building Material*, India, Vol. 2, pp. IV.63–IV.73, 1994.

[SSW77] V.A. Sastry, D.E. Seborg and R.K. Wood, 'An application of a self-tuning regulator to a binary distillation column', *Automatica*, Vol. 13, pp. 417–424.

[SW77] A.P. Sage and C.C. White, *Optimum System Control*, Prentice Hall, Englewood Cliffs, NJ, USA, 1977.

[TC88] T.T.C. Tsang and D.W. Clarke, 'Generalized predictive control with input constraints', *Proceedings of the IEE*, Vol. 135(D), pp. 451–460, 1988.

[Tho75] Y.A. Thomas, 'Linear quadratic optimal estimation and control with receding horizon', *Electronic Letters*, Vol. 11, pp. 19–21, 1975.

[TO90] R. Trujillo and V. Orti, 'SCAP optimization systems: a new era in the operation of processes in the energy field' (in Spanish), *Energía*, Vol. 5, pp. 81–86, 1990.

[WB73] R.K. Wood and M.W. Berry, 'Terminal composition control of a binary distillation column', *Chemical Engineering Science*, Vol. 28, pp. 1707–1717, 1983.

[WCTG94] Q. Wang, G. Chalaye, G. Thomas and G. Gilles, 'An industrial application of predictive control to glass process', *Proc. of the 3rd IEEE Conference on Control Applications*, Glasgow, UK, Vol. 3, pp. 1891–1896, 1994.

[WF76] W.A. Wolovich and P.L. Falb, 'Invariants and canonical forms under dynamic compensation', *SIAM Journal on Control*, Vol. 14(6), pp. 996–1008.

[WMT94] D.J. Wilkinson, A.J. Morris and M.T. Tham, 'Multivariable constrained predictive control (with application to high performance distillation)', *International Journal of Control*, Vol. 59(3), pp. 841–862, 1994.

[Yao72] J.T.P. Yao, 'Concept of structural control', *Journal of Structural Division*, ASCE, Vol. 98(7), pp. 1567–1574, 1972.

[YDL92] J.N. Yang, A. Danielians and S.C. Liu, 'Aseismic hybrid control of nonlinear and hysteretic structures', *Journal of Engineering Mechanics*, ASCE, Vol. 118(7), pp. 1423–1440, 1992.

[Yds84] B.E. Ydstie, 'Extended horizon adaptive control', *Preprints of the 9th IFAC World Congress*, Budapest, Hungary, Vol. VII, pp. 133–137, 1984.

[YKS85] B.E. Ydstie, L.S. Kershenbaum and R.W.H. Sargent, 'Theory and application of an extended horizon self-tuning controller', *AICHE Journal*, Vol. 31, pp. 1771–1780. 1985.

[ZHDM94] R. Zbikowski, K.J. Hunt, A. Dzieliński, R. Murray–Smith and P.J. Gawthrop, 'A review of advances in neural adaptive control systems', Report of the ESPRIT III Project 8039: NACT, European Commission, 1994.

Index

ADAPTIVE PREDICTIVE CONTROL:

From the concepts to plant optimization

Juan M. Martín Sánchez

Department of Energy Systems
School of Mining Engineering
Technical University of Madrid

and SCAP Europa SA
Madrid, Spain

José Rodellar

Department of Applied Mathematics III
School of Civil Engineering
Technical University of Catalonia
Barcelona, Spain

Prentice Hall
London New York Toronto Sydney Tokyo Singapore
Madrid Mexico City Munich

First published 1996 by
Prentice Hall International (UK) Limited
Campus 400, Maylands Avenue
Hemel Hempstead
Hertfordshire, HP2 7EZ
A division of
Simon & Schuster International Group

Printed and bound in Great Britain by
Hartnolls Limited, Bodmin, Cornwall

Library of Congress Cataloging-in-Publication Data

Available from the publisher

British Library Cataloguing in Publication Data

A catalogue record for this book is available from the British Library

ISBN 0-13-514861-8

1 2 3 4 5 00 99 98 97 96

To Refu, Isaías and Julia

– J.M.M.S.

To Anna, Laura and Silvia

– J.R.

Contents